T0144529

RISK ANALYSIS AND MANAGEMENT – TRENDS, CHALLENGES AND EMERGING ISSUES

PROCEEDINGS OF THE 6TH INTERNATIONAL CONFERENCE ON RISK ANALYSIS AND CRISIS RESPONSE (RACR-2017), OSTRAVA/PRAGUE, CZECH REPUBLIC, 5–9 JUNE 2017

Risk Analysis and Management – Trends, Challenges and Emerging Issues

Editors

Aleš Bernatik

Faculty of Safety Engineering, VŠB – Technical University of Ostrava, Czech Republic

Chongfu Huang

Academy of Disaster Reduction and Emergency Management, Beijing Normal University, Beijing, China

Olivier Salvi

INERIS Développement SAS, Verneuil-en-Halatte, France

CRC Press is an imprint of the
Taylor & Francis Group, an **informa** business
A BALKEMA BOOK

CRC Press/Balkema is an imprint of the Taylor & Francis Group, an informa business

Typeset by V Publishing Solutions Pvt Ltd., Chennai, India
Printed and bound in Great Britain by CPI Group (UK) Ltd, Croydon, CR0 4YY

Published by: CRC Press/Balkema
P.O. Box 11320, 2301 EH Leiden, The Netherlands
e-mail: Pub.NL@taylorandfrancis.com
www.crcpress.com – www.taylorandfrancis.com

ISBN: 978-1-138-03359-7 (Hbk)
ISBN: 978-1-315-26533-9 (eBook)

Risk Analysis and Management – Trends, Challenges and Emerging Issues – Bernatik, Huang & Salvi (Eds)
© 2017 Taylor & Francis Group, London, ISBN 978-1-138-03359-7

Table of contents

Risk Analysis and Management – Trends, Challenges and Emerging Issues – Bernatik, Huang & Salvi (Eds)
© 2017 Taylor & Francis Group, London, ISBN 978-1-138-03359-7

Preface

RACR-2017 is the sixth in the series of international conferences on Risk Analysis and Crisis Response, which was launched by the Risk Analysis Council of China Association for Disaster Prevention in 2007, taken over by SRA-China in 2011, and organized by VŠB – Technical University of Ostrava, Czech Republic, in 2017.

The first RACR was held in Shanghai, China in 2007, the second in Beijing, China in 2009, the third in Laredo, Texas, USA in 2011, the fourth in Istanbul, Turkey in 2013, and the fifth in Tangier, Morocco in 2015.

The overall theme of the sixth international conference on risk analysis and crisis response is risk analysis and management – trends, challenges and emerging issues, highlighting science and technology to improve risk analysis capabilities and to optimize crisis response strategy.

In the framework of the conference a specific symposium on the heritage of the ARAMIS project is organized. ARAMIS is the Accidental Risk Assessment Methodology for IndustrieS in the context of the Seveso II Directive developed within the 5th EC Framework Programme. The session title is: ARAMIS project heritage, 10 years after the end of the project.

RACR-2017 will be organized and hosted by VSB-Technical University of Ostrava, which is a modern European university, whose pillars are built on more than 165 years of history. It provides tertiary education in technical and economic sciences across a wide range of degree programs and courses at the Bachelor's, Master's and Doctoral level. Education is organized within 7 faculties and 3 all-university degree programs.

Risk Analysis and Management – Trends, Challenges and Emerging Issues – Bernatik, Huang & Salvi (Eds)

Organisation

ORGANIZED BY:

VŠB – Technical University of Ostrava, Czech Republic

CO-ORGANIZERS:

Association of Fire and Safety Engineering (SPBI), Czech Republic

SCIENTIFIC GUARANTORS

Aleš Bernatík – *VŠB – Technical University of Ostrava, Czech Republic*
Chongfu Huang – *Academy of Disaster Reduction and Emergency Management, Beijing Normal University, China*
Olivier Salvi – *INERIS Développement SAS, France*

Risk Analysis and Management – Trends, Challenges and Emerging Issues – Bernatik, Huang & Salvi (Eds)
 ISBN 978-1-138-03359-7

Sponsor

SPONSORED BY:

INERIS Development SAS, France

RACR

Risk Analysis and Management – Trends, Challenges and Emerging Issues – Bernatik, Huang & Salvi (Eds)
 ISBN 978-1-138-03359-7

Application of slope unit for rainstorm-induced shallow landslide hazard zonation in South China

Qinghua Gong, Guangqing Huang & Jun Wang
Guangdong Open Laboratory of Geospatial Information Technology and Application, Guangzhou, China
Guangzhou Institute of Geography, Guangzhou, China

Junxiang Zhang
Tourism College, Huangshan University, Huangshan, China

ABSTRACT: In South China, shallow landslides result in enormous casualties and huge economic losses in mountainous regions. The landslide hazard zonation is to offer planners with overview information of landslide prone areas which is very important in disaster mitigation since both disasters drastically increase in recently years. The slope, rock types and land use type were the main controlling factors in the shallow landslide formation process. Based on landslide formation mechanism and characteristics in small watershed region; this paper puts forward a method for landslide hazard zonation. Firstly, this paper selected the slope-unit which contains a set of ground features same in the unit and different from the adjacent units as the basic cell for landslide zonation. The partition and distinguish of slop units were made by the Digital Elevation Model (DEM) and GIS system. Secondly, the index system of hazard assessment of landslide was established by analyzing the factors that affect landslide hazard. Nine variables was imposed to create the landslide hazard risk evaluation map including topography relief, slope, slope aspect, slopeshape, soil type, thickness of weathering layer, road density, construction intensity and NDVI are chosen for landslide hazard. Finally, Landslide hazard zonation by means of GIS automatically was overlaid on the nine data layers of hazard. Through the hazard assessment model for the calculation of degree of hazard, the hazard map was deduced.

Keywords: Slope unit, Rainstorm-induced Shallow Landslide, hazard

1 INTRODUCTION

Extreme rainfall-induced landslides usually gathered in a small watershed caused great harm in South China. Those landslides formation mechanism are essentially different from general rainfall pattern of landslides. Geological environment has a unique regional characteristic in South China. Hazard assessment and management has been recognized as an important component of disaster risk reduction strategy. The risk management of landslides began in the late 1980s and became a popular approach in landslide management in the late 1990s. And the risk zoning has become a hotspot in the field of disaster prevention and control in recent years. Experts and scholars have made many exploratory achievements in landslide risk zoning. Landslide hazard risk zoning developed gradually from qualitative to quantitative, from macro to micro, from two-dimensional to three-dimensional (Jin, et al. 2007; Zhou, et al. 2008; Zhu, et al. 2003; Gao, et al. 2011; Qiao, et al. 2008; Tang, et al. 2011). Most of landslide disasters are small shallow landslides, but the landslide hazards are densely distributed in a small watershed in south China. Because of the unique formation mechanism and characteristics of landslide hazard in South China, the existing landslide zoning method cannot meet the spatial accuracy requirements of landslide hazard zoning in South China.

Table 1. Advantages and disadvantages of spatial division unit used in risk mapping.

Cell type	Advantages	Disadvantages
Grid cell	Easy to operate, simple and efficient operation	The grid unit does not reflect the disaster mechanism. The correlation between grid cells and hydrology, geology, topography and other environmental factors is poor.
Administrative zones	The use of administrative zones for risk zoning is more conducive to administrative management	The use of administrative zones to carry out risk zoning ignores the natural attributes of disasters
River basin	Using the river basin as a unit for risk zoning can reflect the characteristics and laws of the geological environment	The spatial scale of risk zoning based on river basin is large, which cannot meet the spatial accuracy requirement of small—scale shallow landslide in South China.
Slope cell	Theoperation process is complex, and requires a higher data accuracy using the slope as the smallest unit for risk division.	The slope cell can reflect the disaster mechanism of the landslide and can meet the precision requirements of the small landslide risk zoning.

There are four spatial division cells used in landslide hazard zonation including grid cell, administrative zones, river basin and slope cell (Guzztti, et al. 2007). Each of the spatial units has its advantages and disadvantages as shown in Table 1.

Each landslide is associated with the slope, and each slope has a clear terrain border (Ermini, et al. 2005). The slope has a different terrain feature (terrain, streamline, and slope) with its adjacent slopes (Tian, et al. 2013). And terrain and geological conditions within the slope unit are basically same. Based on the mechanism, the slope cell is identified as the basic cell of disaster risk zoning. Based on the analysis of meteorological, hydrological and human processes and mechanism of rainstorm-induced shallow landslides in small watershed in south china, this paper established a risk assessment index system for rainstorm-induced shallow landslides. The risk evaluation index of each slope unit is calculated, and the database of rainstorm-induced shallow landslide risk analysis based on slope unit was constructed. Finally, the hazard grade of each slope unit is determined by comprehensive hazard index calculation. On the one hand, both the index system and the spatial unit are based on the mechanism of rainstorm-induced shallow landslide formation, and the result of zoning is more objective and scientific. On the other hand, it provides spatial precision for disaster prevention.

2 STUDY AREA AND METHODS

2.1 *Study area*

The landslide hazard zonation was tested on the Magui town which is located at the southeast of china, with a total area 162 km^2 and the elevation ranging from 186 to 1627.3 m. (Figure 1). There are widespread mountains in the study area with complex geological conditions and strenuous structural movement, an area with serious landslides disasters which had a bad effect on the economic development in mountain area. Considering the hazard from landslides, there is a violent need to identify area which may be prone to landslide.

The study area possesses the typical surf-layer and bedrock geologic feature of the south China. The climate is typical subtropical and monsoonal, with hot and humid summers but mild and dry winters. Rainfall is heavy and occasionally intense during rainstorms and typhoons, with average annual precipitation varying from 1175 to 2090 mm, over the period of 1990–2015. induced by the "Van Asia than" typhoon, a major geological disaster resulting

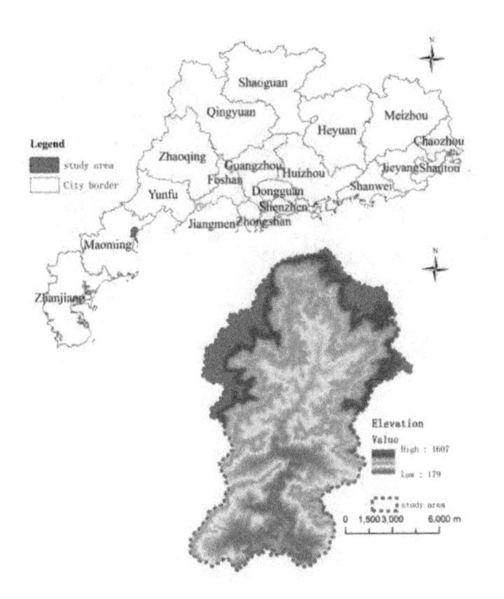

Figure 1. The study area.

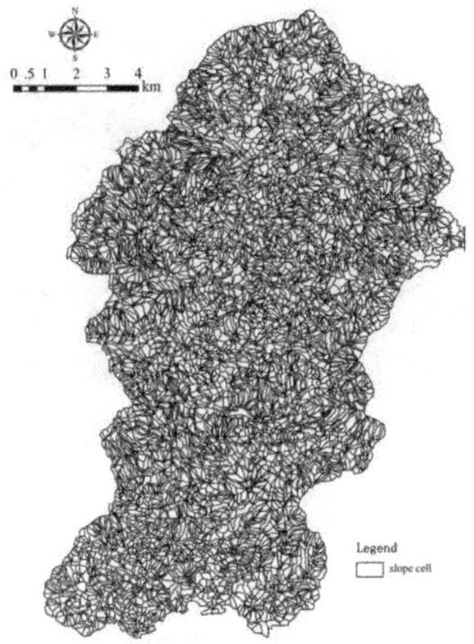

Figure 2. Slope unit division in study area.

in 73 deaths and 8337 houses collapsed occurred in the MaGui River Basin in September 21, 2010. Most of the traffic and water conservancy facilities were destroyed. Landslide type is dominated by small shallow landslide (Figure 3). Slip body thickness of about 1 m or so, most of the landslides from the ridge line to the direction of the valley slip. Landslide edge directly cut watershed. And the sliding surface of the landslide is bedrock. The loose layer of the mountain surface slides into the valley with the landslide. This is a typical small-scale regional rainstorm-induced landslide disaster.

2.2 *Methods*

2.2.1 *Hazard assessment index system*

Landslide hazard assessment is calculating the possibility of the occurrence of landslide hazard under the combined effect of various hazard factors. Based on the spatial distribution and geological environment characteristics of more than a thousand landslide hazards in the study area and surrounding areas in recent years, the principal component analysis is used to study the relationship between landslide hazard and geology, geomorphology, meteorology, hydrology and ecological humanities. The results show that the formation conditions of landslide disasters include geological structure, topography, and vegetation and human activities conditions. The topographyis the external condition of the formation of geological disasters, which restricts the development form and scale of the landslide disaster. Geological conditions are the inherent factors of the formation of geological disasters, which determine the degree and type of development of landslide disasters. Rainfall and human engineering activities are the triggering factors of landslides, which determine the speed and time of occurrence of landslide hazards. In the mountains of southern China, slopes are the main sites of human engineering activities. Human activities have changed the slope of the landforms, hydrological activities and the structural characteristics of rock and soil, destroyed the original slope of the balance, leading to the occurrence of landslide disasters. The main form of human activity in mountainous areas is the construction of houses and roads. So the intensity of construction intensity and the cutting slope by road was chosen for landslide hazard. There is no significant difference in rainfall and rock types due to the smaller area of the study area, so the rainfall and rock types are not listed separately. According to the comprehensive analysis, the paper chose the topographic conditions (topography relief, slope, slope aspect, slope shape), geological conditions (soil type, thickness of weathering layer) and human conditions (road density, construction) nine factor as risk evaluation indexes

Figure 3. Landslide disaster induced by the "Van Asia than" typhoon.

Table 2. Evaluation index system of shallow landslide hazard in mountain area.

Conditions	Indicator	Data source	Scale or resolution
Topography	Relief	DEM	1:10000
	Slope gradient	DEM	1:10000
	Slope aspect	DEM	1:10000
	Slope shape	DEM	1:10000
Geological structure	Soil type	Soil map	1:100000
	Weathering layer thickness	Field investigation and projections	1:10000
Human activities	Construction intensity	DOM	1:10000
	Road density	DOM	1:10000
Vegetation	Vegetation index	DOM	1:10000

(Table 2). These parameters were transformed into a spatial vector database, and the thematic layers were converted into a raster grid for application in the hazard assessment.

2.2.2 *Slope cell division*

The primary task of landslide hazard risk zoning is to determine the basic spatial units of the zoning. The slope cell was chosen the basic unit of the hazard zoning in this paper. Slope unit division method is to use the source cut method. Based on the DEM generated by 1:10000 topographic maps, the valley line and the ridge line are respectively extracted to generate the slope units (Figure 4). The division results are shown in Figure 2. The whole study area is divided into 13601slope units, and the landslide hazard assessment was carried out according to the different geological environment elements of every the slope unit.

2.2.3 *Landslide hazard assessment*

According to the classification and statistical results of landslide hazard assessment factors, the attribute table based on the 13601slope units was established in ARCGIS. The hazard value for each slope cell was obtained by weighted the nine factors of the risk analysis.

$$H_i = \sum WH_j P_{ij} (i = 1,2,\ldots 13601, j - 1,2,\ldots 10) \tag{1}$$

H_i the hazard value of i slope cell; WH_j the weight of the jindicator; P_{ij} the value of the j indicator of the i slope cell. In the evaluation of landslide hazard risk, it is necessary to determine the weight of each indicator (WH_j) in the risk assessment. Analytic hierarchy process was used to determine the weight of each factor in the paper. Firstly, the hierarchical mode was established according to the interrelationship of each factor. The evaluation indicators were divided into three levels. The first level is the general level of landslide hazard. The second layer includes topographic conditions, geological conditions, human conditions

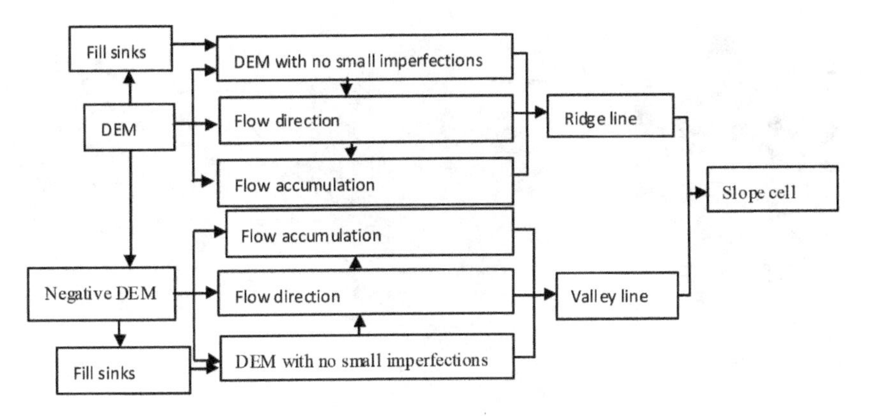

Figure 4. Slope cell partition flow chart.

Table 3. The weight of each indicator.

Indicator	Relief	Slope gradient	Slope aspect	Slope shape	Soil type	Weathering layer thickness	Construction intensity	Road density	Vegetation index
Weight	0.1	0.15	0.06	0.11	0.12	0.16	0.09	0.11	0.1

and land cover conditions. The third level for the sub-system specific nine evaluation indicators. The weight of each factor (Table 3) was determined according to the basic principles of analytic hierarchy process.

3 RESULTS

3.1 *Evaluation index system*

3.1.1 *Topographic conditions*

The terrain condition index was calculated based on DEM data to calculate. The slope value (slope aspect, relief or terrain curvature) of the study area was calculated using the spatial analysis tool in ARCGIS platform. Then the average slope (slope aspect, relief or terrain curvature) of each slope unit was calculated using the regional statistical function. The slope shape of each slope was identified based on the profile curvature of surface in direction of slope. The positive curvature value represents the convex slope, and the negative curvature value represents concave slope. The results were shown in the Figures 5–9.

3.1.2 *Geological structure*

The greater the thickness of the weathering layer, the worse the slope stability. The thickness of the weathering layer is related to the slope of the terrain, the lithology of the formation and the geological structure. According to the geologic lithologic structure of geological map, the study is divided into several regions. Accordance with the direction of the water, field sampling points was applied. The thickness of the weathered layer of each sampling point was measure during the manual auger and tape in the field. And the slope of each sampling point was record. Finally, the relationship between the slope and the thickness of the weathering layer was obtained based on regression analysis by using the data of the sampling point.

$$y = -2.864\ln(x) + 10.719$$

y is the thickness of the $J = \frac{A_J}{A}$ weathering and x is the slope value. The weathering layer thickness distribution can be deduced base the average slope of the unit. The thickness of weathering value is 0.5–16.3 m in the study area (Figure 10).

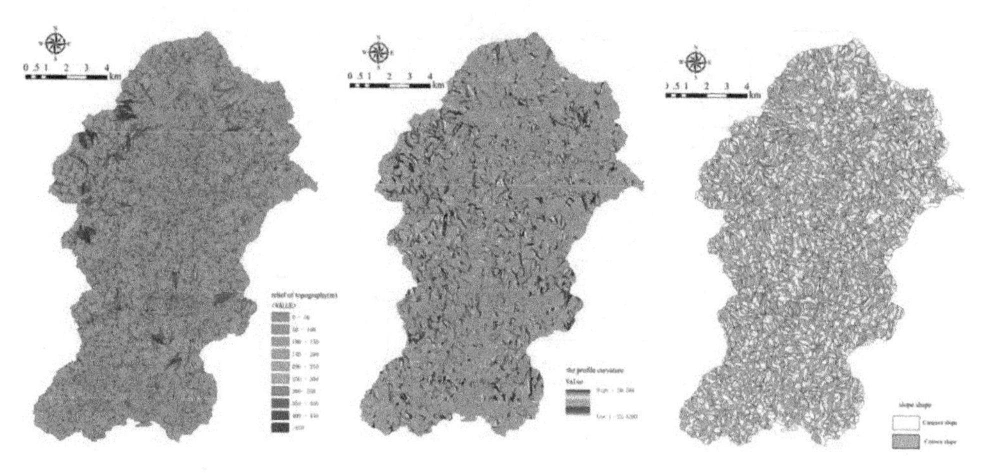

Figure 5. The terrain relief map.

Figure 6. Terrain curvature map.

Figure 7. Slope shape map.

3.1.3 *Human activities*

In the mountains of southern China, people must cut the slope to build house subject to terrain restrictions. The cutting slope affects the groundwater cycle and destroys the stability of the slope. Landslides hazard was highly correlated with the cutting height. The formula of construction intensity is:

$$J = \frac{A_J}{A}$$

J is construction intensity, A_J is the building area in the slope unit. A is the area of the slope unit (Figure 12). Similarly, road density was deduced as follow.

$$R = \frac{A_R}{A}$$

R is road density, A_R is the road area in the slope unit, and A is the area of the slope unit (Figure 11).

3.1.4 *Vegetation*

The vegetation index can reflect the vegetation coverage in the study area. The vegetation coverage is directly related to the stability of the mountain. The NDVI normalized vegetation index was extracted by remote sensing map in the study area. Then the vegetation coverage of each slope unit was calculated based on the regional statistics function in ARCGIS platform.

3.2 *Hazard zonation results*

The hazard value of 13601slope units was calculate during map algebra function on the ARCGIS platform. The hazard value range is 0–1. The slope units were divided into very high hazard degree zones, high hazard degree zones, medium hazard degree zones, low hazard degree zones and very low hazard degree zones at intervals of 0.2. The hazard zonation of each slope unit in study area is shown in Figure 7. In 162.16 km^2 of the studied area, the areas of the very high, high, medium, low and very low hazard degree area respectively were 47.18 km^2, 33.25 km^2, 22.65 km^2, 32.79 km^2 and 26.29 km^2. There are more slope units in very high-grade and high-grade, which is consistent with the disaster investigation and remote sensing identification in the study area (Table 4).

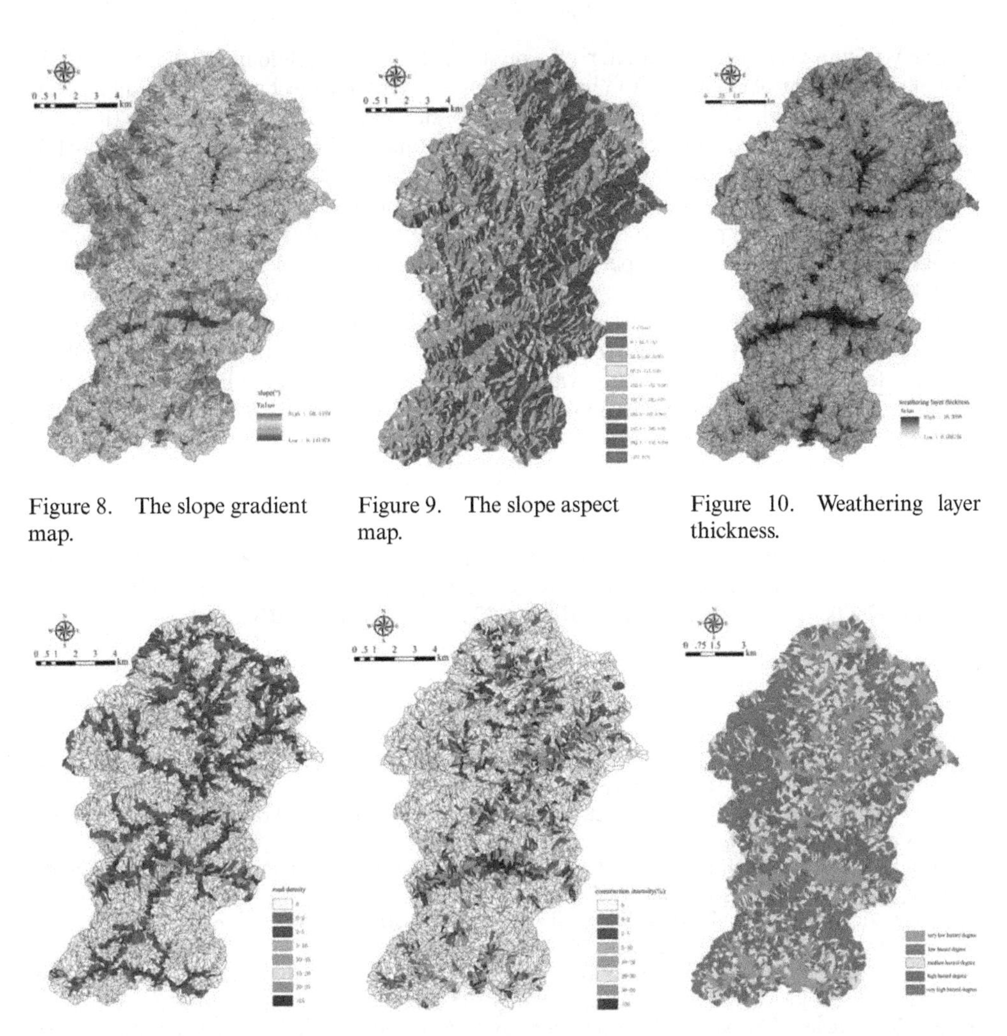

Figure 8. The slope gradient map.

Figure 9. The slope aspect map.

Figure 10. Weathering layer thickness.

Figure 11. Road density map.

Figure 12. Construction intensity map.

Figure 13. Landslide hazard zoning.

Table 4. Statistics on the proportion of different hazard degree.

Hazard degree	Area (km^2)	Proportion (%)
Very low	26.29	16.21
Low	32.79	20.22
Medium	22.65	13.97
High	33.25	20.51
Very high	47.18	29.09
Total	162.16	100.00

4 DISCUSSION AND CONCLUSION

The mountainous area in southern China is a heavy area with landslide hazard caused by heavy rain. It is characterized by lack of basic data and weak foundation of research. The aim of this paper is to study the hazard zonation method of shallow landslide hazard induced by heavy rain in view of the characteristics of small shallow landslide in southern China. In view of the geological environment characteristics and disaster mechanism of small shallow landslide in South China, this paper introduced the slope unit as the basic unit for the small shallow landslide hazard zonation. The use of slope units as a basic spatial unit for landslide hazard zoning can ensure the integrity of the disaster process and provide appropriate spatial scale and accurate spatial guidance for disaster prevention and emergency work.

Nine variables was imposed to create the landslide hazard risk evaluation map including topography relief, slope, slope aspect, slopeshape, soil type, thickness of weathering layer, road density, construction intensity and NDVIare chosen for landslide hazard zonation. The index system is scientific, systematic, operable and wide applicable. Based on the slope unit landslide hazard risk division method can be applied to mountain land use decision-making and disaster prevention and mitigation work. This methodhas broad application prospects in the field of disaster prevention decision-making.

ACKNOWLEDGMENTS

This paper is supported by Chinese National Natural Science Foundation (41671506) and the Science and Technology Planning Project of Guangdong Province (2013B020314003, 2013B030700005, 2014 A020218013, 2014 A020219006, 2015B070701020).

REFERENCES

Ermini, Leonardo, Catani, Filippo, Casagli, Nicola. Artificial Neural Networks applied to Landslide susceptibility assessment [J]. Geomorphology, 2005: 327–343.

Gao, Huaxi, Yinkun-Long. GIS-based spatial prediction of landslide hazard risk [J]. *Journal of Natural Disasters*, 2011, 20(01): 31–36.

Guzztti, F., Peruccacci, S., Rossi, M., et al. Rainfall thresholds for the initiation of landslides [J]. Meteorology and Atmospheric Physics, 2007, 98(3/4): 239–267.

Jin, Jiangjun, Pan, Mao, Li, Tiefeng. Regional Landslide Disaster Risk Assessment Methods [J]. *Journal of Mountain Science*, 2007, 25(02): 197–201.

Qiao, Jian-Ping, Shi, Li-Li, Wang, Meng. Landslide risk zoning based on the contributing weight stack method [J]. Geological Bulletin of China, 2008, 27(11): 1787–1794.

Tang, Y.M., Zhang, M.S., Xue, Q. Landslide risk assessment methods and flow on a large scale-A case study of loess landslides risk assessment in Yan'an urban districts, Shaanxi, China [J]. Geological Bulletin of China, 2011, 30(1): 166–172.

Tian, Shujun, Kong, Jiming. Risk Assessment of Landslide Based on Slope Unit and Highway Function [J]. *Journal of Mountain Science*, 2013, 31(05): 580–587.

Zhou, Guanhua, Zhuang, Wei, Chen, Yunhao, etc. RS/GIS-based regional evaluation of landslide hazard risk: A case study of GuangChang County, Jiangxi Province [J]. *Journal of Natural Disasters*, 2008, 17(06): 68–72.

Zhu, Liangfeng, Wu, Xincai, Yin, Kunlong, etc. Risk assessment of landslide in China using GIS technique [J]. Rock and Soil Mechanics, 2003, 24(S2): 221–224+230.

Risk Analysis and Management – Trends, Challenges and Emerging Issues – Bernatik, Huang & Salvi (Eds)
 ISBN 978-1-138-03359-7

A study on operational risk factors of hotels and building a risk management system in mountain scenic resorts

Yurong He
School of Tourism, Huangshan University, Huangshan, China

ABSTRACT: Hotels are important parts of tourist infrastructure in mountain scenic resorts. Because of the characteristics of the tourism industry and the geological and meteorological conditions of mountain-type scenic spots, hotels will face more unpredictable operational risks in the process of operation and management. Operating risk factor analysis is an important part of operation and management of hotels in mountain-type scenic spots. On the basis of the operating characteristics of hotels in mountain-type scenic spots, this paper puts some management measures to minimize the operational risk loss.

Keywords: Risk, operating risks, risk management, mountain scenic resorts, hotels

1 INTRODUCTION

Mountain-type scenic areas have a high ecological, ornamental, and scientific value. Mountain is a major landscape of resources and constitutive elements. There are 97 mountain scenic resorts in 208 state-level scenic spots in China. Hotels are important parts of tourist infrastructure in the mountain scenic resorts. Because of the characteristics of the tourism industry and the geological and meteorological conditions of mountain-type scenic spots, hotels will face more unpredictable operational risks in the business management process. Risk factors are a necessary condition of the risk formation. This paper identifies and analyzes the operating risk factors of hotels in mountain scenic resorts and then takes some measures to reduce the risk of loss. It is necessary for the sustainable development of hotels.

2 BASIC THEORY AND LITERATURE REVIEW

2.1 *Basic theory*

Risk refers to the possibility of the occurrence of a particular danger and the consequences of a combination of threats, and it is the uncertainty between the production purpose and the work achievement. Risk management originated in the United States in the 1930s. After the economic crisis, in 1929, people gradually realized the importance of risk management. The term "risk management" was coined by American scholar Glaller in 1952. The purpose of risk management is to minimize the adverse impact of the inevitable risks.

2.2 *Literature review*

Not only hotels but also other firms face risks. Since the 1990s, European and American scholars have studied the investment in hospitality industry and operational risk. These studies have focused on management strategy, management system, marketing, human resources, financial management, and other factors affecting the performance of hotels. Hsieh and

Oliveira (Hsieh et al., 2010; Oliveira et al., 2013) pointed out that products and services of hotels bring uncertainty to the business performance of the hospitality industry. Köseoglu et al. (2013) analyzed the influence of the corporate strategy of hotel-to-hotel performance and its operation. Sin (Sin LY et al., 2005) studied the relationship between market orientation and hotel performance. These studies mainly focused on the quantitative analysis of the factors concerned using data. Chinese scholars mainly focused on the qualitative analysis of hotel group consolidation's merger and acquisition risk, financial risk, cultural risk, liability risk, manpower risk, system risk, legal risk, property-type hotel investment risk, budget hotels, and so on (Diao, 2011; Li, 2005; Li et al., 2015; Lili, 2008; Luo, 2013; Yi, 2006). Studies conducted on the operation risk analysis and the risk management of hotels in mountain scenic resorts are very scarce.

3 OPERATING RISK FACTORS OF HOTELS IN MOUNTAIN-TYPE SCENIC AREAS

Operational risk of hotels refers to the uncertainty and the associated loss that hotels will face in the business management process (Luo, 2013), where the factors of economy, environment, market, and the enterprise itself will have a great influence on the hospitality industry. In mountain scenic resorts, hotels face risk factors, which may come from the external environment and the internal management. The analysis of the operating risk source is the premise and foundation of the operational risk management. On the basis of the source of the operational risk factors for classification, this paper analyzes the natural environment, the policy environment, the market environment, and the internal control risk factors of hotels in mountain-type scenic spots, as shown in Figure 1.

3.1 *Risk factors of the natural environment*

In mountain scenic resorts, hotels operate depending highly on the natural environment. Because of specific geological and weather conditions, there are many natural disaster risks in mountain scenic resorts. In mountain-type resorts, the factors of the natural disaster risk, which hotels will face in the process of operation, include the meteorological disaster risk (such as droughts, storms, floods, fog, lightning, wind, typhoon), the geological disaster risk (such as earthquake, landslide, collapse, debris flow, rock fall), and the biological disaster risk.

3.2 *Risk factors of the policy environment*

Risk factors of policy environment refer to the political and economic environment change and the change of exchange rate, which have an adverse impact on the operation of hotels in mountain scenic resorts. These adverse effects include the change of the country's financial, fiscal, and taxation policies, national policy factors, and the adjustment of tourism policy. All of these will bring uncertainty to hotels. For example, "eight rules," "six ban," and "no meeting in scenic area" measures proposed by the government bring "traffic decline"

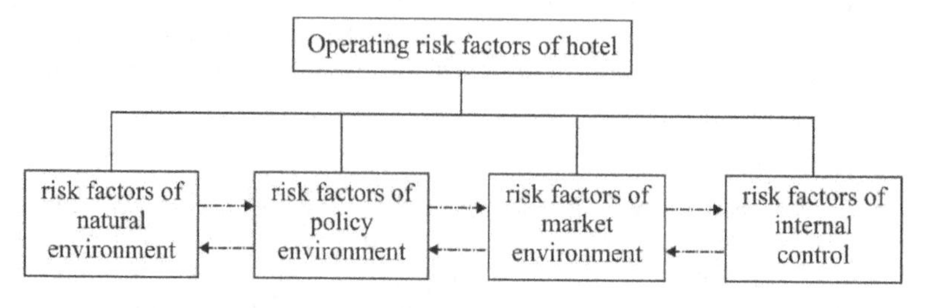

Figure 1. Operating risk of hotels in mountain-type resorts.

in different degrees to hotels in mountain scenic resorts as well as a great uncertainty to high-end reception investment of hotels.

3.3 *Risk factors of the market environment*

Market environment risk factors refer to the change of market positioning error and the main market segment and marketing decision-making errors that bring uncertainty and associated loss to hotels. The uniqueness of tourism resources of mountain-type resorts provides hotels a unique market advantage. For over a long period of time, hotels are in a great need of tourists in mountain-type resorts, where the probability of market environment risk factors is small.

3.4 *Risk factors of the internal control*

Changes of internal management factors of hotels, such as finance, organizational structure, organizational culture, and human resource, will directly affect business activities. The organizational structure of hotels is complex. Many different types of buildings and facilities are affected by careless management. The following types of risks are brought by improper internal controls. The first is the fire risk. Once fire occurs, the difficulty of rescue is higher in an average hotel because of special geographical environment conditions and the lack of social corresponding infrastructure. Because the forest coverage rate of mountain-type scenic forest is high, forest fire can occur easily. The second risk is the risk of human. Hotel is a labor-intensive industry. Personnel moral quality is not equal level. Moral factors may bring operation risk. The third risk is the risk of liability. The illegal stay and activities of hotel guests bring liability risk to hotels. In the era of information network, the hotel responsibility causes guest complaints, exposure, and fermentation by media, which bring loss to the hotel reputation. The fourth risk is the financial risk. In mountain scenic resorts, hotels often undertake more hospitality business. On the one hand, loose internal control will cause financial corruption crime. On the other hand, investment budget mistakes, improper accounting methods, and current accounting treatment caused by bad debt bring financial loss to hotels.

4 ANALYSIS OF HOTEL OPERATING RISKS IN MOUNTAIN SCENIC RESORTS

4.1 *Dual characteristics of risk*

Operating risks of hotels have the risk characteristics of mountain scenic resorts and the characteristics of general hotel operational risk. Because of mountain's fragile geological conditions, steep terrain, and weather conditions, hotels encounter the threat of natural risks in the business management process. Because of the effect of the fast development of tourism, hotels also face all kinds of uncertainty of judgment of physical assets risk, financial risk, human risk, and liability risk. Changeful policy, emergency, and disaster events bring risks to hotels.

4.2 *Natural risk is a great threat and has the characteristics of transitivity*

In mountain scenic resorts, natural disaster risks in the event that results are very serious. Natural disaster risk will also lead to the occurrence of other risk factors and cause a chain reaction. For example, drought is a common risk disaster in mountain scenic resorts. Drought is prone to fire risk and the lack of water to residents and tourists. It also brings down the scenic attraction, thereby reducing the traffic seriously. In mountain freeze–thaw areas, after melting, secondary disasters, such as aggravation of the weathering phenomenon, collapse of rocks, loose pour pile, stone gap, and increase of rift fracture, seriously exist. A serious pest destruction will affect the ecological environment and the natural landscape of mountain scenic resorts. After natural disaster risk, transportation difficulties cause shortages of raw materials, and hotels could not provide some services. Delay or cancellation of the reservation by customers brings losses to hotels. Sometimes, the occurrence of natural disaster risk will lead to the risk of devastation of hotels. For example, on 18 January 2017, a hotel was buried by avalanche in Italy.

4.3 *Seasonal and cyclical characteristics of risk factors*

In mountain scenic resorts, risk factors of hotels are seasonal and cyclic. These characteristics come from the cyclic nature of risks and the timeliness and seasonal products of hotels. Market demand is obviously seasonal, and there is a serious shortage of room occupancy rate in the off-season. During the peak season, there is a shortage in the availability of rooms, and average room price is much higher. Products and services of a hotel are almost confirmed. If the consistency of products and services of a hotel cannot be guaranteed, higher prices may cause guest complaints or bad review. Network issues may also bring loss to a hotel's brand and reputation.

5 BUILDING A MANAGEMENT SYSTEM OF OPERATIONAL RISK OF HOTELS

Risk management is the process to avoid and reduce the loss through effective monitoring, early warning, and risk, that is, decision-making risk. It also sums the rules of hotel operational risk prevention and finds treatment of operational risk to hotels in the mountain-type scenic spot. On the basis of the characteristics of mountain-type resorts and hotels, the following operational risk management system is built, as shown in Figure 2.

5.1 *Building up awareness of operational risk*

In mountain scenic resorts, operational risks of hotels have characteristics of long persistence, objectivity, and inevitability, so daily operational risk management must be paid attention. First, a full range of operational risk defense and early warning should be set up by all employees. Employees should be trained in operational risk management. Second, emergency management training should be strengthened. The training system of employees on special duty should be strictly implemented. Third, carrying out regular drills of emergency management plans and adjusting it with the real-time situations are necessary.

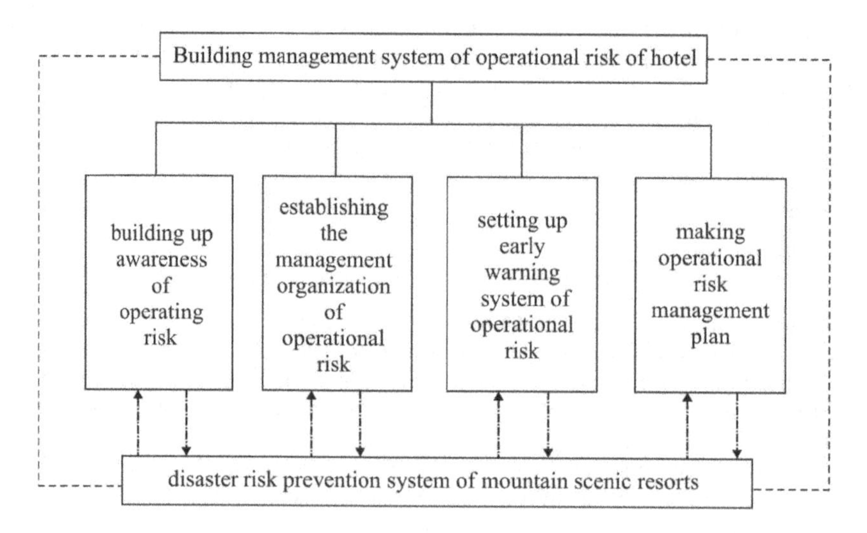

Figure 2. Management system of operational risks of hotels.

5.2 *Establishing the management organization of operational risk*

Management of hotel operational risk involves the aspects of natural disaster prevention and control department of mountain scenic resorts and hotels. In hotels, these aspects involve internal departments of hotel, grass-roots staff, senior management, and the board of directors. Each management main body should have certain responsibility. The management organization of operational risk needs to be set up. The management system of controlling risk will be formulated by it. The organization is responsible for the early warning of operational risk and the prevention and treatment of the planning and implementation.

5.3 *Setting up an early warning system of operational risk*

The process of operational risk early warning system is as follows. First, the management organization of operational risk collects the data of past operational risk loss of hotels and the meteorological and geological risk data of mountain-type resorts by using information technology. Second, the organization of operational risk analyzes and predicts the operational risk, which will occur or has occurred, and prepares charts of all kinds of operational risk loss. Third, the organization of operational risk should calculate the occurrence probability and loss degree of all kinds of risk and evaluate the normally expected loss, possible maximum loss, and maximum possible loss of all kinds of risks (Luo, 2014). Finally, the organization of operational risk should determine the level of all kinds of operating risk of hotels.

5.4 *Making an operational risk management plan*

After mastering the type and characteristics of different operational risk of hotels in mountain scenic resorts, a scientific, standardized, and feasible operating risk management plan will be set up on the basis of risk identification and rank. It would be specially mentioned that the management plan of operating risk of hotels must be included in the prevention and control planning of geological and meteorological disaster risk in mountain-type resorts. In addition, in the era of information network, when hotels treat some operational risk, which may cause a great social impact, the communication between the well and the media must be paid attention.

6 CONCLUSIONS

With the rapid expansion of tourism and the frequent change of the internal and external environments of hotels, more operational risk factors are faced by hotels in mountain scenic resorts. In the process of business, strengthening the internal management of hotels and paying more attention to the change of market and policy environment are of utmost importance to achieve the ultimate goal of operational risk management.

REFERENCES

Diao, Z. 2011. Risk Control and Management of Hotel Industry in China [J]. *Humanities & Social Sciences Journal of Hainan University*, 2011, 29(3): 61–66.

Hsieh, L.F., Lin L.H. 2010. A performance evaluation model for international tourist hotels in Taiwan-An application of the relational network DEA [J]. *International Journal of Hospitality Management*, 2010, 29(1): 14–24.

Köseoglu, M.A., Topaloglu, C., Parnell, J.A., et al. 2013. Linkages among business strategy, uncertainty and performance in the hospitality industry: Evidence from an emerging economy [J]. *International Journal of Hospitality Management*, 2013, 34(1): 81–91.

Li, H. 2005. Reunderstanding of the merging of tourist hotels in China [J]. *Hebei Normal University of Science & Technology* (Social Science), 2005(1): 36–37.

Li, Y., Qingning, He, Y. 2015. Internal influencing factors of hospitality operating risk [J]. *Journal of Beijing second foreign studies college*, 2015, 237(1): 41–49.
Lili, Chen X. 2008. Study on the Risk of Expansion Hotel and Avoiding Strategy [J]. *Hotel Modernization*, 2008(9): 34–36.
Luo, X. 2013. An analysis of probability estimation about hotel operational risk [J]. *Journal of Chifeng Univerity natural science*, 2013, 29(11): 80–81.
Luo, X. 2014. Comprehensive evaluation of the hotel operational risk [J]. *Journal of Hunan First Normal University*, 2014, 14(3): 107–109.
Oliveira, R., Pedro, M.I., Marques, R.C. 2013. Efficiency and its determinants in Portuguese hotels in the Algarve [J]. *Tourism Management*, 2013, 36(1): 641–649.
Sin LY, M., TseAC, B., Heung, V., et al. 2005. An analysis of the relationship between market orientation and business performance in the hotel industry [J]. *International Journal of Hospitality Management*, 2005, 24(4): 555–577.
Yi, A. 2006. Study on our domestic hotel group integration based on the risk management [J]. Market modernization, 2006 (8): 92–94.

Risk Analysis and Management – Trends, Challenges and Emerging Issues – Bernatik, Huang & Salvi (Eds)
© 2017 Taylor & Francis Group, London, ISBN 978-1-138-03359-7

Local betweenness for risk analysis in network systems

Xiao-Bing Hu, Ming-Kong Zhang & Hang Li
Academy of Disaster Reduction and Emergency Management, Beijing Normal University, Beijing, China

Jain-Qin Liao
MiidShare Technology Ltd., Chengdu, China

ABSTRACT: Betweenness is a fundamental, important network property for risk analysis of network systems. Basically, node/link betweenness indicates how many times a node/link appears as an intermediate node/link in all of those shortest paths between all OD (origin-destination) pairs of interest in a given network, and it is useful to assess how likely to what extent a disruptive event might impact on the transmitting efficiency of the network system. In this paper, we propose a new concept, local betweenness, which is fundamentally different from the traditional betweenness from two aspects: (i) Rather than the shortest paths between all OD pairs, it is all of those suitable paths whose length is within certain given ranges between a given OD pair that are considered; (ii) It is not how many times to appear in those suitable paths, but the ratio of how many times of appearance divided by the number of all suitable paths between a given OD pair that is calculated. As demonstrated by the preliminary experimental results, the proposed local betweenness can reveal more details about network topology, and therefore helps to make a more realistic and more precise analysis on the risk of a network system confronted with disruptive events.

1 INTRODUCTION

Network systems, as widely observed in both the nature and the man-made world, have a role of paramount importance to play in our daily life (Albert and Barabasi, 2002; Boccaletti, et al., 2006; Ball, 2012). The complexity of a network system is by no means the linear sum of the complexity of those components, and the safety and risk of a network system are far beyond the study and understanding of the safety and risk of each individual component. For example, when we try to assess the performance of a network system against some disruptive events (such as natural disasters, terrorist attacks, and component malfunction), we often find it difficult to deal with the amplifying effect, cascading effect and/or ripple effect between nodes (e.g., see Lerner, 2005; Mazzocchi, 2010), because there truly lack effective theories, models and methods to study such interactions between nodes (OECD, 2011; Helbing, 2013; Hu, et al., 2016a). Therefore, besides the traditional study on components of network systems, there is an urgent demand for research attention to be put on the performance of network systems against disruptive events, in order to more effectively address the safety and risk issues of network systems. This study attempts to shed a little more light on the risk analysis of network systems by revisiting a traditional network property, betweenness, which provides an important measure to assess how likely a disruptive event might impact on the transmitting efficiency of a given network system (Albert and Barabasi, 2002; Boccaletti, et al., 2006).

Basically, the traditional betweenness of a node/link in a given network system indicates how many times the node/link appears as an intermediate node/link in all of those shortest paths between all OD (origin-destination) pairs of interest in the network (Albert and

Barabasi, 2002; Boccaletti, et al., 2006). Apparently, betweenness plays a crucial role of defining the criticality of components in a network system. Roughly speaking, the larger the betweenness of a node/link, the more important the node/link to the transiting efficiency of the network. In other words, if a disruptive event stands a high chance to fail some nodes and/or links with large betweenness values, then the network system will have a high risk to have its transmitting efficiency degraded significantly by the disruptive event. However, is the traditional betweenness enough to measure the importance of a node/link to the transmitting efficiency of a network system? Or, it the traditional betweenness truly effective to assess the risk of the network system confronted with disruptive events? Unfortunately, as will be disclosed in this study, the concept of traditional betweenness has some obvious drawbacks: (i) it only considers the extreme situation of the shortest paths between OD pairs being impacted by disruptive events; (ii) the usage of the absolute number of how many times of appearance makes it difficult to conduct comparison between network systems of different scales. In Section 2, we will further explain why the traditional betweenness is not enough and therefore needs modification for the sake of risk analysis in network systems. In Section 3, we will mathematically propose the concept of local betweenness, and then discuss it in more depth. The usefulness of local betweenness will be demonstrated by some preliminary experimental results in Section 4. This paper will end with some conclusions and discussions for future research in Section 5.

2 THREE SITUATIONS FOR RISK ANALYSIS IN NETWORK SYSTEMS

With respect to risk analysis for network systems against disruptive events, relevant research results reported in literature often focus on two extreme situations: Situation 1, the network degrades into some separated sub-graphs because of disruptive events; Situation 2, the 1st best paths between those OD pairs of interest are cut off by disruptive events.

To study Situation 1, many theories, models and methods have been developed based on the concept of "degree of connectedness" (CND), which is defined as the number of nodes directly connected to a given node. Then, by analyzing the distribution of CND, it is found that most real-world network systems can be roughly classified into two categories: random networks with Poisson CND distribution and scale-free networks with power-law CND distribution (Albert and Barabasi, 2002; Boccaletti, et al., 2006). In terms of the possibility of network degradation, scale-free networks are robust to random disruptive events but vulnerable to intentional attacks to hub nodes, while random networks exhibit opposite performance against different disruptive events.

To study Situation 2, the concept of "betweenness" plays a crucial role. As discussed in Section 1, traditionally, node/link betweenness is defined as how many times the node/link appear as an intermediate node/link in all of those shortest paths (to be more precise, all of those 1st shortest paths) between all OD pairs of interest in the network. In general, the failure of a node/link with a higher betweenness means that more OD pairs of interest cannot be connected in the most cost-efficient way (but they are usually still connected by the network).

Then, are Situation 1 and Situation 2 enough to cover all of the demands for risk analysis in network systems? To answer this question, let us first look at a real life scenario. Assume a researcher needs to go from his hotel to a conference venue to give a presentation about his/her research work scheduled at 10:00 am. Here are three concerns s/he might have in mind before s/he sets off. Concern 1: How likely will the route network degrade due to disruptive events, so that there would be no path available for her/him to travel from the hotel to the conference venue? Concern 2: How likely will the 1st shortest path from the hotel to the conference venue be cut off by disruptive events, so that s/he could not travel at the most cost-efficient way? Concern 3: How likely will the route network be disrupted so that s/he could not get to the conference venue in time to given her/his presentation at 10:00 am? Clearly, Concern 1 is associated with Situation 1 largely addressed by CND based methods, and Concern 2 with Situation 2 by the traditional concept of betweenness. Compared the first two concerns, Concern 3 has much less been discussed. However, Concern 3 is apparently much more common and more important to the daily life of normal people. Therefore, besides Situation 1 and Situation 2, we need to study a third situation of risk analysis for network systems to address Concern 3.

To introduce the third situation of risk analysis for network systems in a more intuitive way, Figure 1 gives an illustration about all the three situations. From Figure 1, one can see clearly that the new situation, denoted as Situation 3 hereafter, is a more general situation than Situation 1 and Situation 2. Basically, Situation 3 can be described as that how likely all of those suitable paths whose length is within certain given length ranges between all OD pairs of interest might be cut off by disruptive events. If such given length ranges cover all possible paths between all OD pairs of interest, then Situation 3 becomes Situation 1. If such given length ranges are defined based on the lengths of shortest paths between OD pairs of interest, then Situation 3 becomes Situation 2.

As a more general situation of risk analysis for network systems, Situation 3 deserves at least the same amount of attentions as the other two extreme situations from researchers. Unfortunately, this is not the case, and there actually even lack relevant basic network properties which are especially defined for or explicitly linked to Situation 3. By extending the traditional concept of betweenness, we have recently proposed a generalized betweenness particularly for the study of Situation 3 (Hu, et al., 2016c). Basically, the generalized betweenness of a node/link indicates how many times the node/link appears as intermediate node/link in all of those suitable paths whose length is within certain given length ranges between all OD pairs of interest. If a disruptive event stands a high chance to shut down more nodes and/or links with higher values of generalized betweenness, then the system has a higher risk of having its suitable paths between OD pairs been impacted by disruptive events.

Thus, is generalized betweenness enough to study Situation 3 for risk analysis in network systems? Now, let us look at the illustration of Figure 2.(a), where we are concerned about the first 2 shortest paths (highlighted as red lines) between two given OD pairs. In the example of Figure 2.(a), node 1 and node 2 have the same value of generalized betweenness, i.e., value of 2. Then, here comes the question: which node, node 1 or node 2, is more important in terms of ensuring each of the two given OD pairs to be connected by at lease a path whose length is shorter than the 3rd shortest path between the associated OD pair? Node 1 has a generalized betweenness of 2, because both the first 2 shortest paths between O1 and D1 pass node 1. If node 1 is shut down by disruptive events, O1 and D1 can no longer be connected by any path with a satisfactory length. Node 2 also has a generalized betweenness of 2,

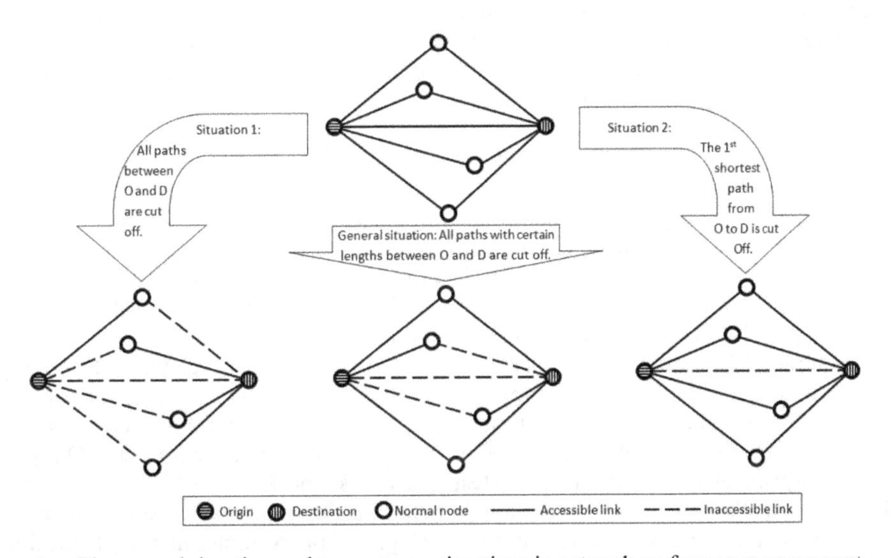

Figure 1. The general situation and two extreme situations in network performance assessment.

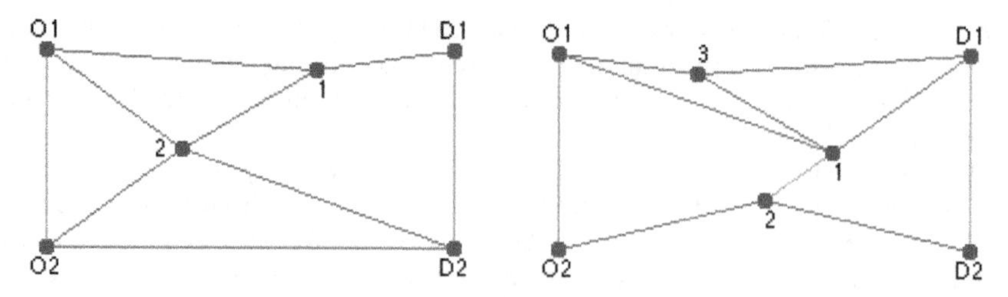

Figure 2. The difference (a) between GB and LB (left subplot), and (b) between ALB and RLB (right subplot).

because node 2 is passed by one the first 2 shortest paths between O1 and D1 and one of the first 2 shortest paths between O2 and D2. If node 2 is shut down by disruptive events, both of the two given OD pairs will still be connected by at least one path with a satisfactory length. Therefore, node 1 is more important and deserves more resources for reinforcement than node 2 in terms of ensuring each of the two given OD pairs to be connected by at lease one path with a satisfactory length. Clearly, the Generalized Betweenness (GB) proposed in Hu, et al., 2016c, fails to distinguish the importance of node 1 and node 2 in the example of Figure 2.(a). This paper will further propose a new betweenness, so called *Local Betweenness* (LB), which will fairly well tell the difference in importance of node/link in Situation 3 for risk analysis in network systems.

3 LOCAL BETWEENNESS

3.1 *The basic idea of local betweenness*

As explained in Section 2, particularly as illustrated in Figure 1 and Figure 2.(a), (i) the concept of traditional betweenness cannot address the issue of risk analysis for network systems in Situation 3, which is however a more general and more realistic situation in our daily life; (ii) the concept of generalized betweenness reported in Hu, et al., 2016c, fails to distinguish the importance of nodes/links in terms of ensuring every OD pair of interest to be connected by at least one path with satisfactory length after disruptive events.

The failure of the generalized betweenness of Hu, et al., 2016c, roots in the fact that it is defined based on all those suitable paths between all OD pairs of interest. In other words, the generalized betweenness, just like the traditional betweenness, is a global definition/concept, which inherently ignores local details such as whether a specific OD pair could have all of its suitable paths cut off by disruptive events. The concept of local betweenness proposed in this paper, as a local definition/concept, will pay a particular attention to such local details which the generalized betweenness ignores.

Definition 1: For a given OD pair, the *local betweenness* of a node/link indicates how many times the node/link appears as intermediate node/link in all of those suitable paths whose length is within certain given ranges between the given OD pair.

Clearly, for a specific node/link, it may have a different local betweenness value for each OD pair of interest in the network system. In other word, assuming the network system has N_N nodes, then a node may have up to $(N_N - 1) \times (N_N - 2)$ local betweenness values, and a link may have up to $N_N \times (N_N - 1)$ local betweenness values.

With the local betweenness given by Definition 1, we can clearly distinguish the importance of node 1 and node 2 in the example of Figure 2.(a). The maximal local betweenness value of node 1 is 2, because both the first 2 shortest paths between O1 and D1 pass node 1, while the maximal local betweenness value of node 2 is just 1, because for either OD pair of interest, only one of the 2 associated shortest paths passes node 2.

However, the number of suitable paths between an OD pair may be different from the number of suitable paths between another OD pair. For example, in the illustration of

Figure 2.(b), node 1 has a maximal local betweenness value of 2, because 2 of the 3 suitable paths (highlighted as red lines) between O1 and D1 pass node 1, and node 2 has a maximal local betweenness value of 1, because only 1 of the 1 suitable path between O2 and D2 passes node 2. Therefore, in terms of ensuring each OD pair of interest to be connected by at least one suitable path after disruptive events, node 2 is clearly more important than node 1, although node 2 has a smaller maximal local betweenness value than node 1. Therefore, we have the following modified definition of local betweenness, which is called *Relative Local Betweenness* (RLB). For distinguishing purposes, Definition 1 is called *Absolute Local Betweenness* (ALB) hereafter.

Definition 2: For a given OD pair, suppose a node/link has an ALB value of b_{ALB}. Suppose the number of all suitable paths between the given OD pair is N_{SP}. Then, the *Relative Local Betweenness* (RLB) of a node/link for the given OD pair is defined as:

$$b_{RLB} = b_{ALB} / N_{SP} \tag{1}$$

Based on Definition 2, we know that, in the example of Figure 2.(b), node 1 has a maximal RLB value of 0.5, while node 2 has a maximal RLB value of 1. Thus, we can tell that, although node 2 has a smaller ALB value than node 1, node 2 is actually more important than node 1 in terms of ensuring every OD pair of interest to be connected by as least one suitable path after disruptive events.

3.2 *Mathematical description of local betweenness*

For the purpose of calculation, we further give the mathematical description of Generalized Betweenness (GB), Absolute Local Betweenness (ALB) and Relative Local Betweenness (RLB). Assuming a route network $G(V,E)$ is composed of node set V and link set E. V has N_N different nodes, and E has N_L links between nodes. This route network can be recorded as an $N_N \times N_N$ adjacent matrix A. The matrix entry $A(i,j) = 1$, $i = 1, \ldots, N_N$ and $j = 1, \ldots, N_N$, defines a link from node i to node j. Otherwise, $A(i,j) = 0$ means no link. There is a cost, i.e., $C(i,j)$, associated with each link $A(i,j)$, and $C(i,j)$ will be used to calculate the length of a path between an OD pair. Suppose there are N_{OD} OD pairs to be considered, and O_h and D_h are the origin and destination of the hth OD pair, $h = 1, \ldots, N_{OD}$. Theoretically, for a directed network with N_N node, there could be up to $N_{OD} = N_N \times (N_N - 1)$ OD pairs of interest. Between an OD pair, say, the hth OD pair, suppose a feasible/candidate path is recorded as an integer vector whose element $P(i) = j$ means node j is the ith node in the path, $i = 1, \ldots, N_P$, and $j = 1, \ldots, N_N$, where N_P tells how many nodes, including O_h and D_h, are included in the path. Thus, the length of path P is:

$$f(P) = \sum_{i=1}^{N_P - 1} C(P(i), P(i+1)) \tag{2}$$

Suppose there are $N_{GR,h}$ given ranges for the hth OD pair, $h = 1, \ldots, N_{OD}$, let $[R_{GR}(h,i,1), R_{GR}(h,i,2)]$ denote the ith given range, $i = 1, \ldots, N_{GR,h}$ and:

$$R_{GR}(h,i,2) < R_{GR}(h,i+1,1) \tag{3}$$

If a path connects an OD pair of interest and the length of the path is within one of such given ranges, i.e., if there exist a $1 \le h \le N_{OD}$ and $1 \le i \le N_{GR,h}$, such that:

$$P(1) = O_h \; and \; P(N_P) = D_h \tag{4}$$

$$R_{GR}(h,i,1) \le f(P) \le R_{GR}(h,i,2) \tag{5}$$

then we call the path a suitable path.

Suppose for the hth OD pair, the number of all suitable paths is $N_{SP}(h)$. And suppose we know a node/link i appears as intermediate node/link for n_{ALB} times in all of the $N_{SP}(h)$ suitable paths of the hth OD pair. Then, according to Definition 1, we record the ALB value of node/link i for the hth OD pair as:

$$b_{ALB}(h,i) = n_{ALB} \tag{6}$$

Thus, the GB value of node/link i can be calculated as:

$$b_{GB}(i) = \sum_{h=1}^{N_{OD}} b_{ALB}(h,i) \tag{7}$$

And, the RLB value of node/link i can be calculated as:

$$b_{RLB}(h,i) = b_{ALB}(h,i)/N_{SP}(h) \tag{8}$$

Based on $b_{RLB}(h,i)$, we can further calculate GB in a relative sense as following:

$$b_{RGB}(i) = \sum_{h=1}^{N_{OD}} b_{RLB}(h,i) \tag{9}$$

Since each node/link i has N_{OD} values of ALB or RLB, it is often useful to conduct statistic analysis on those N_{OD} values, in order to better understand the importance of node/link i. Basically, we can have the mean and maximal of those N_{OD} values:

$$b_{ALB}^{-}(i) = \sum_{h=1}^{N_{OD}} b_{ALB}(h,i)/N_{OD}, b_{ALB}^{*}(i) = \max_{h=1,\ldots,N_{OD}} b_{ALB}(h,i) \tag{10}$$

$$b_{RLB}^{-}(i) = \sum_{h=1}^{N_{OD}} b_{RLB}(h,i)/N_{OD}, b_{RLB}^{*}(i) = \max_{h=1,\ldots,N_{OD}} b_{RLB}(h,i) \tag{11}$$

In general, the larger the $b_{ALB}^{-}(i), b_{RLB}^{-}(i), b_{ALB}^{*}(i)$, or $b_{RLB}^{*}(i)$ value, the more important node/link i.

3.3 *Methods to calculate local betweenness*

To calculate the local betweenness of a node/link, we must find out all of those suitable paths between a given OD pair in the first place. This can be converted into the k shortest paths problem (k-SPP) with an uncertain k value. Basically, the traditional k-SPP aims to find out the first k best paths between a given OD pair under a specified k value. Based on the traditional k-SPP, we conduct an iteration process as follows, where the value of k keeps increasing until all suitable paths between a given OD pair have been found.

For $h = 1{:}1{:}\ N_{OD}$
  $L_{\max} = 0$; $k = 1$
  While $L_{\max} \le R_{GR}(h, N_{GR,h}, 2)$ or not all paths between O_h and D_h have been explored.
  Calculate the k shortest paths between O_h and D_h; Let P^*_k denote the kth shortest path.
  $L_{\max} = f(P_k^*); k = k + 1$
  End
Calculate the local betweenness of each node/link for the hth OD pair;
End

The traditional k-SPP has long been studied and many effective methods have been reported (Yen, 1971; Eppstein, 1998; Aljazzar and Leue, 2011; Mohanta, 2012; Hu, et al.,

2016b). Therefore, we can simply borrow such a method for the traditional k-SPP, and integrate it into the core loop of While-End.

Resolving a traditional k-SPP is often time-consuming, particularly when the network has a large scale. Calculating local betweenness for all OD pairs of interest makes the thing even worse, because we need to resolve not one, but N_{OD} k-SPPs. Therefore, it is important choose or design a computationally efficient method for the traditional k-SPP, and readers may refer to relevant studies, such as the ripple-spreading algorithm reported in Hu, et al., 2016b, which has a good potential of performing parallel computation to find out all suitable paths between all OD pairs by just a single run. However, the computational efficiency for the traditional k-SPP is beyond the scope of this paper.

4 A PRELIMINARY CASE STUDY

Here we give a simple case study to illustrate how different the proposed Local Betweenness (LB) is from either the Classical Betweenness (CB) in Albert and Barabasi, 2002, or the Generalized Betweenness (GB) in Hu, et al., 2016c. In this case study, we generate a simple network with $N_N = 20$ and $N_L = 80$ (see the left subplot in Figure 3). Here, a suitable path between an OD pair is a path whose length is no more than 20% larger than the length of the 1st shortest path between the OD pair. The normalized values of 3 kinds of betweenness for each node and link are given in the right subplot of Figure 3, which clearly shows that LB is different from either CB or GB. For example, the maximal normalized values of CB, GB and LB appear at 3 different nodes, which means, if we can invest on only 1 node due to limited resources and we need to reinforce the most important node, then, CB, GB and LB will suggest different nodes. Therefore, the proposed local betweenness can provide a new angle for risk analysis in network systems.

5 CONCLUCIONS

Different from the traditional betweenness in Albert and Barabasi, 2002, or the generalized betweenness in Hu, et al., 2016c, this paper proposes a new network property, so called local betweenness, in order to address the issue of risk analysis for network systems in a general situation where all of those suitable paths whose length is within certain given ranges between all node pairs of interest are concerned with. The difference between the traditional betweenness, the generalized betweenness and the new local betweenness is discussed in depth, some

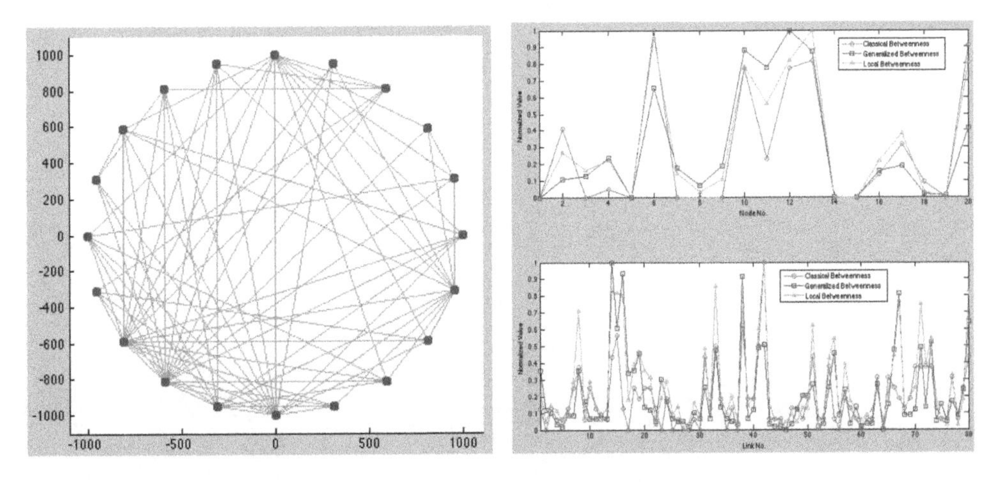

Figure 3. Experimental results of preliminary case study.

mathematical descriptions of local betweenness are given, and the method of calculating the local betweenness is described. A preliminary case study is presented to demonstrate the usefulness of the proposed local betweenness for risk analysis in network systems.

ACKNOWLEDGEMENTS

This work was supported in part by the National Key Research and Development Programme (Grant No. 2016YFA0602404), and the National Natural Science Foundation of China (Grant No. 61472041).

REFERENCES

Albert, R., Barabasi, A.L. (2002). "Statistical mechanics of complex networks", *Reviews of Modern Physics*, 74: 47.

Aljazzar, H., Leue, S. (2011) "K: A heuristic search algorithm for finding the k shortest paths", *Artificial Intelligence*, 175(18): 2129–2154.

Ball, P. (2012). Why Society is a Complex Matter, Springer.

Boccaletti, S., Latora, V., Moreno, Y., Chaves, M., Hwang, D.U. (2006). "Complex networks: Structure and dynamics", *Phys. Rep.*, 424: 175.

Eppstein, D. (1998). "Finding the k shortest paths," *SIAM Journal on Optimization*, 28(2): 652–673.

Helbing, D. (2013). "Globally networked risks and how to respond", Nature, doi:10.1038/nature12047.

Hu, X.B., Gheorghe, A.V., Leeson, M.S., Leng, S.P., Bourgeois, J. (2016a). "Risk and Safety of Complex Network Systems," *Mathematical Problems in Engineering*, vol. 2016, Article ID 8983915.

Hu, X.B., Wang, M., Leeson, M.S., Di Paolo, E., Liu, H. (2016b). "Deterministic agent-based path optimization by mimicking the spreading of ripples," *Evolutionary Computation*, 24(2): 319–346.

Hu, X.B., Zhang, M.K., Liao, J.Q., Zhang, H.L. (2016c). "A Ripple-Spreading Algorithm for Network Performance Assessment", *2016 IEEE Symposium Series on Computational Intelligence (IEEE SSCI 2016)*, 5–9 Dec 2016, Athens, Greece.

Lerner, E.J. (2005). What's wrong with the electric grid? Gravitational, Electric, and Magnetic Forces: An Anthology of Current Thought, 41.

Mazzocchi, M., Hansstein, F., Ragona, M. (2010). "The 2010 volcanic ash cloud and its financial impact on the European airline industry," *CESifo Forum*, 92–100.

Mohanta, K. (2012). "Comprehensive Study on Computational Methods for K-Shortest Paths Problem," *International Journal of Computer Applications*, 40(14): 22–26.

OECD (2011). Future Global Shocks-Improving Risk Governance, OECD Reviews of Risk Management Policies.

Yen, J.Y. (1971). "Finding the k shortest paths in a network," *Management Science*, 17(11): 712–716.

Risk Analysis and Management – Trends, Challenges and Emerging Issues – Bernatik, Huang & Salvi (Eds)
 ISBN 978-1-138-03359-7

Towards a general risk assessment on the transmitting efficiency of network systems

Xiao-Bing Hu, Ming-Kong Zhang & Hang Li
Academy of Disaster Reduction and Emergency Management, Beijing Normal University, Beijing, China

Jain-Qin Liao
MiidShare Technology Ltd., Chengdu, China

ABSTRACT: Basically, the transmitting efficiency of a network system means how efficiently the network can facilitate the travelling between nodes. When disturbances (such as external attacks and/or internal malfunction) are concerned, the transmitting efficiency might drop significantly due to possible failures of nodes and links in the network. Therefore, risk assessment on the transmitting efficiency is crucial to design, develop and improve network systems. Traditionally, such risk assessment can be classified into two streams: (i) how likely the network might degrade into some separated sub-graphs under disturbances, and (ii) how likely the shortest paths between nodes might be cut off by disturbances. Differently, this study introduces a general risk assessment focusing on how likely all those paths whose length is within certain specified ranges might be impacted by disturbances. To this end, a general situation of risk assessment on the transmitting efficiency of network systems is discussed, a new network property, i.e., generalized betweenness, is proposed, and a general methodology is developed. Actually, risk assessments on network degradation and the shortest paths are just two special cases of the new situation, which therefore has more potentials of applying to study various real world network systems.

1 INTRODUCTION

Many complex systems in the nature and our society can be described as network composed of many nodes and links. Nowadays, network systems play an unprecedented important role in our daily life, and network performance assessment against disturbances (such as natural disasters, terrorist attacks, and component malfunction) has long been a hot topic in various research and engineering communities (Albert and Barabasi, 2002; Boccaletti, et al., 2006; Ball, 2012). The challenge of network performance assessment largely roots in the fact that the impact of a local disturbance event is usually not limited to the local area where the disturbance event occurs, but will easily spread out through the network due to the amplifying effect, cascading effect and/or ripple effect at nodes (e.g., see Lerner, 2005; Mazzocchi, 2010). As emphasized by OECD (2011), Helbing (2013) and Hu, et al., (2016a), studying the performance of network systems against disturbances is a big challenge as well as a big opportunity, many traditional approaches can hardly address this challenge, and therefore, to address network related risk, there is an urgent demand for new theories, models and methods to assess the performance of a network system against disturbances.

One of the most fundamental functions of network systems is to facilitate the travelling between nodes (Albert and Barabasi, 2002; Boccaletti, et al., 2006). For example, a route network enables people to travel between cities, a communication network allows information to transmit among cell phones, and a decision tree illustrates how input states evolve to output states. Therefore, transmitting efficiency has long been a topic of paramount importance to design and analysis of network systems. Basically, it is always desirable that a network could connect all nodes in the most efficient way. Achieving a high transmitting efficiency is not all the story about network topology. An established network system is always subject to various disturbances, which could shut down certain nodes and/or links. If not property designed, the transmitting efficiency of a network system, although initially at a satisfactorily high level, might drop significantly after a few nodes and/or links are malfunctioned by disturbances. Therefore, in reality it is crucial to analyze how the transmitting efficiency of a network system could change due to disturbances. In other words, achieving a robust transmitting efficiency against disturbances makes a realistic sense. Risk assessment on the transmitting efficiency of network systems aims to develop effective theories and methods to answer how likely the transmitting efficiency might drop to what extent due to what kind of disturbances.

Basically, research on risk assessment of network transmitting efficiency against disturbances mainly focuses on two extreme situations. Situation 1, the network degrades into some separated sub-graphs because of disturbances; Situation 2, the 1st best paths between those pairs of Origin and Destination (OD) of interest are cut off by disturbances. "Degree of connectedness" (CND), indicating how many nodes are connected to a given node, is a key concept for addressing the risk assessment in Situation 1. Based on CND, many theories, models and methods have been developed, which have promoted unprecedented advances in the study on network performance against disturbances in the last few decades (Albert and Barabasi, 2002; Boccaletti, et al., 2006). For instance, one of the most important findings in system science is that the CND distribution of most real-world complex networks significantly deviates from a Poisson distribution but has a power-law tail or a scale-free property, and a scale-free network is robust to random disturbances but vulnerable to intended attacks to hub nodes, i.e., attacking a few hub nodes can easily degrade a scale-free network into some separated sub-graphs, which means some OD pairs of interest might be isolated from each other. "Betweenness" is an important concept to study the risk in Situation 2, which indicates how many times a node/link appears as an intermediate node/link in all of those 1st shortest paths between all OD pairs of interest. Obviously, the failure of a node/link with a higher betweenness means that more OD pairs of interest cannot be connected in the most cost-efficient way (but they are usually still connected by the network). Differently, this paper will discuss a more general situation for risk assessment on network transmitting efficiency against disturbances, and also propose an effective methodology in order to accomplish the risk assessment in such a general situation.

2 A GENERAL SITUATION

As discussed in Section 1, when risk assessment on the transmitting efficiency of a network system is concerned, existing studies often focus on two extreme situations. Situation 1: how likely might the network degrade into some separated sub-graphs? In other words, how likely might all paths between some node pairs of interest be cut off by disturbances? As illustrated by the left subplot of Figure 1, if all of the 5 paths between the given origin and destination nodes (O and D) are cut off, then the network will fail the purpose of connecting the given OD pair. The risk assessment of Situation 1 aims to analyze how likely all of the 5 paths in Figure 1 might be cut off. In the other extreme situation, i.e., Situation 2, it is concerned about how likely the shortest paths between all OD pairs of interest might be cut off due to disturbances. For example, in the right subplot of Figure 1, if the shortest path between the given OD pair is cut off, then the most cost-efficient way of connecting the given OD pair will become unavailable. The risk assessment of Situation 2 focuses on the possibility of the most cost-efficient way being impacted.

Different from either Situation 1 or Situation 2, this paper is concerned with a more general situation: between all OD pairs of interest, how likely might all of those paths whose length/cost is within certain given ranges be cut off by disturbances. For example, in the middle subplot of Fig. 1, there are 3 paths between the given OD pair satisfying the length

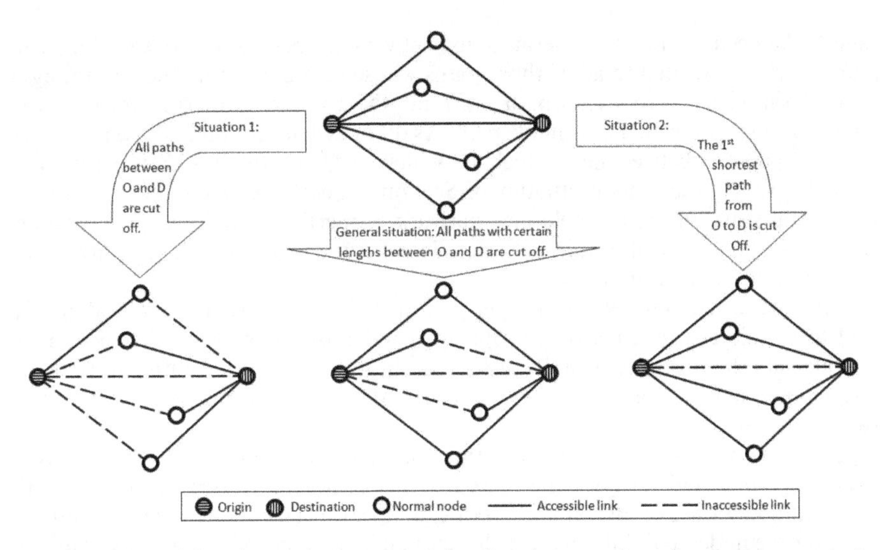

Figure 1. The general situation and two extreme situations in network performance assessment.

requirements, and then we would like to know the possibility for such 3 paths to be cut off simultaneously by disturbances.

So far, this general situation has barely been touched in existing studies on risk assessment of network transmitting efficiency against disturbances. One might argue that this general situation might be less important than those two well-studied situations, i.e., Situaton1 and Situation 2. However, this argument is largely doubtable if we take the following scenario as an example. For a traveller who needs to go from O to D for a scheduled meeting, there are 3 questions to concern about. Question 1: How likely will the route network become apart so that there is no way to travel from O to D? Question 2: How likely will the 1st shortest, or the 1st most cost-efficient path between O and D become unavailable? Question 3: How likely will it turn out, due to disturbances to route network, there is no way to get to D in time, say, before 10:00am? Obviously, it is Question 3 that is often more concerned by normal people in our daily life. The above scenario and Question 3 can be extended to many areas of daily life, e.g., how likely is there no investment plan to achieve the minimal return ratio due to market uncertainties? Obviously, the general situation introduced in this paper is, if not more important than, at least as important as Situation 1 and Situation 2. Actually, Situation 1 and Situation 2 are just two special cases of the general situation. Figure 1 clearly illustrates the differences and relationships between the two extreme situations and the general one.

Then, why has this general situation been barely studied in existing literatures on risk assessment of network systems? This is largely because it is very difficult, if not impossible, for existing methods, particularly those methods for Situation 1 and Situation 2, to find out all (not just some) of those paths whose length/cost is within the given ranges. For example, in the middle subplot of Figure 1, if we have found out all of those 3 paths which satisfy the given length requirements, then it will actually be easy to analyze the possibility for such 3 paths to be cut off simultaneously by disturbances. The true difficulty is: In the first place, how can we find out such 3 paths, and how can we know for sure such 3 paths are exactly all of those which satisfy the given length requirements?

3 A GENERAL METHODOLOGY

3.1 *A general framework*

To assess the risk on the transmitting efficiency of a network system against disturbances in the general situation, there are basically three stages.

Stage 1: We need to choose or define some network properties or indexes which can well reflect or be closely related to all of those paths whose length is within the given ranges. For the sake of simplicity, hereafter, a path satisfying the given length requirements between a node pair of interest is called as suitable path. As discussed in Section 2, those network properties, e.g., CND and betweenness, which are widely used in Situation 1 and Situation 2, are not suitable to study the general situation of Section 2. Therefore, to conduct risk assessment in the general situation, reasonable new network properties or indexes are demanded. In the next subsection, we will propose a generalized betweenness to serve the purpose of risk assessment in the general situation.

Stage 2: Once a relevant network property or index is chosen, we need to find an effective method to calculate such a network property or index on the scale of the entire network, i.e., to calculate the property or index value for all of those suitable paths. In particular, if we design a new network property, then we have to develop a new method to accomplish the relevant calculation.

Stage 3: Once we have found out all of those suitable paths, and have got their associated network property or index values, then we can easily conduct risk assessment by referring to various traditional ways of risk analysis. For example, we can conduct hazard simulation to analyze how the nodes and links in all of those suitable paths could be impacted, and how the associated network property or index values might change under hazards.

Clearly, the challenge largely lies in the above Stage 1 and Stage 2, and introducing innovative ideas and concepts in Stage 1 and Stage 2 holds the key to address the risk assessment in the general situation.

3.2 *A generalized betweenness*

Here, we introduce a novel network property which is closely related to all of those suitable paths in a network system. This new network property is called as generalized betweenness. As discussed in Section 1, the traditional betweenness is used to study Situation 1. Here, the generalized betweenness is particularly proposed for studying the general situation of Section 2. The detailed definition of generalized betweenness is given as following.

Definition: The generalized betweenness of a node/link indicates how many times the node/link appears as an intermediate node/link in all of those paths whose length is within certain given ranges between all OD pairs of interest in the network.

We further give a mathematical description of generalized betweenness. Assuming a route network $G(V,E)$ is composed of node set V and link set E. V has N_N different nodes, and E has NL links between nodes. This route network can be recorded as an $N_N \times N_N$ adjacent matrix A. The matrix entry $A(i,j) = 1$, $i = 1,\ldots, N_N$ and $j = 1,\ldots, N_N$, defines a link from node i to node j. Otherwise, $A(i,j) = 0$ means no link. There is a cost, i.e., $C(i,j)$, associated with each link $A(i,j)$, and $C(i,j)$ will be used to calculate the length of a path between a pair of Origin and Destination (OD). Suppose there are N_{OD} OD pairs to be considered, and O_h and D_h are the origin and destination of the hth OD pair, $h = 1,\ldots, N_{OD}$. Theoretically, for a directed network with N_N node, there could be up to $N_{OD} = N_N \times (N_N - 1)$ OD pairs of interest. Between an OD pair, say, the hth OD pair, suppose a feasible/candidate path is recorded as an integer vector whose element $P(i) = j$ means node j is the ith node in the path, $i = 1,\ldots, N_P$, and $j = 1,\ldots, N_N$, where N_P tells how many nodes, including O_h and D_h, are included in the path. Thus, the length of path P is:

$$f(P) = \sum_{i=1}^{N_P - 1} C(P(i), P(i+1)). \tag{1}$$

Suppose there are $N_{GR,h}$ given ranges for the hth OD pair, $h = 1,\ldots, N_{OD}$, let $[R_{GR}(h,i,1), R_{GR}(h,i,2)]$ denote the ith given range, $i = 1,\ldots, N_{GR,h}$, and:

$$R_{GR}(h,i,2) < R_{GR}(h,i+1,1) \tag{2}$$

If a path connects an OD pair of interest and the length of the path is within one of such given ranges, i.e., if there exist a $1 \le h \le N_{OD}$ and $1 \le i \le N_{GR,h}$, such that:

$$P(1) = O_h \text{ and } P(N_P) = D_h \tag{3}$$

$$R_{GR}(h,i,1) \le f(P) \le R_{GR}(h,i,2) \tag{4}$$

then we call the path a suitable path. Thus, the generalized betweenness of a node needs to count how many times the node appears, not as $P(1)$ or $P(N_P)$, in all of those paths satisfying Conditions (3) and (4), and the generalized betweenness of a link, how many times the link appears in all of those suitable paths.

If for every $1 \le h \le N_{OD}$, $N_{GR,h} = 1$ and:

$$R_{GR}(h,1,1) = 0, \text{ and } R_{GR}(h,1,2) = \infty \tag{5}$$

then, the general situation becomes Situation 1, i.e., all paths between all OD pairs are concerned.

If for every $1 \le h \le N_{OD}$, $N_{GR,h} = 1$ and $R_{GR(h,1,1)} = -\infty$,

$$f(P^*_{h,1}) \le R_{GR}(h,1,2) < f(P^*_{h,2}) \tag{6}$$

where $P^*_{h,1}$ and $P^*_{h,2}$ are the 1st and 2nd shortest paths from O_h to D_h, respectively, then, the general situation becomes Situation 2, i.e., only the 1st shortest paths between all OD pairs are concerned.

3.3 *To find out all of those suitable paths*

To calculate the generalized betweenness, we must find out all of those suitable paths in the first place. The methods for the traditional betweenness only calculate the 1st shortest paths, and they are obviously useless in the calculation of generalized betweenness. In this study, we have to turn to those methods for the *k* shortest paths problem (*k*-SPP) (Yen, 1971; Eppstein, 1998; Aljazzar and Leue, 2011; Mohanta, 2012; Hu, et al., 2016b). Basically, the *k*-SPP aims to find out the first *k* best paths between a given OD pair. Here, we develop a method based on the *k*-SPP, in order to find out all of those suitable paths.

For $h = 1{:}1{:}\, N_{OD}$
  $L_{\max} = 0$; $k = 1$
  While $L_{\max} \le R_{GR}\left(h, N_{GR,h}, 2\right)$ or not all paths between O_h and D_h have been explored.
  Calculate the k shortest paths between O_h and D_h; Let P^*k denote the kth shortest path.
  $L_{\max} = f(P^*_k)$; $k = k + 1$
  End
End

Calculating the *k* shortest paths for a given OD pair in the loop of "While-End" part above is the core. Figure 2 gives an example of finding all suitable paths from the left-bottom node to the right-top node by calculating the 4 shortest paths between the OD pair. As discussed in (Hu, et al. 2016b), classical *k*-SPP methods are not computationally efficient, as they have to keep reconstructing route networks in order to find the *k*th shortest path based on the (*k*–1) shortest paths found. For such classical *k*-SPP methods, finding the *k* shortest paths between a single OD pair is already time-consuming, let alone for all of the $N_{OD} = N_N \times (N_N - 1)$ OD pairs. Therefore, more effective methods to find all suitable paths are still needed.

A recently reported nature-inspire method, so called *Ripple-Spreading Algorithm* (RSA), could possibly bring an innovative solution. The natural ripple spreading phenomenon reflects an optimization principle: a ripple spreads at the same speed in all directions, so it always reaches the closest spatial point first. This very simple principle can be easily applied to accomplish Path Optimization Problems (POPs) effectively, and several Ripple-Spreading

Algorithms (RSAs) were developed to resolve various POPs, including the k-SPP (Hu, et al., 2016b). Most existing POP methods are centralized, top-down, logic-based search algorithm. Differently, RSAs are actually decentralized, bottom-up, agent-based simulation model. By defining the behavior of individual nodes, optimality will automatically emerge as a result of the collective performance of the model. Simply speaking, RSA simulates a ripple relay race in route network. The race starts with an initial ripple at the origin; As the ripple spreads and reaches a node, a new ripple may be triggered at the node under certain condition; a ripple will become inactive when it has reached all the ends of links of its epicenter node; All active ripples keep spreading and more and more ripples are triggered at spatially farther away nodes; the relay race will terminate when certain criteria (e.g., the destination has been reached by a ripple) is met. As an agent-based model, RSA has a good capability of performing parallel computation, which means a computational efficiency. Actually, for a given OD pair, the RSA can find the k shortest paths by a single run on the original network, no need of reconstructing any route networks at all (Hu, et al., 2016b). In this paper, we modify the RSA simply by letting N_{OD} initial ripples start from all origin nodes simultaneously, and then, by a single run of RSA, all suitable paths between all OD pairs will be identified. Due to limited space, here we only give an example in Figure 3 to illustrate how the modified RSA works efficiently. Readers may refer to Hu, et al., (2016c) for more details about the modified RSA.

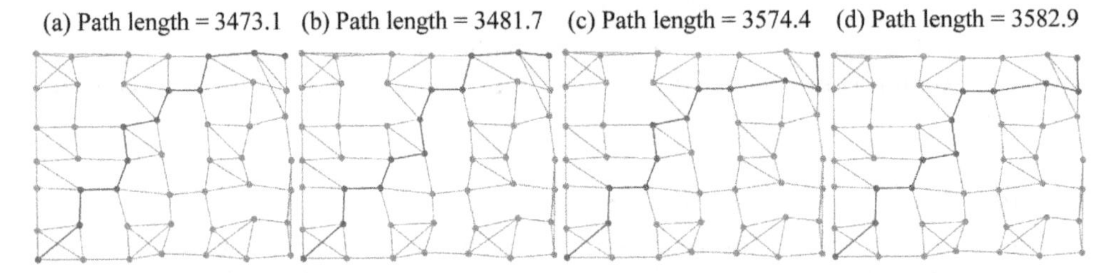

Figure 2. All of the 4 paths whose length is within the range [0, 3583] in a network.

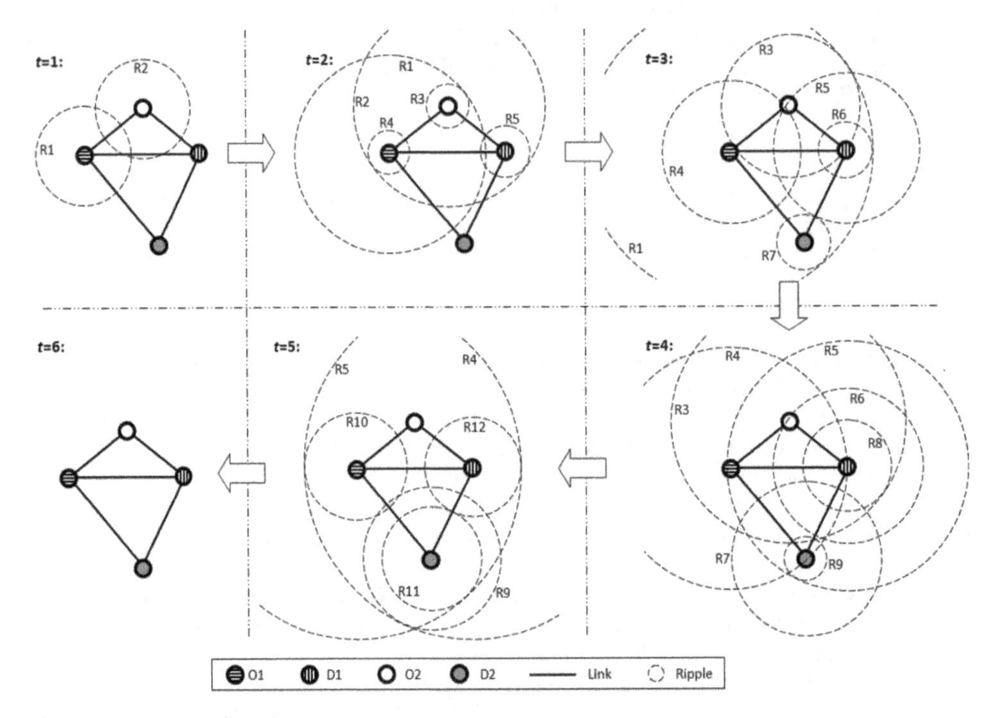

Figure 3. Finding all suitable paths between all OD pairs by RSA.

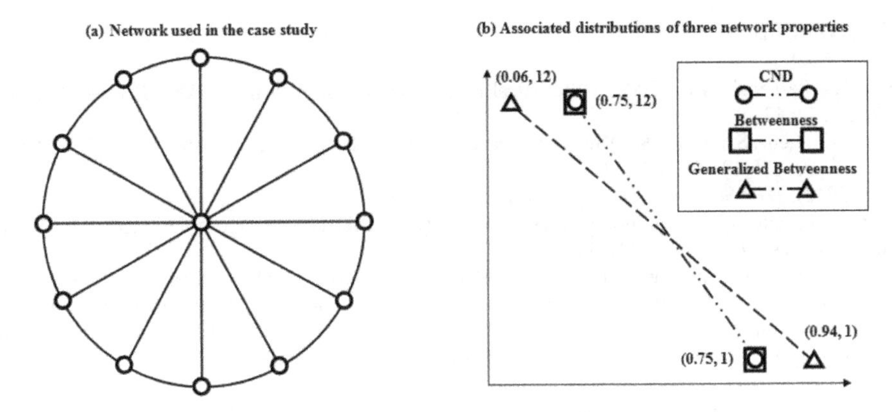

Figure 4. The distributions of 3 different network properties in the case study.

4 A PRELIMINARY CASE STUDY

Here we give a simple case study to illustrate how different the proposed generalized betweenness is from traditional network properties, i.e., CND and betweenness, used in Situation 1 and Situation 2, respectively. The network system used is plotted in the left subplot of Figure 4. Obviously, it has a scale-free topology. The distributions or CND, betweenness and generalized betweenness are plotted in the right subplot of Figure 4 as circle points, square points and triangle points, respectively. The distributions of three network properties are normalized along x axis (the axis of network properties) for comparative purposes. In this case study, when the generalized betweenness is concerned, the first 2 shortest paths between each node pair are considered for the sake of simplicity. From Figure 4, one can see that the central hub node has the same importance in terms of either CND or betweenness, while a greater importance in terms of the generalized betweenness, which means, if standby or alternative paths are needed, then the central hub node is worth of investing more resources to reinforce it against disturbances. This case study clearly shows that the general situation and the generalized betweenness offer a new angle which may help to deepen our understanding of network performance against disturbances, and thus, to improve relevant risk assessment.

5 CONCLUCIONS

This paper focuses on the risk assessment of network transmitting efficiency in a general situation, i.e., how likely those paths whose length is within certain given ranges between all node pairs of interest might be impacted to what extent by disturbances. The general situation of risk assessment is discussed in depth, a new network property, i.e., generalized betweenness, is introduced to gain a better understanding of the vulnerability of network systems, a general methodology, particularly based on ripple-spreading algorithm, is developed to identify all suitable paths in the general situation, and a preliminary case study is presented to show that the proposed general risk assessment makes good sense, and therefore is worth further efforts in future research, including theoretical improvements and real-world case studies.

ACKNOWLEDGEMENTS

This work was supported in part by the National Key Research and Development Programme (Grant No.2016YFA0602404), and the National Natural Science Foundation of China (Grant No. 61472041).

REFERENCES

Albert, R., Barabasi, A.L. (2002). "Statistical mechanics of complex networks", *Reviews of Modern Physics*, 74: 47.

Aljazzar, H., Leue, S. (2011) "K: A heuristic search algorithm for finding the k shortest paths", *Artificial Intelligence*, 175(18): 2129–2154.

Ball, P. (2012). *Why Society is a Complex Matter*, Springer.

Boccaletti, S., Latora, V., Moreno, Y., Chaves, M., Hwang, D.U. (2006). "Complex networks: Structure and dynamics", *Phys. Rep.*, 424: 175.

Eppstein, D. (1998). "Finding the k shortest paths," *SIAM Journal on Optimization*, 28(2): 652–673.

Helbing, D. (2013). "Globally networked risks and how to respond", Nature, doi:10.1038/nature12047.

Hu, X.B., Gheorghe, A.V., Leeson, M.S., Leng, S.P., Bourgeois, J. (2016a) "Risk and Safety of Complex Network Systems," *Mathematical Problems in Engineering*, vol. 2016, Article ID 8983915.

Hu, X.B., Wang, M., Leeson, M.S., Di Paolo, E., Liu, H. (2016b). "Deterministic agent-based path optimization by mimicking the spreading of ripples," *Evolutionary Computation*, 24(2): 319–346.

Hu, X.B., Zhang, M.K., Liao, J.Q., Zhang, H.L. (2016c). "A Ripple-Spreading Algorithm for Network Performance Assessment", *2016 IEEE Symposium Series on Computational Intelligence (IEEE SSCI 2016)*, 5–9 Dec 2016, Athens, Greece.

Lerner, E.J. (2005). What's wrong with the electric grid? Gravitational, Electric, and Magnetic Forces: An Anthology of Current Thought, 41.

Mazzocchi, M., Hansstein, F., Ragona, M. (2010). "The 2010 volcanic ash cloud and its financial impact on the European airline industry," *CESifo Forum*, 92–100.

Mohanta, K. (2012). "Comprehensive Study on Computational Methods for K-Shortest Paths Problem," *International Journal of Computer Applications*, 40(14): 22–26.

OECD (2011). Future Global Shocks-Improving Risk Governance, OECD Reviews of Risk Management Policies.

Yen, J.Y. (1971). "Finding the k shortest paths in a network," *Management Science*, 17(11): 712–716.

Risk Analysis and Management – Trends, Challenges and Emerging Issues – Bernatik, Huang & Salvi (Eds)
 ISBN 978-1-138-03359-7

An approach to assess an integrated risk caused by two hazards

Chongfu Huang
State Key Laboratory of Earth Surface Processes and Resources Ecology, Beijing Normal University, Beijing, China
Academy of Disaster Reduction and Emergency Management, Ministry of Civil Affairs and Ministry of Education, Beijing, China
Faculty of Geographical Science, Beijing Normal University, Beijing, China

ABSTRACT: In this paper, we define an integrated probability risk of multihazards as the expected value of disaster, which is determined by a joint probability distribution and a disaster function. To overcome the associated difficulty, that is, with a small sample, it is impossible to theoretically construct a valid distribution and a valid function, we develop the information diffusion technique for their construction. The two-dimensional normal diffusion and the three-dimensional normal diffusion were employed to construct a discrete joint probability distribution and a discrete disaster function, respectively.

1 INTRODUCTION

In the latest United Nations' Framework for Disaster Risk Reduction 2015–2030 (UNISDR, 2015), international financial institutions, such as the World Bank and regional development banks, proposed to consider the priorities for providing financial support and loans for *integrated disaster risk reduction*. The pursued goal is to prevent new and reduce existing disaster risk through the implementation of *integrated* and inclusive economic, structural, legal, social, health, cultural, educational, environmental, technological, political, and institutional measures that prevent and reduce hazard exposure and vulnerability to disaster, increase the preparedness for response and recovery, and thus increase the resilience. The framework continues the idea appeared over the past 20 years to integrate multidisciplines (Munns et al. 2003; Sekizawa and Tanabe 2005) and the integrated databases (Fedra, 1998) to provide a systematic overview of the sources of risks or hazards. In the framework, "integrated risk assessment" has been developed to be *integrated disaster risk reduction*. In a sense, the bureaucrats are more interested in how to get more resources to reduce disaster risk rather than to know what is integrated risk caused by multihazards such as earthquake and flood.

It is impossible to find the better method between the integrated risk assessment and the classical risk assessment before application. Few studies are conducted to distinguishing between *integrated "risk assessment"* and *"integrated risk" assessment*. Hence, there is no common approach to assess the integrated risk of multihazards.

In this paper, from the viewpoint of probability risk, we define the integrated risk of multihazards. Reviewing some efforts in theoretically combining random variables, we develop the information diffusion technique to construct a discrete probability distribution and a discrete disaster function to assess an integrated risk caused by two hazards.

2 INTEGRATED RISK CAUSED BY MULTIHAZARDS

Risk, similar to a ghost, is a scene in the future that is associated with some adverse incident (Huang and Ruan, 2008). If we can accurately predict the scene, it is called a pseudo-risk. In the case, it is not a wandering ghost but a familiar thing. For example, if a person falls to the ground from a plane flying at a height of 500 m without parachute, he will die. Evidently, it is a pseudo-risk. About 77% of the risk definitions are suggested with the probability, which is employed to measure random uncertainty. It implies that most risk analysts are interested in probability risks, which can statistically predict by using probability models with more data. For example, there are powerful probability models and more data to study traffic incidents. For accident insurance, the traffic incident is a probability risk.

There are three methods to profile a probability risk: the first is risk = (event, loss, probability) generally in a tabular format; the second is risk matrix, which can be constructed in two ways; and the third is risk curve (real probabilistic loss distribution). Particularly, when the connotation of a probability risk is defined as the expected value of disaster, the risk is represented as:

$$Risk = \int_{u_0}^{u_1} f(x)p(x)dx \tag{1}$$

Where $p(x)$ is the probability distribution about an index x measuring adverse event, such as earthquake or flood, $f(x)$ is the disaster function caused by x, u_0 is the minimum of x that could cause a disaster, and u_1 is the maximum of x that would occur. For example, if x is employed to measure the magnitude of earthquake occurred in China, u_0 is 4.5 and u_1 is 8.5.

Extending equation (1), we formally define an integrated probability risk of multihazards as follows:

Definition 1 Let $x_1, x_2, \ldots, x_n$ be n random variables for measuring n hazards. Let $p(x_1, x_2, \ldots, x_n)$ be the joint probability distribution of the n random variables, $f(x_1, x_2, \ldots, x_n)$ be the disaster function caused by the n hazards, and $u_0^{(xi)}$ and $u_1^{(xi)}$ be the minimum and maximum of the ith hazard, respectively, $i = 1, 2, \ldots, n$. The expected value of disaster is called the integrated risk caused by n hazards, which is represented as:

$$Risk = \int_{u_0^{(x_1)}}^{u_1^{(x_1)}} \int_{u_0^{(x_2)}}^{u_1^{(x_2)}} \cdots \int_{u_0^{(x_n)}}^{u_1^{(x_n)}} f(x_1, x_2, \cdots, x_n)p(x_1, x_2, \cdots, x_n)dx_1 dx_2 \cdots dx_n \tag{2}$$

It is necessary to verify whether a probability distribution and a disaster function are reliable before they are employed to assess a probability risk. For example, we suppose that a set of observations of earthquake magnitude x and loss y is recorded in the past T years. The set as a sample provides statistical data to estimate the probability distribution of the magnitude of the earthquake, x, occurred, denoted as $p(x)$, and to estimate the loss function as a relationship between x and loss y, denoted as $y = f(x)$. Several models are recommended for estimating probability distributions, and much of them are suggested for estimating input–output relation functions. Some have been demonstrated with practical effects, and others would be proven by using mathematical theory.

However, when the type of the population from which the sample is drawn is unknown, it is impossible to precisely estimate the underlying probability distribution. When the size of the sample of input–output observations is small, it is difficult to reasonably estimate the underlying input–output relation function.

Particularly, to multivariate random variables $x_1, x_2, \ldots, x_n$, it is more difficult to estimate $p(x_1, x_2, \ldots, x_n)$ and $f(x_1, x_2, \ldots, x_n)$ with traditional statistical methods.

Despite the difficulties to estimate joint probability distributions, many researchers are working hard to do that. A key limitation of the existing parametric methods is that the distribution needs to be assumed a priori. Evidently, this is a strong assumption, because the form of the distribution is frequently unknown. Even the nonparametric models still make some assumptions regarding the form of the distribution. In addition, the nonparametric methods have other limitations (Alghalith, 2016; Talamakrouni et al, 2016).

The current popular approach is "copulas". Copulas are functions that fully define the multivariate distribution of a random vector or a set of random variables (Nelsen, 2006). They link or join multivariate distributions functions of random variables to their univariate marginal distributions. According to Sklar's theorem (Sklar, 1959; Salvadori et al., 2007), any multivariate joint distribution can be written in terms of a copula function and marginal distribution functions. However, the existence of a copula function does not mean that we can get it. There are several copula models to construct a copula function with given marginal distribution functions, such as Gaussian copulas and Archimedean copulas. There are many families of Archimedean copulas, such as Frank, Gumbel, Clayton, which are uniparametric (Montes-Iturrizagaand Heredia-Zavoni, 2016). Estimating marginal distribution functions and finding an applicable copula function are extremely difficult.

To estimate a disaster function, $f(x_1, x_2, \ldots, x_n)$, the coarsest method is multiple linear regression, and the more popular method is bio-inspired computing, such as neural networks and the leaping frog and bat algorithms. All of these algorithms try to replicate the way biological organisms and suborganism entities operate to achieve a high level of efficiency, even if sometimes the actual optimal solution is not achieved (Kar, 2016). However, no bio-inspired algorithm is a practical universal algorithm. For example, it has been theoretically proved that multilayer neural networks using arbitrary squashing functions can approximate any continuous function to any degree of accuracy, provided enough hidden units are available (Hornik et al., 1989). However, the neural networks has the problem of becoming trapped in local minima, which may lead to a failure in finding a global optimal solution (Marco and Alberto, 1992). Besides, the convergence rate of the algorithm is still too low even if learning can be achieved. When a trained neural network is performing as a mapping from input space to output space, it is a black box. This means it is not possible to understand how a neural system works, and it is very difficult to incorporate human a priori knowledge into a neural network.

It is clear that an integrated probability risk caused by multihazards can be regarded as the expected value of disaster excited by a multivariate random variable and a multivariate disaster function. Also, it is not easy to estimate a joint probability distribution and a disaster function with given sample.

There is no loss in generality when we suppose that $n = 2$ in equation (2). In this case, the integrated risk caused by two hazards could be represented as:

$$Risk = \int_{u_0(x_1)}^{u_1(x_1)} \int_{u_0(x_2)}^{u_1(x_2)} f(x_1, x_2) p(x_1, x_2) dx_1 dx_2 \tag{3}$$

When we discuss a bivariate random variable with two components x_1, x_2, their joint probability distribution:

$$P(x_1, x_2) = \Pr(X_1 \le x_1, X_2 \le x_2) \tag{4}$$

is defined by the probability of a product event. In a random trial, events $X_1 \le x_1$ and $X_2 \le x_2$ simultaneously occur.

Obviously, the probability that the two types of disasters, earthquake and flood, simultaneously occur in a region is almost zero. In other words, there is no bivariate random sample in the strict sense in terms of respective two hazards.

To understand the phenomena, let us consider an experiment of tossing a dice and a coin at the same time. We will assign an indicator random variable to the result of tossing the coin. If it turns head, we assign 1 to the variable, and if it turns tail, we assign 0 to the variable. Consider the following random variables:

X_1: The number of dots appearing on the dice.
X_2: The sum of the number of dots on the dice and the indicator for the coin.

Strictly speaking, the probability that the dice and the coin land at the same time is almost zero. In the strict sense of "simultaneously occur", the experiment cannot give any bivariate random sample. In practice case of a random experiment, X_1 and X_2 are recorded after a trial. Extending the time length of a trial to a year, we can regard the disaster events occurred in a year that occurred simultaneously.

Let W be a sample of observations on disaster D caused by two hazards S and Z in T years. We write the sample as follows:

$$W = \{(s_1, z_1, d_1), (s_2, z_2, d_2), \cdots, (s_T, z_T, d_T)\} \quad (5)$$

where s_i and z_i represent the magnitudes of hazards S and Z in ith year and d_i is the disaster caused by the hazards in the year.

An integrated probability risk caused by S and Z is considered as the expected value of disaster caused by the hazards. According to equation (3), we know that it is necessary to estimate a joint probability distribution of S and Z, $p(s,z)$, and a disaster function, $d = f(s,z)$. If the given sample, W, does not come from too many years, we employ the information diffusion technique to estimate $p(s,z)$ and $f(s,z)$ with the sample.

3 INFORMATION DIFFUSION TECHNIQUE

The concept of information diffusion (Huang, 1997) was proposed in function learning from a small sample of data (Huang and Moraga, 2004). The approximate reasoning of information diffusion was used to estimate probabilities and fuzzy relationships from scant, incomplete data for grassland wildfires (Liu et al., 2010). The simplest models of the information diffusion technique are the linear information distribution and the normal diffusion. The latter is more convenient to use. Mathematically, the normal diffusion can be illustrated in fuzzy set as follows (Huang, 2002).

Let $W = \{w_i \mid i = 1,2,\ldots,m\}$ be a given sample and let $U = \{u\}$ be its universe. The function in equation (6) is called a normal diffusion function:

$$\mu(w.u) = \exp\left[-\frac{(w-u)^2}{2h^2}\right], w \in W, u \in U. \quad (6)$$

The diffusion coefficient h can be calculated by using equation (7) (Huang, 2012):

$$h = \begin{cases} 0.8146\,(b-a), & m = 5; \\ 0.5690\,(b-a), & m = 6; \\ 0.4560\,(b-a), & m = 7, \\ 0.3860\,(b-a), & m = 8; \\ 0.3362\,(b-a), & m = 9; \\ 0.2986\,(b-a), & m = 10; \\ 2.6851(b-a)/(m-1), & m \geq 11. \end{cases} \quad (7)$$

where $b = \max\{w_i\}$ and $a = \min\{w_i\}$:

Using a diffusion function, $\mu(w,u)$, we can change a given sample point w into a fuzzy set with membership function $\mu_w(u) = \mu(w,u)$ on universe U. The principle of information diffusion guarantees that there are reasonable diffusion functions to improve the nondiffusion estimates when the given samples are incomplete.

When we employ the normal diffusion to estimate a probability distribution, it is just the same as the Gaussian kernel estimate. It implies that the Gaussian kernel connects to some simple diffusion without the birth–death phenomenon. However, the coefficient from the kernel theory is both nonexplanatory and rough. When the size m of a given sample is small, the method of normal diffusion is superior with respect to almost any distribution.

When we employ the information diffusion technique to estimate an input–output relation function with a given sample, the first requirement is to construct an information matrix with the sample.

Let:

$$X = \{(x_i, y_i) \mid i = 1, 2, \cdots, m\} \tag{8}$$

Be an r-dimensional random sample with input x_i and output y_i, and the input and output monitoring spaces of X be U and V, respectively, which are denoted as:

$$\begin{cases} U = \{u_j \mid j = 1, 2, \cdots, J\} \\ V = \{v_k \mid k = 1, 2, \cdots, K\} \end{cases} \tag{9}$$

The monitoring spaces serve for diffusing with some steps. For a sample point (x_i, y_i) and a monitoring point (u_j, v_k), with a diffusion function μ, we can obtain a value $\mu((x_i, y_i), (u_j, v_k))$, called *diffused information* on (u_j, v_k) from (x_i, y_i), denoted as q_{ijk}.

Let:

$$Q_{jk} = \sum_{i=1}^{m} q_{ijk} \tag{10}$$

where

$$Q = \begin{matrix} & v_1 & v_2 & \cdots & v_K \\ u_1 & Q_{11} & Q_{12} & \cdots & Q_{1K} \\ u_2 & Q_{21} & Q_{22} & \cdots & Q_{2K} \\ \cdots & \cdots & \cdots & \cdots & \cdots \\ u_J & Q_{J1} & Q_{J2} & Q_{J3} & Q_{JK} \end{matrix} \tag{11}$$

is called an information matrix of X on $U \times V$.

Then, according to the characteristic of the information matrix, we can change it to be a fuzzy relationship matrix. With an appropriate approximate reasoning operator, we can estimate the input–output relation function with respect to the given sample in equation (8).

4 JOINT DISTRIBUTION CONSTRUCTED BY TWO-DIMENSIONAL NORMAL DIFFUSION

From the sample in equation (5), we can have a sample of observations on hazards S and Z in T years. We describe the sample in equation (12):

$$W_1 = \{(s_1, z_1), (s_2, z_2), \ldots, (s_T, z_T)\} \tag{12}$$

We employ S and Z to denote universes of the two hazards, respectively:

$$\begin{cases} S = \{u_j \mid j = 1, 2, \cdots, J\} \\ Z = \{v_k \mid k = 1, 2, \cdots, K\} \end{cases} \tag{13}$$

The two-dimensional normal diffusion of W_1 on $S \times Z$ is defined in equation (14):

$$\mu((s_i, z_i), (u_j, v_k)) = \exp\left[-\frac{(S_i - u_j)^2}{2h_s^2}\right] \exp\left[-\frac{(z_i - v_k)^2}{2h_z^2}\right], \quad i = 1, 2, \ldots, T;\ j = 1, 2, \ldots, J;\ k = 1, 2, \ldots, K \tag{14}$$

The diffusion coefficient h_g, $g \in \{s,z\}$, can be calculated by using equation (15):

$$h_g = \begin{cases} 0.8146\,(b_g - a_g), & T=5 \\ 0.5690\,(b_g - a_g), & T=6 \\ 0.4560\,(b_g - a_g), & T=7 \\ 0.3860\,(b_g - a_g), & T=8 \\ 0.3362\,(b_g - a_g), & T=9 \\ 0.2986\,(b_g - a_g), & T=10 \\ 2.6851(b_g - a_g)/(T-1), & T \geq 11. \end{cases} \tag{15}$$

where $b_g = \max_{1\leq i\leq T}\{g_i\}$ and $a_g = \min_{1\leq j\leq T}\{g_i\}$; for $g = s$, $g_i = s_i$; for $g = z$, $g_i = z_i$.

Then, using equation (16):

$$P(u_j, v_k) = \frac{\sum_{i=1}^{T} \mu((s_i, z_i),(u_j, v_k))}{\sum_{j=1}^{J}\sum_{k=1}^{K}\sum_{i=1}^{T} \mu((s_i, z_i),(u_j, v_k))} \tag{16}$$

we can obtain a joint probability distribution $P(u_j, v_k)$ which is a discrete function.

In the above algorithm to estimate $P(u_j, v_k)$, we ignore the normalizing to unity $\mu((s_i, z_i),(u_j, v_k))$ on $S \times Z$ to guarantee that every diffused observation $\mu((s_i, z_i),(u_j, v_k))$ is equally important for constructing a joint distribution, because we can set discrete points (u_j, v_k) so much to reduce the difference.

5 DISASTER FUNCTION CONSTRUCTED BY THREE-DIMENSIONAL NORMAL DIFFUSION

The sample W of observations in equation (5) is a three-dimensional random sample with hazard input (s_i, z_i) and disaster output d_i, $i = 1, 2, \ldots, T$. We employ the three-dimensional normal diffusion to estimate a disaster function.

First, we employ S, Z, and D to denote universes of the two hazards and the disaster, respectively:

$$\begin{cases} S = \{u_j \mid j = 1,2,\ldots,J\} \\ Z = \{v_k \mid k = 1,2,\ldots,K\} \\ D = \{O_l \mid l = 1,2,\ldots,L\} \end{cases} \tag{17}$$

The three-dimensional normal diffusion of W on $S \times Z \times D$ is defined in equation (18):

$$\mu((s_i, z_i, d_i),(u_j, v_k, O_l)) = \exp\left[-\frac{(s_i - u_j)^2}{2h_s^2}\right]\exp\left[-\frac{(z_i - v_k)^2}{2h_z^2}\right]\exp\left[-\frac{(d_i - O_l)^2}{2h_d^2}\right], \tag{18}$$

where the diffusion coefficient h_g, $g \in \{s, z, d\}$, can also be calculated by using equation (15).

Second, let:

$$Q_{jkl} = \sum_{i=1}^{T} \mu((s_i, z_i, d_i),(u_j, v_k, O_l)),\ j = 1,2,\ldots,J;\ k = 1,2,\ldots,K;\ l = 1,2,\ldots,L \tag{19}$$

We obtain an information matrix of W on $S \times Z \times D$ as shown in equation (20):

$$Q = \left(\begin{array}{c} \begin{array}{cc} & \begin{array}{cccc} d_1 & d_2 & \cdots & d_L \end{array} \\ v_1 \begin{array}{c} u_1 \\ u_2 \\ \vdots \\ u_J \end{array} & \left[\begin{array}{cccc} Q_{111} & Q_{112} & \cdots & Q_{11L} \\ Q_{211} & Q_{212} & \cdots & Q_{21L} \\ \vdots & \vdots & \vdots & \vdots \\ Q_{J11} & Q_{J12} & \cdots & Q_{J1L} \end{array} \right] \end{array} \\ \\ \begin{array}{cc} v_2 \begin{array}{c} u_1 \\ u_2 \\ \vdots \\ u_J \end{array} & \left[\begin{array}{cccc} Q_{121} & Q_{122} & \cdots & Q_{12L} \\ Q_{221} & Q_{222} & \cdots & Q_{22L} \\ \vdots & \vdots & \vdots & \vdots \\ Q_{J21} & Q_{J22} & \cdots & Q_{J2L} \end{array} \right] \end{array} \\ \\ \vdots \quad \vdots \qquad \qquad \vdots \\ \\ \begin{array}{cc} v_K \begin{array}{c} u_1 \\ u_2 \\ \vdots \\ u_J \end{array} & \left[\begin{array}{cccc} Q_{1K1} & Q_{1K2} & \cdots & Q_{1KL} \\ Q_{2K1} & Q_{2K2} & \cdots & Q_{2KL} \\ \vdots & \vdots & \vdots & \vdots \\ Q_{JK1} & Q_{JK2} & \cdots & Q_{JKL} \end{array} \right] \end{array} \end{array} \right) \tag{20}$$

$\forall l \in \{1, 2, \cdots, L\}$, let:

$$H_l = \max_{\substack{1 \le j \le J \\ 1 \le k \le K}} \{Q_{jkl}\} \tag{21}$$

$$r_{jik} = Q_{jld} / H_l, j = 1, 2, \cdots, J; k = 1, 2, \cdots, K \tag{22}$$

then

$$R = \{r_{jkl}\}_{J \times K \times L} \tag{23}$$

is a fuzzy relationship between input (s, z) and output d.

For a fuzzy input A with membership function $\mu_A(u_j, v_k)$, $u_j \in S$, $v_k \in Z$ employing the approximate reasoning operator represented in equation (24), we can obtain a fuzzy output B with membership function $\mu_B(o_l)$, $o_l \in D$:

$$\mu_B(o_l) = \max_{\substack{1 \le j \le J \\ 1 \le k \le K}} \min\{\mu_A(u_j, v_k), r_{jkl}\} \tag{24}$$

Finally, using the center-of-gravity method, we obtain a crisp value $d(u_j, v_k)$:

$$d(u_j, v_k) = \left(\sum_{l=1}^{L} \mu_B(o_l) o_l \right) \Big/ \left(\sum_{l=1}^{L} \mu_B(o_l) \right) \tag{25}$$

where $d(u_j, v_k)$ is the disaster function constructed by a three-dimensional normal diffusion with the given sample W in equation (5). It is a discrete function.

6 INTEGRATED RISK CALCULATED BY JOINT DISTRIBUTION AND DISASTER FUNCTION

Recalling equation (2) and considering the joint distribution and disaster function we obtained are discrete, we calculate the integrated risk by using formula (26):

$$Risk = \sum_{j=1}^{J} \sum_{k}^{K} d(u_j, v_k) P(u_j, v_k) \tag{26}$$

where $P(u_j, v_k)$ and $d(u_j, v_k)$ are given in equations (16) and (25), respectively. The physical meaning of "*risk*" in (26) is the expected value of disaster, which is assessed by using the sample of observations on disaster D caused by two hazards S and Z in T years, as shown in equation (5).

As already mentioned, a risk is a scene in the future associated with some adverse incident. When we use a random sample to assess a risk, we are in fact using the past to judge the future, because the observations of the sample record historical events occurred in a stochastic system. The reliability of the assessment depends on the assumption that the stochastic system evolving over time in the study period is a stationary Markov process.

In the probability theory, a Markov process is a stochastic process that has the property that the next value of the Markov process depends on the current value, but it is conditionally independent of the previous values of the stochastic process. A Markov process is a stationary Markov process if the moments are not affected by a time shift. In other words, the stochastic behavior of the system in the future is the same as the stochastic behavior in the past.

In fact, most of disaster systems are changing, and the corresponding stochastic processes do not satisfy the stationary Markov process hypothesis. If we collect observations of historical disasters across 100 years to assess disaster risks, the reliability of the assessments will be low.

When we use a traditional statistical method, such as the parameter estimation method, to assess a risk caused by a hazard, if the observations are across 30 years, the result of the risk assessment would have some degree of reliability. However, to an integrated probability risk caused by two hazards, the size of a sample with observations across 30 years is too small. To estimate the joint distribution in equations (16), we need about 900 (i.e. 30×30) observations if we desire the result to have some degree of reliability.

We believe that, for most regions, the observations used for assessing an integrated risk should not span more than 50 years. They are incomplete information to estimate the joint distribution and the disaster function. It is quite good for us to obtain a discrete joint distribution, $P(u_j, v_k)$, and a discrete disaster function, $d(u_j, v_k)$, to estimate an integrated risk. When we have to make a choice between theoretical perfection and respect for reality, we should respect the reality to carry out the risk assessment.

7 CONCLUSION AND DISCUSSION

It is important to distinguish *integrated "risk assessment"* and "*integrated risk assessment*". An integrated probability risk of multihazards is the expected value of disaster, which is determined by a joint probability distribution and a disaster function.

When the type of the population from which the sample is drawn is unknown, it is difficult to estimate the distribution and the function. There are several copula models to construct a copula function to be a joint probability distribution with given marginal distribution functions. It is unknown that which constructed distribution from "copula" is better to assess the integrated probability risk.

The information diffusion technique regards small samples as fuzzy information. It can be employed to construct a discrete joint probability distribution and a discrete disaster function for assessing integrated probability risks.

The proposed approach does not need to know the distribution type of the population from which the given sample is drawn nor need to know the functional form of the causal relationship, which can construct joint probability distribution and disaster function with a clear physical meaning. Therefore, although the assessed risk is inaccurate, it is more credible.

Because of the limited length of this paper, a case study is omitted, which will be provided in an extended version.

ACKNOWLEDGMENTS

This work was supported by the National Natural Science Foundation of China (41671502).

REFERENCES

Alghalith, M. 2016. Novel and simple non-parametric methods of estimating the joint and marginal densities. *Physica A: Statistical Mechanics and its Applications*, 454: 94–98.

Fedra, K. 1998. Integrated risk assessment and management: overview and state of the art. *Journal of Hazardous Materials* 61(1–3): 5–22.

Hornik, K., Stinchcombe, M., and White, H. 1989. Multilayer feedforward networks are universal approximators. *Neural Networks* 2(5): 359–366.

Huang, C.F. 1997. Principle of information diffusion. *Fuzzy Sets and Systems* 91(1): 6–90.

Huang, C.F. 2002. Information diffusion techniques and small sample problem. *International Journal of Information Technology and Decision Making* 1(2): 229–249.

Huang, C.F. 2012. *Risk Analysis and Management of Natural Disaster*. Beijing: Science Press. (in Chinese).

Huang, C.F. and Ruan, D. 2008. Fuzzy risks and an updating algorithm with new observations. *Risk Analysis* 28(3): 681–694.

Huang, C.F. and Moraga C. 2004. A diffusion-neural-network for learning from small samples. *International Journal of Approximate Reasoning* 35(2): 137–161.

Kar, A. K. 2016. Bio inspired computing—A review of algorithms and scope of applications. Expert Systems with Applications 59: 20–32.

Liu, X.P. J., Zhang, Q., Cai, W.Y. and Tong, Z.J. 2010. Information diffusion-based spatio-temporal risk analysis of grassland fire disaster in northern China. *Knowledge-Based Systems* 23(1): 53–60.

Marco, G. and Alberto. T. 1992. On the problem of local minima in back-propagation. *IEEE Transactions on Pattern Analysis and Machine Intelligence* 14(1): 76–86.

Montes-Iturrizaga, R. and Heredia-Zavoni, E. 2016. Reliability analysis of mooring lines using copulas to model statistical dependence of environmental variables. *Applied Ocean Research* 59: 564–576.

Munns, W.R., Kroes, R., Veith, G., Suter, II G.W., Damstra, T. and Waters, M.D. 2003. Approaches for integrated risk assessment. *Human and Ecological Risk Assessment* 9(1): 267–272.

Nelsen, R.B. 2006. *An Introduction to Copulas*. New York: Springer.

Salvadori, G., De Michele, C., Kottegoda, N.T., Rosso, R. 2007. *Extremes in Nature an Approach Using Copulas*. Dordrecht: Springer.

Sekizawa, J. and Tanabe, S. 2005. A comparison between integrated risk assessment and classical health/environmental assessment: emerging beneficial properties. *Toxicology & Applied Pharmacology* 207(2, Sup 1): 617–622.

Sklar, A. 1959. Fonction de re'partition a' n dimensions et leurs marges, vol. 8. *Publications de L'Institute de Statistique, Universite, de Paris*, Paris, pp. 229–231.

Talamakrouni, M., Keilegom, I. and Ghouch, A. 2016. Parametrically guided nonparametric density and hazard estimation with censored data. *Computational Statistics and Data Analysis*, 93 (C): 308–323.

UNISDR. 2015. *The Sendai Framework for Disaster Risk Reduction 2015–2030*. Available at: http://www.preventionweb.net/files/43291_sendaiframeworkfordrren.pdf (accessed December 27, 2016).

Risk Analysis and Management – Trends, Challenges and Emerging Issues – Bernatik, Huang & Salvi (Eds)
 ISBN 978-1-138-03359-7

Risk assessment of $PM_{2.5}$-bound Polycyclic Aromatic Hydrocarbons (PAHs) during wintertime in Beijing, China

Sihong Chao, Yanjiao Chen, Yanxue Jiang & Hongbin Cao
College of Resource Science and Technology, Beijing Normal University, Beijing, China

ABSTRACT: Sixteen Polycyclic Aromatic Hydrocarbons (PAHs) in $PM_{2.5}$ were monitored at one site in the central area of Beijing during wintertime. Their levels and health risks to local residents were evaluated. Results showed that the average total concentration of PAHs in $PM_{2.5}$ was 150.01 ng/m^3. They were dominated by high-molecular-weight PAHs. The health risks of local residents exposed to $PM_{2.5}$-bound PAHs via inhalation were quantitatively calculated by BaP equivalent concentration (BaPeq) using Toxic Equivalence Factors (TEFs) and the Incremental Lifetime Cancer Risk (ILCR). The mean value of BaP equivalent concentration (BaPeq) of the total PAHs is 33.4 ng/m^3. The mean ILCRs of PAHs of local residents exposed to PAHs via inhalation are 3.93×10^{-5} and 3.11×10^{-3}, using the unit risk value provided by California Environmental Protection Agency of U. S. America (CalEPA) and the World Health Organization (WHO), respectively. Both values exceeded the recommended safety standard 1×10^{-6} and one even exceeded 1×10^{-4}, indicating that the $PM_{2.5}$-bound PAHs may pose potential and even serious cancer risk to local residents with lifetime exposure.

1 INTRODUCTION

Fine Particulate Matter ($PM_{2.5}$) with an aerodynamic diameter less than 2.5 μm has become one of the most significant pollutants of ambient air and received much attention (Li et al., 2014; Ma et al., 2016). $PM_{2.5}$ has been linked to increased adverse health effects on human health, especially respiratory and cardiovascular diseases according to epidemiological studies (Lu et al., 2015). The adverse health risk effects of $PM_{2.5}$ are relevant to its complex compositions with inorganic elements, water soluble ions and organic compounds.

Polycyclic Aromatic Hydrocarbons (PAHs) are one of the most common pollutants in $PM_{2.5}$. PAHs are a group of compounds that composed of two or more fused aromatic rings (Cao et al., 2016). They can be formed through the incomplete combustion of fossil fuels. Traffic emissions and industrial activities and coal combustion are the main sources in urban area (Chen et al., 2016). Due to its potential carcinogenic and mutagenic and persistent properties, sixteen PAHs are included in the list of priority pollutants by the USEPA (USEPA, 2003). These 16 PAHs are naphthalene (NAP), acenaphthylene (ACY), acenaphthene (ACE), fluorene (FLO), phenanthrene (PHE), anthracene (ANT), fluoranthene (FLA), pyrene (PYR), benzo(a)anthracene (BaA), Chrysene (CHR), benzo(b)fluoranthene(BbF), benzo(k)fluoranthene (BkF), benzo(a)fluoranthene (BaP), indeno(1,2,3-cd)pyrene (IcdP), dibenzo(a,h)anthracene (DahA), and benzo[g,h,i]perylene (BghiP). Normally, these 16 PAHs can be divided into two categories according to the different rings. Low-molecular-weight (LMW) PAHs include PAHs with 2 or 3 rings. And the high-molecular-weight (HMW) PAHs contain 4–6 rings, which mainly derive from vehicle emissions (Ravindra et al., 2008). The health risk of PAHs in $PM_{2.5}$ usually is calculated by TEF methods. Considering that BaP is the most studied PAH with corresponding Unit Risk value, other PAHs can be ranked according to their cancer potency relative to BaP by using Toxic Equivalence Factors (TEFs).

Beijing, as the capital of China, has successfully bid to host the 2022 winter Olympics. The air quality is directly related to the health of athletes and tourists. This article aims to investigate the levels of $PM_{2.5}$ and $PM_{2.5}$-bound PAHs in Beijing in winter and calculate the health risk of local residents exposed to $PM_{2.5}$-bound PAHs via inhalation.

2 METHODS AND MATERIALS

2.1 *Site description and sample collection*

Beijing, which is considered as the economic and cultural center of China, occupies an area of 16,411 km^2 with a permanent population of more than 20 million. In this article, the $PM_{2.5}$ samples were collected at the south campus of Beijing Normal University in Haidian District, where numerous universities and schools were located.

The 24h integrated samples (from 9:00 a.m. to 9:00 am of the next day) were collected in January and February 2016. Samples were collected by quartz microfiber filters (20.3 × 25.4 cm,Whatman) with high-volume $PM_{2.5}$ samplers (1.05 m^3 min^{-1}, Wuhan Tianhong Ltd.. China) at the roof of a one-storey building (3.5 m, height). A total of 32 samples (14 samples in January, 18 samples in February) were collected. Each filter was wrapped in aluminum foil and baked at 450°C for at least 6h before sampling. After sampling, the loaded filters were wrapped in aluminum foil and a ziplock bag in a refrigerator at −18°C before analysis.

2.2 *Chemical analysis of PAHs*

Each filter was cut into small pieces, which were then extracted with 25 ml n-hexane and acetone (1:1, v/v). The extraction temperature was 110°C and programmed as follows: ramp to 110°C for 10 min, holding at 10 min. Then the extracts were concentrated using a vacuum rotary evaporator (R-201, IKA, Germany) at 22°C before being transferred to an alumina silica gel column for purification. The alumina silica column was eluted with 20 mL of hexane followed by 50 mL hexane:dichloromethane (1:1, v/v). The eluation was concentrated to near dryness in the vacuum rotary evaporator (R-201, IKA, Germany) using a 22°C water bath. The residue was then transferred with n-hexane and then concentrated by nitrogen blow (Eyela MG-1000) at room temperature (25°C). Finally, the nearly dried residue was rinsed with n-hexane and finally adjusted to 1 mL. The samples were sealed in vials and stored at −4°C before analysis. The 16 PAHs was detected using a gas chromatograph with a mass spectrometer detector (Agilent 6890GC/5973MSD). A HP-5MS capillary column (30 m length, 0.25 mm i.d., 0.25 μm film thickness) was used. A mixture of deuterated surrogate compound, including Anthracene-D10, Chrysene-D12 and Perylene-D12 were added into all samples for the determination of the recovery ratio. The PAH recoveries of the standard spiked matrix range from 76% to 102%. The correlation coefficients (R^2) for linear regressions of the calibration curves were >0.99.

2.3 *Risk assessment*

The health risk of a PAH mixture can be expressed by its total BaPeq concentration. It is calculated by multiplying the mass concentration of specific PAHs species with their corresponding Toxic Equivalent Factor (*TEF*), as Eq. (1) (Nisbet and LaGoy, 1992):

$$BEC = \sum_{i=1}^{n} C_i \times TEF_i \qquad (1)$$

where, *BEC* is BaP equivalent total concentration of PAHs. C_i is the concentration of PAH congener *i* in the soil and TEF_i is the Toxic Equivalent Factor (*TEF*) of PAH congener *i*. *TEF* was provided by Nisbet and Lagoy (Nisbet and LaGoy, 1992).

The incremental lifetime cancer risk (*ILCR*) was calculated to evaluate the cancer risk caused by PAHs. *ILCR* is the average excess risk of cancer for an individual. The value of *ILCR* can be calculated by Eq. (2) (USEPA, 2004):

$$ILCR = BEC \times UR_{BaP} \tag{2}$$

where, the UR_{BaP} (unit risk) is the cancer risk via the inhalation route for a lifetime (70years) PAH exposure to one unit of BaP (1 ng/m^3). The values of UR_{BaP} from World Health Organization (WHO) and California Environmental Protection Agency of U. S. America (CalEPA) are 8.7×10^{-5} and 1.1×10^{-6}, respectively (CalEPA, 1994; WHO, 2000).

Crystal Ball software was used to estimate the probabilistic distribution of ILCR value by Monte-Carlo technique. The best-fit distribution was determined for BEC by assessing the goodness-of-fit for a number of parametric distributions using Anderson-Darling test. The software was run for 10,000 iterations to achieve a stable distribution of ILCR. When ILCR is greater than 10^{-4}, it suggested that the potential risk of cancer is serious; when the ILCR is less than 10^{-6}, it suggested that the cancer risk is negligible.

3 RESULTS AND DISCUSSION

3.1 *Levels of $PM_{2.5}$ and PAHs*

The average concentration of $PM_{2.5}$ during the sampling days is 135.6 ± 66.22 μg/m^3 with the range of 55.12–332.51 μg/m^3, which exceeds Grade II value (75 μg/m^3) of the ambient air quality standard of China (China Ministry Of Environmental Protection, 2012). The average concentration of PAHs is 150.01 ± 144.45 ng/m^3 with the range of 10.35–742.28 ng/m^3. The levels of $PM_{2.5}$-bound PAHs observed in this study are lower than those in Harbin (377 ± 228 ng/m^3; (Mohammed et al., 2016)), Qingdao (263 ng/m^3; (Guo et al., 2009)) and higher than Taiyuan (119.8 ng/m^3; (Li et al., 2014)), Italy (13 ng/m^3; (Martellini et al., 2012)) in winter. The species with the highest concentration is BbF. The HMW PAHs are dominant in $PM_{2.5}$ with the percentage of 80.96% in the total $PM_{2.5}$-bound PAHs.

Table 1. The concentration, *TEF*, *BEC* values of 16 PAHs in $PM_{2.5}$.

Rings	PAH	Mean (ng/m^3) N = 3 2	TEF	BEC (ng/m^3)
2	NAP	0.10	0.001	1.04×10^{-4}
2	ACY	0.45	0.001	4.53×10^{-4}
2	ACE	0.03	0.001	3.45×10^{-5}
2	FLO	0.47	0.001	4.66×10^{-4}
3	PHE	7.91	0.001	7.91×10^{-2}
3	ANT	1.65	0.01	1.65×10^{-2}
3	FLA	17.94	0.001	1.79×10^{-2}
4	PYR	15.27	0.001	1.53×10^{-2}
4	BaA	11.60	0.1	1.16×10^{0}
4	CHR	14.25	0.01	1.43×10^{-1}
4	BbF	23.40	0.1	2.34×10^{0}
4	BkF	8.98	0.1	8.98×10^{-1}
5	BaP	16.14	1	1.61×10^{1}
5	IcdP	14.21	0.1	1.42×10^{0}
5	DahA	2.21	5	1.11×10^{1}
6	BghiP	15.39	0.01	1.54×10^{-1}
	Σ16PAHss	150.01		3.34×10^{1}

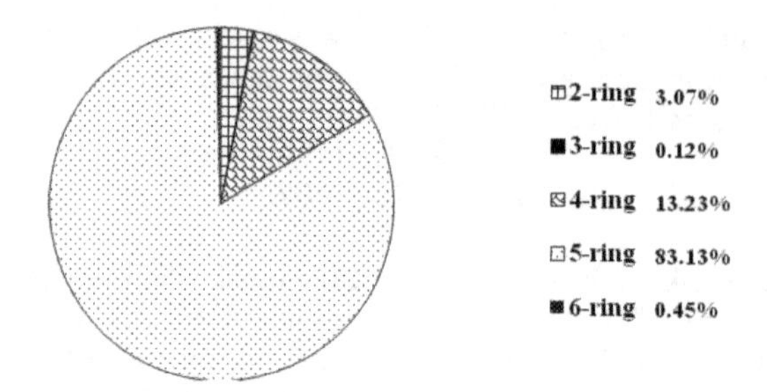

Figure 1. The BEC percentages of different rings in $PM_{2.5}$.

Table 2. The mean and percentiles of ILCR for PAHs.

	Risk (10^{-6})			
ILCR	Mean	10%	50%	90%
CalEPA	39.33	3.52	17.77	89.98
WHO	3110.93	278.29	1405.70	7116.98

3.2 *PAHs risk assessment*

The mean value of *BEC* is 33.4 ± 45.53 ng/m^3 with the range of 1.73–192.75 ng/m^3. The level of *BEC* is much higher than Guangzhou (9.24 ng/m^3; (Liu et al., 2015)) and lower than Harbin (51 ng/m^3; (Mohammed et al., 2016)). BaP has the highest *BEC* value of 16.1 ng/m^3. The second highest PAH is DahA (11.1 ng/m^3) with the highest *TEF* value of 5. Figure 1 shows the composition of *BEC* value. The largest component of BEC value is 5-ring PAHs. Therefore, the HMW PAHs bring the highest health risk with the percentage of 86.81%.

The distribution of *BEC* value follows a lognormal distribution with a mean of 36.18 ng/m^3. And the distribution of *ILCR* of local residents exposed to PAHs via inhalation also follows a lognormal distribution. The mean and percentiles of *ILCR* are show in Table 2. The mean values of *ILCR* are 3.93×10^{-5} (CalEPA) and 3.11×10^{-3} (WHO), respectively and exceeded the recommended safety standard 1×10^{-6}. The ratios of people with risks exceeding 10^{-6} are 98.90% (CalEPA) and 100% (WHO). However, the ratios of people with severe cancer risks exceeding 10^{-4} drop to 8.62% (CalEPA) and 98.16% (WHO). It should be noted that these results are based on the assumption that the local residents exposed to this level of PAHs all their lives. The value of *ILCR* varies largely depending on the different URBaP values. Much more toxicological evidences for deriving URBaP via inhalation route should be collected to attain a concensus on URBaP value.

4 CONCLUSIONS

The HMW PAHs corresponding to $PM_{2.5}$ in wintertime had the highest concentration (80.96%) and health risks (86.81%), which may be linked to vehicle emissions. The mean ILCRs of PAHs of local residents exposed to PAHs via inhalation are 3.93×10^{-5} (CalEPA) and 3.11×10^{-3} (WHO), respectively. In conclusion, the ILCR results indicate that the local residents might have potential serious cancer risk via $PM_{2.5}$-bound PAHs lifetime inhalation.

ACKNOWLEDGEMENT

This work was supported by the Key Technologies R&D Program for the 12th Five-Year Plan (No. 2012BAJ24B04) of the Ministry of Science and Technology of People's Republic of China.

REFERENCES

CalEPA, 1994. Benzo[a]pyrene as a toxic air contaminant., Berkeley, California, USA.

Cao, H., Chao, S., Qiao, L., Jiang, Y., Zeng, X., Fan, X., 2016. Urbanization-related changes in soil PAHs and potential health risks of emission sources in a township in Southern Jiangsu, China. Sci Total Environ.

Chen, Y., Chiang, H., Hsu, C., Yang, T., Lin, T., Chen, M., Chen, N., Wu, Y., 2016. Ambient $PM_{2.5}$-bound polycyclic aromatic hydrocarbons (PAHs) in Changhua County, central Taiwan: Seasonal variation, source apportionment and cancer risk assessment. Environ Pollut. 218, 372–382.

China, M.O.E.P., 2012. Ambient air quality standards.

Guo, Z., Lin, T., Zhang, G., Hu, L., Zheng, M., 2009. Occccurrence and sources of polycyclic aromatic hydrocarbons and n-alkanes in $PM_{2.5}$ in the roadside environment of a major city in China. J Hazard Mater. 170, 888–894.

Li, R., Kou, X., Geng, H., Dong, C., Cai, Z., 2014. Pollution characteristics of ambient $PM_{2.5}$-bound PAHs and NPAHs in a typical winter time period in Taiyuan. Chinese Chem Lett. 25, 663–666.

Liu, J., Man, R., Ma, S., Li, J., Wu, Q., Peng, J., 2015. Atmospheric levels and health risk of polycyclic aromatic hydrocarbons (PAHs) bound to $PM_{2.5}$ in Guangzhou, China. Mar Pollut Bull. 100, 134–143.

Lu, F., Xu, D., Cheng, Y., Dong, S., Guo, C., Jiang, X., Zheng, X., 2015. Systematic review and meta-analysis of the adverse health effects of ambient $PM_{2.5}$ and PM10 pollution in the Chinese population. Environ Res. 136, 196–204.

Ma, Y., Cheng, Y., Qiu, X., Lin, Y., Cao, J., Hu, D., 2016. A quantitative assessment of source contributions to fine particulate matter ($PM_{2.5}$)-bound polycyclic aromatic hydrocarbons (PAHs) and their nitrated and hydroxylated derivatives in Hong Kong. Environ Pollut.

Martellini, T., Giannoni, M., Lepri, L., Katsoyiannis, A., Cincinelli, A., 2012. One year intensive $PM_{2.5}$ bound polycyclic aromatic hydrocarbons monitoring in the area of Tuscany, Italy. Concentrations, source understanding and implications. Environ Pollut. 164, 252–258.

Mohammed, M.O.A., Song, W., Ma, Y., Liu, L., Ma, W., Li, W., Li, Y., Wang, F., Qi, M., Lv, N., Wang, D., Khan, A.U., 2016. Distribution patterns, infiltration and health risk assessment of $PM_{2.5}$-bound PAHs in indoor and outdoor air in cold zone. Chemosphere. 155, 70–85.

Nisbet, I.C.T., LaGoy, P.K., 1992. Toxic equivalency factors (TEFs) for polycyclic aromatic hydrocarbons (PAHs). Regul Toxicol Pharm. 16, 290–300.

Ravindra, K., Sokhi, R., Van Grieken, R., 2008. Atmospheric polycyclic aromatic hydrocarbons: Source attribution, emission factors and regulation. Atmos Environ. 42, 2895–2921.

USEPA, 2003. Appendix A to 40 CFR. Part 423–126 Priority Pollutants.

USEPA, 2004. Risk Assessment Guidance for Superfund Volume I: Human Health Evaluation Manual (Part E, Supplemental Guidance for Dermal Risk Assessment). Office of Superfund Remediation and Technology Innovation. U.S. Environmental Protection Agency. Washington, DC.

WHO, 2000. Air quality guidelines for Europe., Copenhagen, Denmark.

Risk Analysis and Management – Trends, Challenges and Emerging Issues – Bernatik, Huang & Salvi (Eds)
 ISBN 978-1-138-03359-7

Earthquake risk assessment for urban planning in Songming, China

Zhou Jian
Institute of Geophysics, China Earthquake Administration, Beijing, China
China Earthquake Administration, Beijing, China

Chen Kun & Wu Jian
Institute of Geophysics, China Earthquake Administration, Beijing, China

Xuchuan Wu
Institute of Engineering Mechanics, China Earthquake Administration, Harbin, China

ABSTRACT: A high-speed and low-cost method is proposed for seismic risk assessment in Songming city, Yunnan Province in this research. Earthquake hazard information is based on the data of Ground Motion Parameters Zonation Map of China and the method of topographic slope as a proxy for seismic site condition and amplification. Remote sensing image is applied to extract building data. A multistorey concentrated mass shear model is introduced to assess building vulnerability. Unlike the traditional method, this new method reduces the time of field investigation and computational workload and thus can be used as a technical support in seismic regions for urban planning.

Keywords: Earthquake, Risk Assessment, Rapid Method, Urban Planning

1 INTRODUCTION

Disaster risk is a synthetic result of interrelating disaster systems. Combining the influence of the environment, hazard and its affected bodies, disaster risk illustrates the expectation of social and economic impact of natural disasters. Unlike the hazard assessment, which focuses

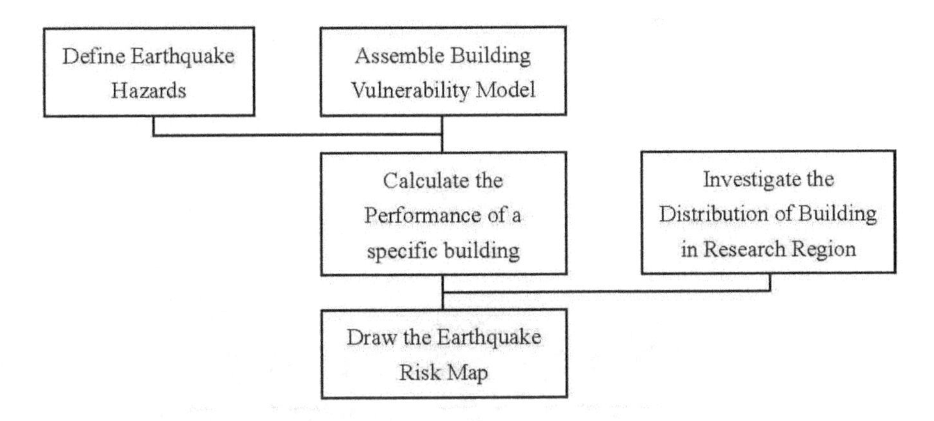

Figure 1. The process of earthquake risk assessment.

on the possibility of hazard intensity, disaster risk assessment comprehensively takes hazard, exposure, and vulnerability into account (Shi Peijun 2002).

With regard to earthquake risk assessment, the hazard can be described as the frequency and intensity of the earthquake, exposure can be measured by the density of the building, population or GDP, while vulnerability can be measured by the correspondent vulnerability model.

If limiting the affected bodies to just the buildings, usually casing the most serious losses in an earthquake, the flowchart below illustrates the relationship between the steps that comprise the risk assessment methodology and risk map drawing.

In earthquake risk assessment, defining earthquake hazards and investigating the distribution of buildings are usually time and money consuming. In the traditional earthquake risk assessment method which was widely used in China, known as earthquake disaster evaluation (GB/T 19428-2014, 2014), defining earthquake hazard needs drillings to be taken in every 1 km to 2 km to obtain site condition distribution and thus get amplification of bedrock ground motion in the research area, that is an amount like over 200,000 dollars in a small city. In the traditional method, investigating the distribution of buildings and getting the building data needs a sampling survey of 8%–11% of the total number of buildings in the study area to get the structure type, material, number of storeys, and for important buildings even the design drawing is needed, which often needs 12 months of work to cover a city.

These lead to unacceptable time and fund requirement in many application areas, including urban planning, especially in some developing cities, like many cities in China, where the urban region is expanding rapidly and building distribution is changing every year. Urban planning in these cities needs earthquake risk assessment catching the step of the changing urban areas; in other words, a rapid earthquake risk assessment method. The method introduced in this paper and applied in Songming urban planning avoids most of the field investigation and reduces the time scale of urban earthquake risk assessment from years to months.

2 THE STUDY AREA—SONGMING

Songming (25.215°N 103.071°E) is located in Yunnan province, southwest China, with the acreage of 1,241 km^2 and a total population of 357,000 (Figure 2). The center of Songming is alluvial plain surrounded by mountains. Due to Kunming Changshui aviation hub located in its southern part and neighboring the economic center of Yunnan province, Songming has a great potential to represent an important hub on which is proposed by the China central government, the 'The Belt and Road Initiative'.

To meet the perspective of its important role, the Songming government plans to renew its urban planning. A plan of quadrupling the population and land use in the construction of

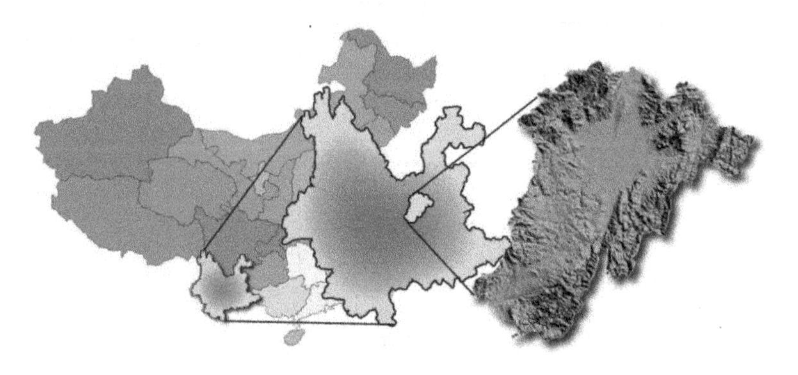

Figure 2. Geographic location of the study area: Songming city, Yunnan Province.

Songming is proposed and due to its history of stroke by a devastating earthquake in 1833 and the high seismic risk background in the whole Yunnan Province, the earthquake issue becomes a critical factor in urban planning.

3 EARTHQUAKE HAZARD EVALUATION

The first step is to get the on-rock ground motion distribution in the study area. The relevant raw data including historical earthquake, potential seismic source, and attenuation model are contributed by the workgroup of 'Seismic Ground Motion Parameter Zonation Map of China (CSGMM) (GB18306-2015, 2015) project. The method of generating the on-rock ground motion map of the study area is also the same as the method introduced in the manual book of CSGMM (Gaomengtan 2016). The process of getting the ground motion map is skipped. The result is shown in Figure 3, which uses 50a10% Ground Peaking Acceleration (GPA) as parameters.

To estimate the strong ground motion, the character of surficial materials and site conditions should be considered due to its dramatic effect on amplification of strong ground motion on the rock site. Unlike the traditional method that uses field investigation and geophysical prospecting including drillings to gain the information of site conditions, the method of topographic slope is introduced as a proxy to estimate site conditions. The framework of this method was introduced in the USGS 2007–1357 report (U.S. Geological Survey 2007).

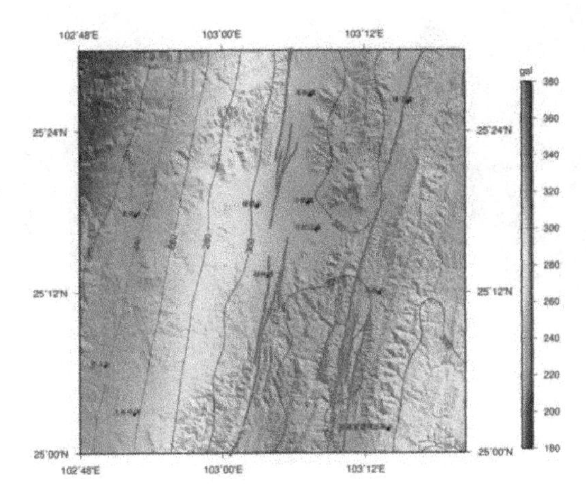

Figure 3. Peak ground motion on rock site (50a10%, gal) in study area and around.

Table 1. Summary of slope ranges for subdivided NEHRP V_s^{30} categories.

Class	V_s^{30} range (m/s)	Slope range (m/m)-Active Tectonic	Slope range(m/m)-Stable Tectonic
E	<180	<1.0E-4	<2.0E-5
	180–240	1.0E-4–2.2E-3	2.0E-5–2.0E-3
D	240–300	2.2E-3–6.3E-3	2.0E-3–4.0E-3
	300–360	6.3E-3–0.018	4.0E-3–7.2E-3
	360–490	0.018–0.050	7.2E-3–0.013
C	490–620	0.050–0.10	0.013–0.018
	620–760	0.10–0.138	0.018–0.025
B	>760	>0.138	>0.025

The method is based on the theory that more competent materials showing high-velocity are more likely to maintain a steep slope whereas deep basin sediments are deposited primarily in environments with low gradients.

In this study, the shuttle Radar Topography Mission (SMRT) data (Farr, Kobrick 2000) is applied to obtain the maximum slope of a specific site. V_s^{30}, the average shear-velocity down to 30 m can be acquired by correlations (Table 1) between V_s^{30} and slopes.

In this study, 92 drillings on 24 sites in or around the study region offer direct shear-wave observations. Among these drillings, 14 are V_s^{20} results. The method of estimating V_s^{30} from V_s^{20} (Kang Chuanchuan, et al. 2015) is applied to transform these V_s^{20} data to V_s^{30} values.

Then, these measured V_s^{30} values are compared with values derived from topographic slope correlations to ensure the correlation is fit for the study region. Referencing with the local geological map in the area of the Mesozoic group and earlier, the trend of measured V_s^{30} is relatively well-recovered using the topographic slope, while in the Cenozoic area, values derived from the topographic slope correlations show a systematic high trend compared with measured V_s^{30}. For improving the result of estimation, the estimated values of V_s^{30} are adjusted by (1).

$$Y = 83 + 0.41X \tag{1}$$

where X is the original estimated V_s^{30} value, and Y is the adjusted value.

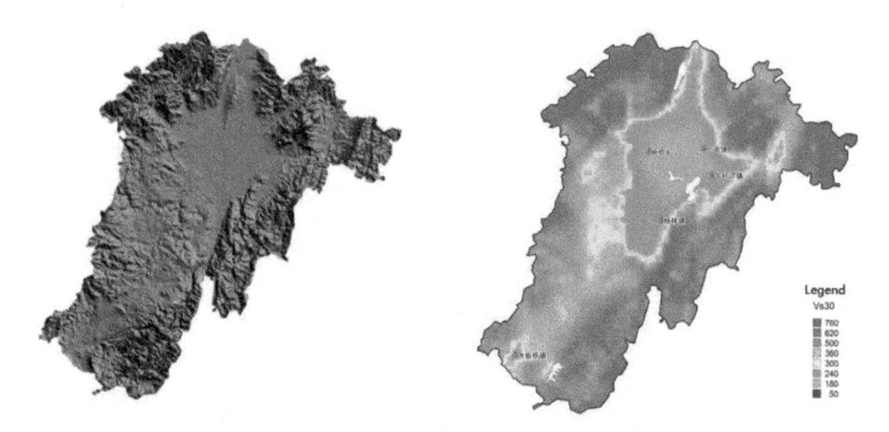

Figure 4. SMRT data and adjusted topographic slope V_s^{30} map for Songming.

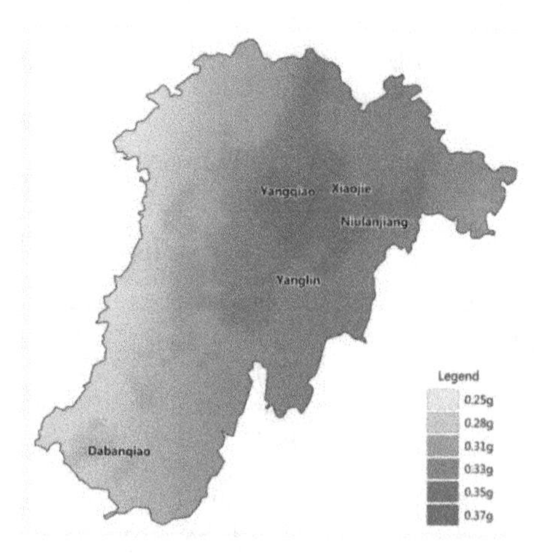

Figure 5. Ground motion map for Songming.

The adjusted topographic slope V_s^{30} map is shown in Figure 4. In the map, lower velocity region in the center is indicated by soft surficial materials whereas in field investigations it is a nearly flat basin. A high-velocity region around the low-velocity region indicated rock wherein field investigations showed mountains and hills.

The short-period site amplification factors were assigned to each site by applying the amplification factors. (Borcherdt 1994) Based on rock site ground motion information, site condition, and amplification factors, the hazard map for Songming, which uses 50a10% Ground Peaking Acceleration (GPA) as a parameter is generated (Figure 5).

4 BUILDING INFORMATION INVESTIGATION

The Yangqiao town and Dabanqiao town, which are the two highest population density towns in the study area are chosen to generate building the earthquake risk map.

The recent development of remote sensing technology gives building information investigation an accessible way to get the roof shape and height without field investigation. Foresight and backsight SPOT images can be used to extract stereo pair images. The digital stereophotogrammetry system is applied to gather elevation and plane coordinates. Fitting with the SPOT pair image, the building distribution map of Yangqiao and Dabanqiao with roof shape and height of each building is generated (Figure 6).

Figure 6. Up-left: Building distribution of Yangqiao town; Up-right: Building distribution of Dabanqiao town; Bottom: Generate building shape model through roof shape and height.

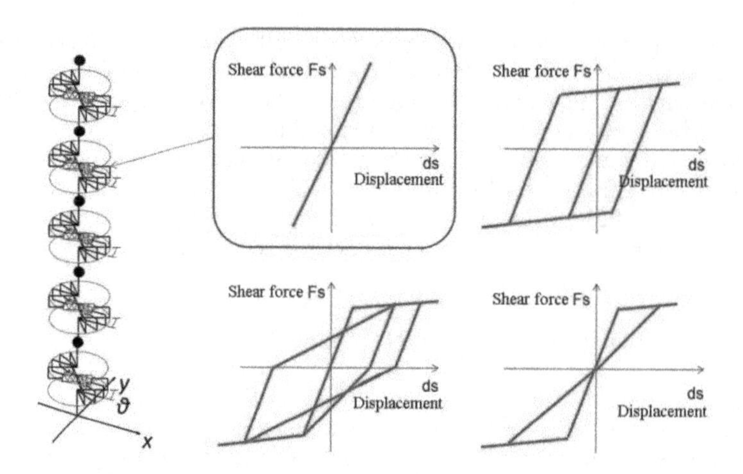

Figure 7. Right: The basic meaning of a multistorey concentrated mass shear model; Up-middle: elastic constitutive model; Up-left: bilinear constitutive model; Bottom-middle: peak point directed constitutive model; Bottom-left: origin directed constitutive model.

5 BUILDING VULNERABILITY MODEL

Multistorey Concentrated mass Shear (MCS) model is adopted to analyze the performance of the buildings. In the MCS model, a multistorey building can be looked as a multi-degree-of-freedom model, each storey concentrating its mass into a point particle, and its performance in the earthquake shaking is assumed to be dominated by the shear mode.

Elastic constitutive model, bilinear constitutive model, peak point directed constitutive model, and origin directed constitutive model were used to represent the response of four dominated types of buildings in the study area, including steel, beam-column frames, masonry-concrete, and wood. The construction type of each building can be classified through height and the guild-line of the Code for Seismic Design of Building of China (Figure 7). (GB50011-2010, 2010)

6 EARTHQUAKE RISK MAP

By comparing the maximum storey drift ratio with limit value defined by the Code for Seismic Design of Building of China, based on the difference, five predicted damage levels are defined including No Damage, Light Damage, Medium Damage, Serious Damage, and Collapse.

Calculating each building in the research area by using the MCS model which relies on the roof shape and height data obtained from remote sensing and ground motion time history generated from earthquake hazard evaluation, the risk map of single buildings was generated. Then multiplying the vulnerability index of each building in a block with its floor area and representing it on the map, the risk map of blocks was generated (Figure 8).

7 DISCUSSION

In the application of this method for city planning of Songming city, the Ground Motion Map for Songming is used to optimize land-use planning, the risk map of blocks for Yangqiao town and Dabanqiao town due to it can be easily deduced from the refugee needed population, and are used in optimizing earthquake shelter planning.

Figure 8. Up-left: Risk map of single buildings for Yangqiao town; Up-right: Risk map of single buildings for Dabanqiao town; Bottom-left: Risk map of blocks for Yangqiao town; Bottom-right: Risk map of blocks for Dabanqiao town.

When using the topographic slope as a proxy for site condition, local modification to the correlations is provided to improve accuracy. Restricted by the database of measured V_s^{30} information, this modification is preliminary. Moreover, other methods to demonstrate and modify the correlations like GIS and geological method will be developed and applied in the future.

When using remote sensing images to obtain building information, in practice, some sort of field investigation to obtain structure type and use of the building can greatly improve the precision of the result. While in this research, the structure type is associated with the height of the building by referencing the code of seismic design of China, the use of the building is not considered. This approach can lead to acceptable precision in block scale, but cannot meet the demands when using the earthquake vulnerability analysis for single buildings, especially for low-rise buildings. Further information extraction technique through remote sensing images should be developed in the future.

REFERENCES

Borcherdt, R.D., 1994. Estimates of site-dependent response spectra for design (methodology and justification): Earthquake Spectra, v. 10, no 4, p. 613–653.

Farr, T.G., Kobrick, M., 2000. Shuttle Radar Topography Mission produces a wealth of data: EOS Trans., v. 81, p. 583–585.

Gaomengtan. Manual book of Seismic ground motion parameters zonation map of China [M]. China Zhijian Publishing House, Standards Press of China. 2016.

GB/T 19428-2014. Code for earthquake loss estimation and its information management system [M]. China Standards Press. 2014.

GB18306-2015. Seismic ground motion parameters zonation map of China [M]. China Standards Press. 2015.

GB50011-2010. Code for Seismic Design of Buildings [M]. China Standards Press. 2010.
Kang, Chuanchuan, Yu, Yanxiang, Ma, Chao, Li, Jianliang, Huang, Chengcheng. 2015. Estimation Methods of V_s^{30} for Drilling with Depth Less than 30 Meters in Sichuan Area [J]. Technology for Earthquake Disaster Prevention, Vol. 10(2): 316–323.
Shi, Peijun. Theory on disaster science and disaster dynamics [J]. *Journal of Natural Disaster*. 2002. Vol. 11, No. 3: 1–9.
U.S. Geological Survey. 2007. Topographic Slope as a Proxy for Seismic Site-Condition (V_s^{30}) and Amplification Around the Globe [R]. Open-File Report 2007–1357.

Risk Analysis and Management – Trends, Challenges and Emerging Issues – Bernatik, Huang & Salvi (Eds)

A comparative study on spatial vulnerability of European and Chinese civil aviation networks under spatial disruptive events

Hang Li & Xiao-Bing Hu
Academy of Disaster Reduction and Emergency Management, Beijing Normal University, Beijing, China

ABSTRACT: The European Civil Aviation Network (ECAN) and the Chinese Civil Aviation Network (CCAN), as two important air transport systems in the world, are facing constantly increasing spatial disruptive events, such as extreme weathers and terrorist attacks. Based on a newly proposed spatial vulnerability model, this paper reports a study on the spatial vulnerability of the ECAN and the CCAN under random and intentional spatial hazards. The simulation results show that both of the ECAN and the CCAN are vulnerable to random and intentional spatial hazards. When comparing the results of the two networks, one can see that the spatial vulnerability of the ECAN is much larger than that of the CCAN under intentional spatial hazards, which suggests that the ECAN needs to pay more attention to such hazards, particularly when the terrorism risk is considerably high in Europe nowadays.

1 INTRODUCTION

The Civil aviation, as one of the most important transportation modes, is not only closely linked with our daily life, but also makes a great contribution to the world economy (Verma et al., 2014; Dunn et al., 2016). In 2014 approximate 3.3 billion passengers and 50 million tons of freight were carried by airplane, which totally makes 758 billion USD revenues for the whole world (International Civil Aviation Organization, 2014). The European Civil Aviation Network (ECAN) and the Chinese Civil Aviation Network (CCAN), as two large air transport systems, play significant roles in the global air transportation business. However, in recent years, air transport systems are facing more and more spatial disruptive events, such as extreme weathers, natural disasters, traffic accidents, and terrorist attacks (Zanin et al., 2013). Usually these events could cause more severe global impacts when they happen to large air transport systems, such as the ECAN and the CCAN. For example, the 2008 snowstorm disaster in southern China led to a large scale airport closure and thousands of stranded passengers (Guo et al., 2015); the 2010 eruption of the Iceland Volcano had a severe impact on the European air transport system, causing more than 10 million passengers delayed and leading to almost 1.7 billion US Dollars loss (Mazzocchi et al., 2010).

This paper is particularly concerned with the vulnerability of the ECAN and the CCAN when facing spatial disruptive events. To this end, a newly proposed spatial vulnerability model (SVM) in Li, et al., (2016a) will be applied. This model emphasizes on the global impact of spatial hazards on network systems by considering hazard location, hazard covered area and global impact of hazard (including direct impact and indirect impact). In this model, two important curves, impact curve and neutral curve were introduced. and two spatial vulnerability indices, Absolute Spatial Vulnerability Index (ASVI) and Relative Spatial Vulnerability Index (RSVI), were developed, in order to study the performance of network systems under spatial hazards. By considering random hazards and intentional hazards as two major categories of disruptive events, this paper applies a modified SVM to analyze the spatial vulnerability of the ECAN and the CCAN under random and intentional spatial hazards.

The rest of paper is organized as follows. The method to calculate the spatial vulnerability of civil aviation networks is described in Section 2. Then we conduct a comparative study on the spatial vulnerability of the ECAN and the CCAN under random and intentional spatial hazards in Section 3. This paper ends with its main conclusions in Section 4.

2 METHOD

2.1 *The introduction of SVM*

In the SVM of Li, et al., (2016a), the spatial vulnerability of a given network system is defined as the degree of the system as a whole to be likely harmed due to its exposure to spatial hazards. The SVM emphasizes on the global impact of spatial hazards on network systems. Based on two important curves, the impact curve and the neutral curve, the SVM proposes two indexes, Absolute Spatial Vulnerability Index (ASVI) and relative spatial vulnerability index (RSVI) to quantify the spatial vulnerability of a network system under spatial hazards (see Figure 1).

Specifically, in the SVM, the impact curve is often obtained by establishing a mathematical relationship between the area covered by hazards and the global impact of hazards. For the sake of calculation, it replaces the area covered by hazards with the percentage of hazard covered area, and the global impact of hazards by the percentage of hazard impacted nodes. It should be noted that the definition of global impact of hazards on a network system is highly problem dependent. For example, we can use the number of impacted nodes to quantify the global impact of hazards. We can also use the volume of function losses/failures in a network system to calculate the global impact of hazards. Here it simply defines the global impact of hazards on a network system as their impact on network nodes, i.e., the percentage of impacted nodes. Moreover, even for the same area covered by a hazard, depending on different hazard locations in a network system, the percentage of impacted nodes may vary largely. Therefore, it further introduces the concept of average percentage of impacted nodes under a specific percentage of hazard covered area, which can be obtained from historical data or hazard simulations. A neutral curve usually reflects an expectation on the resistance capacity of a network system in the face of spatial hazards, which is also problem dependent. Here it is simply defined according to a common sense, i.e., the percentage of impacted nodes are proportional to the percentage of hazard covered area. Based on these two curves, it further calculates the ASVI and RSVI. Their mathematical descriptions are as follows:

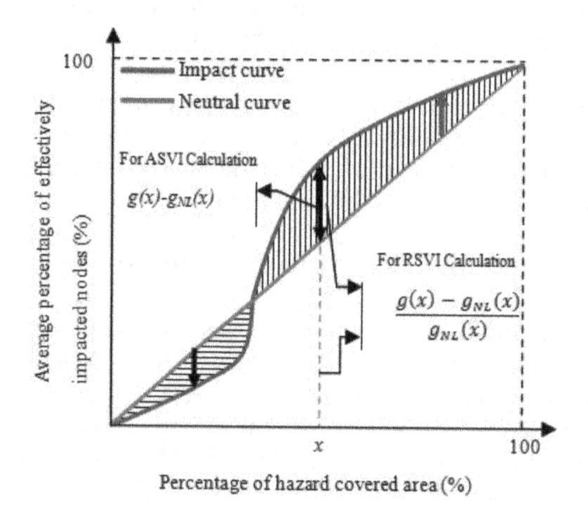

Figure 1. Spatial vulnerability model.

$$V_{ASVI} = \int_0^1 \left(g(x) - g_{NL}(x)\right) dx \quad (1)$$

$$V_{RSVI} = \int_0^1 \frac{g(x) - g_{NL}(x)}{g_{NL}(x)} dx \quad (2)$$

where x is the percentage of hazard covered area, $g(x)$ is the average percentage of impacted nodes for a given x value, and $g_{NL}(x)$ is the associated neutral curve value.

According to the definition of the neutral curve above, one has:

$$g_{NL}(x) = x \quad (3)$$

Basically, a larger value of ASVI/RSVI means a higher spatial vulnerability. Thus, based on the impact curve, the neutral curve, and the ASVI and RSVI values, we can analyze the spatial vulnerability of a network system quantitatively and also compare the spatial vulnerability of different network systems under spatial hazards. For more details of the SVM, readers may refer to Li, et al., (2016a) and Li, et al., (2016b).

2.2 *A modified SVM*

In this study, we aim to using the SVM to analyze the impact of random and intentional spatial hazards on the ECAN and the CCAN. And we are particularly concerned with the impact of spatial hazards on air lines of the two networks. Therefore, some modifications to the SVM are needed.

Firstly, when getting the impact curve, we replace the percentage of hazard covered area by the percentage of hazard covered airports, and the percentage of impacted nodes by the percentage of impacted air lines. Then we establish a mathematical relationship between these two factors and get the impact curve. In this study, we define the neutral curve as follows: the percentage of impacted air lines is proportional to the percentage of hazard covered airports.

Secondly, when calculating the percentage of impacted air lines, we further consider the weight of each air line, WL, as different air lines may have different importance to the system. The mathematical description of the weight of an air line, WL_{ij}, is as follows,

$$WL_{ij} = \frac{F_{ij}}{\sum F} \quad (4)$$

where, WL_{ij} is the weight of the air line between airport i and airport j, F_{ij} is the number of flights between airport i and airport j, $\sum F$ is the total number of flights of the whole system.

Finally, we can use this modified SVM to analyze the spatial vulnerability of the ECAN and the CCAN under random and intentional spatial hazards.

3 THE SPATIAL VULNERABILITY OF THE ECAN AND THE CCAN UNDER RANDOM AND INTENTIONAL SPATIAL HAZARDS

In this study, the airport distributions of the ECAN and the CCAN are shown in Figure 2 and Fig. 3, respectively. The links of the two networks are built based on the flights between their airports. The distribution of random spatial hazards is spatially uniform. The possibility of an intentional spatial hazard at an airport is determined based on the amount of flights at this airport. Then we study the spatial vulnerability of the two networks under four hazard scenarios, random spatial hazards without considering WL, random spatial hazards with considering WL, intentional spatial hazards without considering WL and intentional spatial hazards with considering WL. In each hazard scenario, we change the percentage of

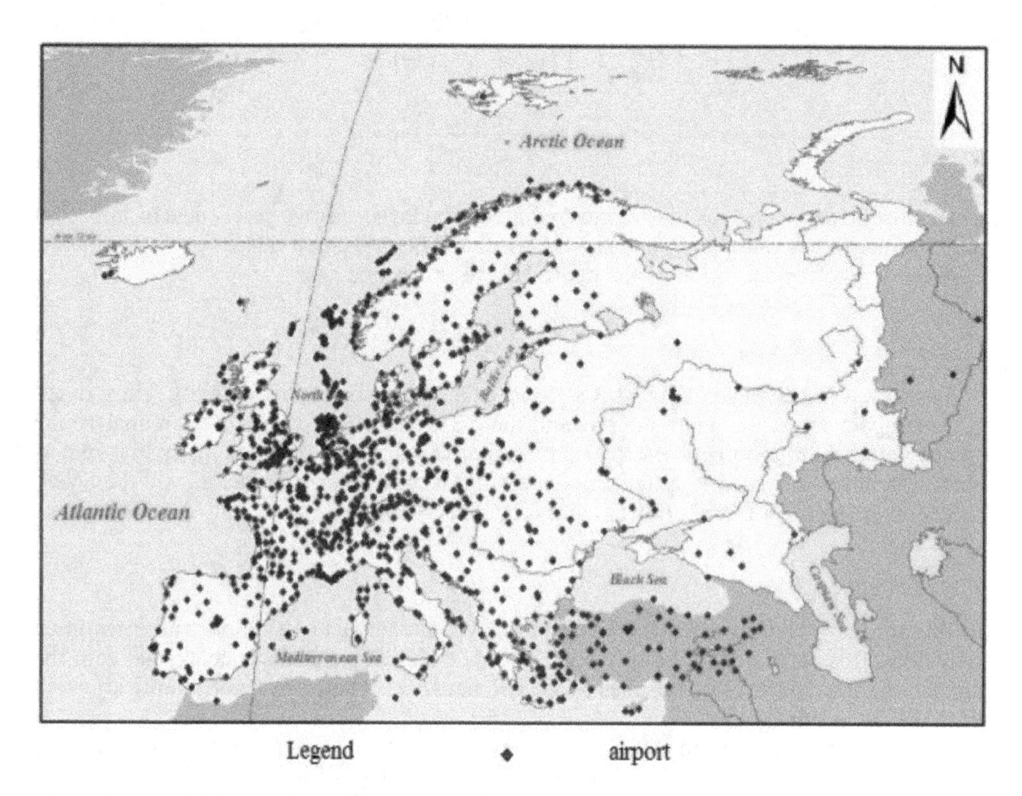

Figure 2. Spatial distribution of European civil airports.

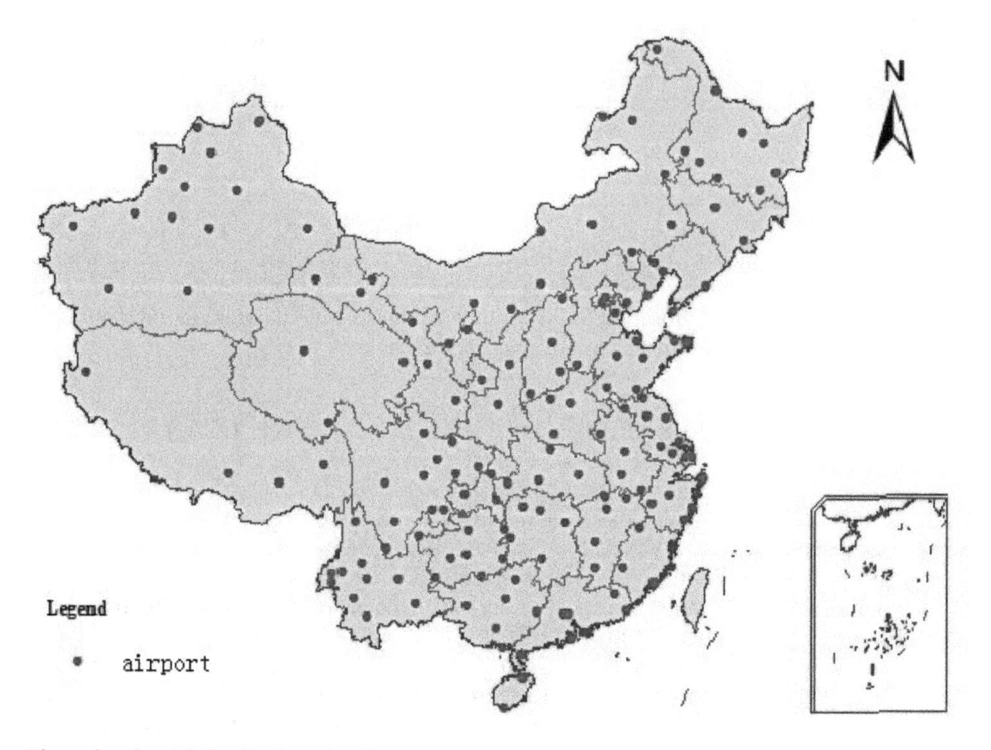

Figure 3. Spatial distribution of Chinese civil airports.

hazard covered airports from 0% to 100% by a step of 0.5%. For each step, we conduct 100 times hazard simulations. The results of spatial vulnerability of the ECAN and the CCAN are shown in Table 1, Figure 4, and Figure 5.

In the four hazard scenarios of the ECAN, all the ASVI and RSVI values are positive, as shown in Table 1 and Figure 4, which means the European civil aviation network is vulnerable to both random and intentional spatial hazards. Furthermore, when comparing the spatial vulnerability of the ECAN under random and intentional spatial hazards, one can see that, no matter whether considering *WL* or not, the ECAN has much larger ASVI/RSVI values under intentional spatial hazards, indicating the ECAN is more vulnerable to intentional spatial hazards. This is reasonable because under random spatial hazards, each airport has equal possibility to be affected by hazards. However, under intentional spatial hazards, airports with more flights are prior to be attacked. And it usually further leads to more impacts on air lines, as airports with more flights also means more possible air lines connected to them. In addition, when analyzing the impact of *WL* on the spatial vulnerability of the ECAN, one

Table 1. Results of spatial vulnerability of the ECAN and the CCAN.

	ECAN				CCAN			
	Without *WL*		With *WL*		Without *WL*		With *WL*	
	ASVI	RSVI	ASVI	RSVI	ASVI	RSVI	ASVI	RSVI
Random hazard	16.6904	49.8130	33.3545	100.1481	16.7117	49.7063	33.4687	100.2332
Intentional hazard	39.9003	191.0863	89.0512	501.0921	35.3069	138.4071	79.2370	357.0290

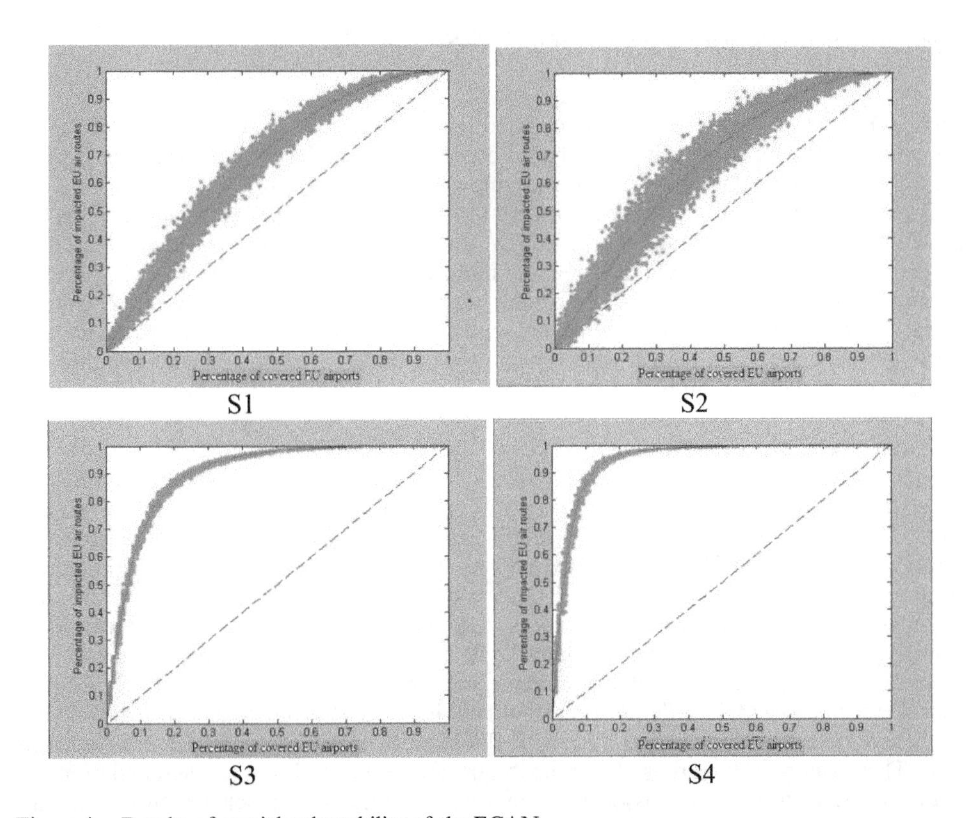

Figure 4. Results of spatial vulnerability of the ECAN.

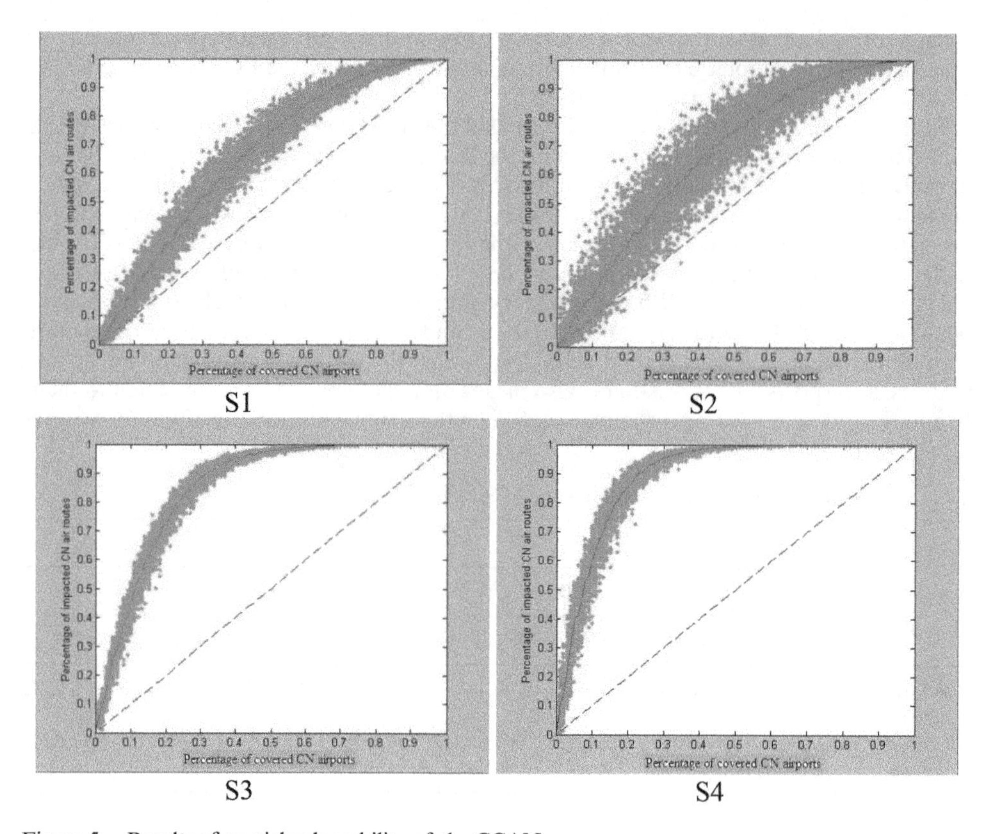

Figure 5. Results of spatial vulnerability of the CCAN.

can see that, no matter under random or intentional spatial hazards, the ASVI/RSVI values are much larger with considering *WL* than those without considering *WL*, which means that the ECAN is more vulnerable when considering *WL*.

For the CCAN, according to the associated results shown in Table 1 and Figure 5, one can get similar conclusions as those of the ECAN. Firstly, the CCAN is also vulnerable to both random and intentional spatial hazards, and more vulnerable to intentional spatial hazards. Meanwhile, when considering *WL*, the system become more vulnerable to random and intentional spatial hazards, especially intentional spatial hazards. Since intentional spatial hazards are largely related to terrorism attacks, which have nowadays become an increasing threat in the real world, it is very important to take *WL* into consideration when assessing the spatial vulnerability.

Furthermore, when comparing the results of the ECAN with those of the CCAN, one can see that, under random spatial hazards, basically these two networks have similar ASVI/RSVI values, no matter whether considering *WL* or not (see Table 1). It means that the ECAN and the CCAN show a similar vulnerability level when facing random spatial hazards. However, under intentional spatial hazards, the ECAN has much larger ASVI/RSVI values than those of the CCAN, which indicates that the ECAN is more vulnerable than the CCAN when facing intentional spatial hazards. This result is explainable. As mentioned before, under intentional spatial hazards, airports with more flights will have larger possibility to be attacked. According to the amount of flights of each airport, we firstly sort airports of the ECAN and the CCAN in descending order, respectively. That is, we get the importance of each airport to the two networks. Meanwhile, we record the number of air lines connected to each airport. Then, based on the order of airport importance, we can obtain a general relationship between covered airports and associated connected air lines of each network by plotting the

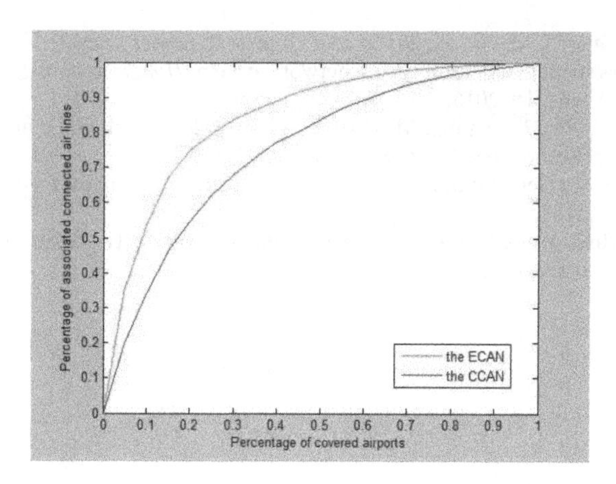

Figure 6. Relationship between covered airports and associated connected air lines of the ECAN and CCAN based on airport importance.

percentage of covered airports against percentage of associated connected air lines, as shown in Figure 6. Obviously, for any specific percentage of covered airports, one can see that, the ECAN will contain more percentage of air lines than that of the CCAN (see Figure 6). That is, if an intentional spatial hazard happens to these two networks simultaneously, more air lines in ECAN will be affected by the hazard, i.e., the ECAN is more vulnerable to intentional spatial hazards than the CCAN.

4 CONCLUSION

This paper applies a modified vulnerability model to analyze the spatial vulnerability of the European civil aviation network (ECAN) and the Chinese civil aviation network (CCAN) under random and intentional spatial hazards. The simulation results show that both of the two network systems are vulnerable to random and intentional spatial hazards, and more vulnerable to intentional spatial hazards. Furthermore, when comparing the spatial vulnerability of the two network systems, the ECAN is more vulnerable than the CCAN to intentional spatial hazards.

ACKNOWLEDGEMENTS

This work was supported in part by the National Key Research and Development Programme (Grant No.2016YFA0602404), and the National Natural Science Foundation of China (Grant No.61472041). The flight data of the ECAN and the CCAN are collected from the websites of EUROCNTROL and Qunar, respectively.

REFERENCES

Dunn, S., Wilkinson, S.M. 2016. Increasing the resilience of air traffic networks using a network graph theory approach. *Transportation Research Part E Logistics & Transportation Review* 90, 39–50.

Guo, X.M., Hu, X.B., Li, H., et al. 2015. A study on spatial-temporal rainstorm risk at civil airports in China. *Risk Analysis and Crisis Response* 5(3): 188–198.

International Civil Aviation Organization. 2014. Annual report of the ICAO council. http://www.icao.int/annual-report-2014/Pages/default.aspx.

Li, H., Hu, X.B., Guo, X.M., et al. 2016a. A new qualitative and quantitative method to study the vulnerability of civil aviation network system. *Disaster Risk Science* 7(3): 245–256.

Li, H., Hu, X.B., Guo, X.M., et al. 2016b. An improved model considering traditional network properties to assess spatial vulnerability of a network system. *Human and Ecological Risk Assessment* published online: August 10, 2016.

Mazzocchi, M., Hansstein, F., Ragona, M. 2010. The 2010 volcanic ash cloud and its financial impact on the european airline industry. *Cesifo Forum* 11(2): 92–100.

Verma, T., Araújo, N.A., Herrmann, H.J. 2014. Revealing the structure of the world airline network. *Scientific Reports* 4, 5638.

Zanin, M., Lillo, F. 2013. Modelling the air transport with complex networks: a short review. *The European Physical Journal Special Topics* 215(1), 5–21.

Risk Analysis and Management – Trends, Challenges and Emerging Issues – Bernatik, Huang & Salvi (Eds)
 ISBN 978-1-138-03359-7

Time-dependent probabilistic seismic hazard assessment for Central Bayan Har Block

Changlong Li
Institute of Geophysics, China Earthquake Administration, Beijing, China

Mengtan Gao
China Earthquake Risk and Insurance Laboratory, Beijing, China

ABSTRACT: Predicting ground motion intensity and compiling seismic hazard maps are important methods for earthquake disaster reduction. This type of work requires scientific and effective Probabilistic Seismic Hazard Assessment (PSHA; Cornell 1968). PSHA involves the determination of seismic hazard models including source models, seismicity models, Ground Motion Prediction Equation (GMPE) models and site condition models. Bayan Har Block is a part of Tibetan Plateau of China, and possesses serious seismic hazard. Seismic hazard models can be divided into time-independent and time-dependent models. Previous PSHA work in China has used time-independent seismic hazard models. In this study, we constructed a time-dependent seismic hazard model for Central Bayan Har Block and the surrounding area based on several major-earthquake seismogenic structures for which historical and paleoseismic event data were available. With the time-dependent model, we calculated the distribution of peak ground acceleration with 10% probability of exceedance in the next 50 years in the area, and compared the results with those calculated using the time-independent model. The results showed that for the time-dependent seismicity model, the seismic hazard is higher on faults that are close to their expected event recurrence times, and is lower on the faults that have recently experienced a major event.

Keywords: Central Bayan Har; Time-dependent probabilistic seismic hazard assessment; paleoearthquake; Jiaocheng Fault

1 INTRODUCTION[1]

Predicting ground motion intensity and compiling seismic hazard maps are important methods for earthquake disaster reduction. This type of work requires scientific and effective Probabilistic Seismic Hazard Assessment (PSHA; Cornell 1968). PSHA involves the determination of seismic hazard models including source models, seismicity models, Ground Motion Prediction Equation (GMPE) models and site condition models.

Seismic hazard models can be divided into time-independent and time-dependent models. Previous PSHA work in China has used time-independent seismic hazard models. That model assumed that the source model (Zhou et al. 2013) is an area source model, that the seismicity model (Pan et al. 2013) follows the Gutenberg-Richter relationship (Gutenberg and Richter 1944), and that earthquakes of any magnitude occur according to a Poisson distribution. The time-dependent model assumes that earthquake probability varies with time, and is dependent on the time elapsed since the last earthquake event. As major earthquakes usually have a

[1]Sponsored by: National Natural Science Fund (51678537); Fundamental Research Specific Fund (DQJB16B19).

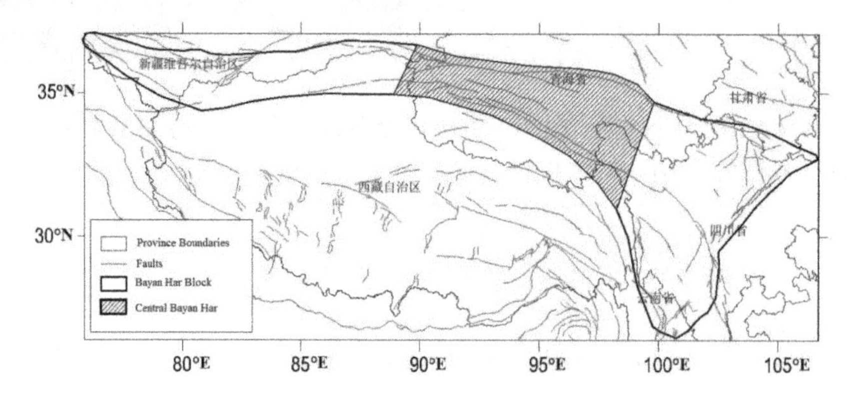

Figure 1. The study area—Central Bayan Har Block.

recurrence cycle of hundreds of years or more, it is more appropriate to use a time-dependent model when considering seismic hazard within a short period of time (e.g. 50 years).

Time-dependent PSHA was developed for the purpose of describing seismogenic structures with time-dependent seismicity, and has been applied in several countries. Meanwhile, Bayan Har Block shows active seismicity in recent years. It is important to make a new Probabilistic Seismic Hazard Analysis (PSHA) in the east part of the block. This paper reevaluates probabilistic seismic hazard of Central Bayan Har Block (Figure 1). In this paper, we construct a new PSHA for Central Bayan Har with time-dependent seismic hazard models based on new data. The new data used included two aspects. (1) Different from the area source model, the source model is considered to be a fault source model. This model is based on the consideration that a major-earthquake rupture surface can significantly affect the ground-motion distribution in the near-field (Aki 1968; Schnabel and Seed 1973; Chen et al. 1998; Hu 1999). (2) The time-dependent seismicity model is based on several major-earthquake seismogenic structures. We produced a distribution map of the Peak Ground Acceleration (PGA) with 10% probability of exceedance in the next 50 years in the area using the time-dependent model, and compared the results with the map obtained from a time-independent model.

2 TIME-DEPENDENT SEISMIC HAZARD MODEL

In the time-dependent model, we considered several major-earthquake seismogenic structures in the area for which historical and paleoseismic event data were available (Figure 2). We used the fault source model for these faults and the time-dependent model for major earthquake probabilities.

2.1 *Fault source model of the major-earthquake seismogenic structures in the area*

Faults in the area for which historical and paleoseismic event data are available are shown in Figure 2. The parameters of these structures are listed in Table 1.

The empirical formula for earthquake rupture length and width is (Wells and Coppersmith 1994):

$$\lg L = -1.43 + 0.88 M_W \tag{1}$$

$$\lg W = -1.01 + 0.32 M_W \tag{2}$$

where L is the length of the rupture surface, W is the width of the rupture surface, and M_W is the moment magnitude.

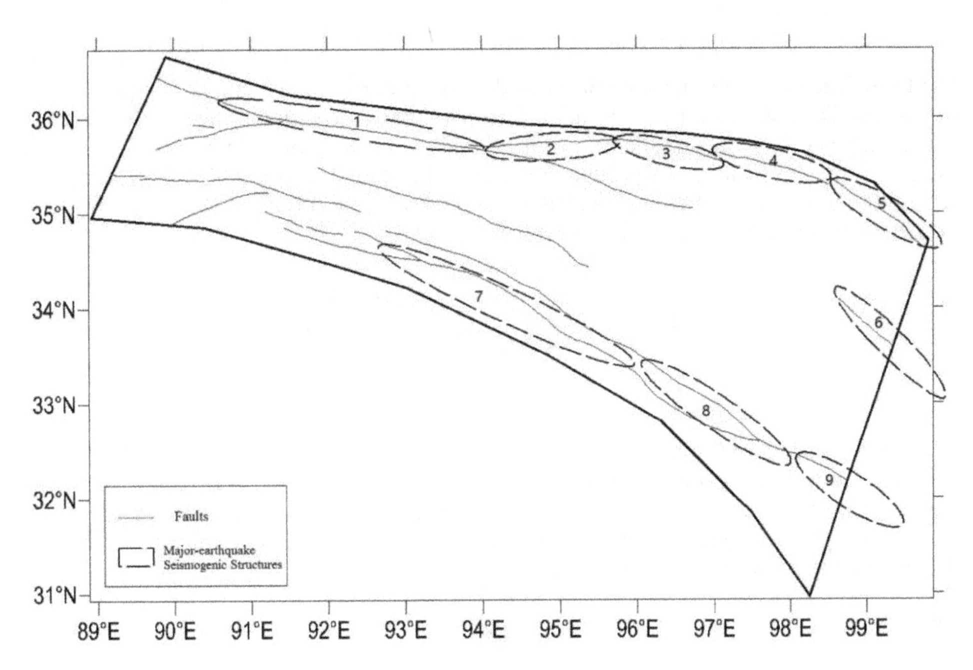

Figure 2. Major earthquake seismogenic structures in East Bayan Har Block, seismogenic structures the numbers are referred to in Table 1.

Table 1. Fault parameters of the major-earthquake seismogenic structures.

No.	Fault name	Magnitude	Dip (°)	Rake (°)	Elapsed time, recurrence interval and standard deviation of major earthquake (aBP)	Annual event probability	Data Source
1	Kusai fault	8.5	11.5	6.4	14 423	0	Xia et al., 2013
2	Xidatan fault	8	12		223 848 500	0.00044	Xia et al., 2013
3	Alake fault	8	6.5		462 1284 300	0	Xia et al., 2013
4	Tuosu fault	8	6	7	78	0	Xia et al., 2013
5	Maqumaqin fault	8	6.3		1070 1041.7 500	0.0019	Fu, 2012
6	Dari fault	8			68	0	
7	Wulanwula fault	7.5			5k 5k 1.3k	0.00068	Du et al., 2012
8	Yushu fault	7.5	6.4		5	0	
9	Dengke fault	7.5 <6.5	7.2 ± 1.2		119 400 133	0.000107	

2.2 *Time-dependent seismicity model of major-earthquake seismogenic structures*

The time-dependent seismicity model considers the recurrence of major earthquakes to be quasi-periodic, and that earthquake annual probability varies with the elapsed time since the last earthquake. For the seismicity model of faults with historical and paleoseismic event data, most previous studies (Wen 1998, Yang and Liu 2000; Petersen et al. 2007; Hebden and Stein 2009) used a lognormal model (NB model; Nishenko and Buland 1987). The probability density function for earthquake recurrence is:

$$f\left(T/\bar{T}\right)=\frac{\bar{T}}{\sigma_D\sqrt{2\pi T}}\exp\left\{\frac{-\left[\ln\left(T/\bar{T}\right)\right]^2}{2\sigma_D^2}\right\} \tag{3}$$

where T is the time elapsed since the last earthquake, $\bar{T}$ is the median recurrence interval, and σ_D is the standard deviation.

In this study, historical and paleoseismic event data were available for major-earthquake seismogenic structures, so the NB model was selected. In the calculation, we considered two aspects of standard deviation, the intrinsic uncertainty of earthquake recurrence series and the uncertainty of paleoearthquake dating:

$$\sigma_D=\sqrt{\sigma_T^2+\sigma_P^2} \tag{4}$$

where σ_T is the uncertainty of the unevenly distributed earthquake recurrence series and σ_P is the uncertainty of paleoearthquake dating.

On the basis of the data provided in Table 1, we can calculate the annual probability of a major earthquake on each fault. The calculation results are also listed in Table 1.

2.3 *GMPE model and site condition model*

From the features of the fault source model, GMPEs must be selected to have equations with the distance parameter R_{rup} (the shortest distance from site to rupture surface). Thus, we chose the GMPEs of Campbell and Bozorgnia (2008). The distance parameter of the GMPEs is R_{rup}, and the distance term f_{dis} is:

$$f_{mag}=\left(c_4+c_5M\right)\ln\left(\sqrt{R_{RUP}^2+c_6^2}\right) \tag{5}$$

where the dimension of R_{rup} is km, and c_4, c_5 and c_6 are constant.

The site condition model is the same as that of time-independent PSHA.

3 PSHA AND RESULTS

The computational formulation of PSHA is developed by Cornell (1968), Esteva (1970), and McGuire (1976):

$$p(a>A)=\sum_i v_i\int_m\int_r\int_\sigma P(a>A\mid m,r,\sigma)f_m(m)f_r(r)f_\sigma(\sigma)dmdrd\sigma \tag{6}$$

where $p(a > A)$ is the annual frequency of exceedance of ground motion amplitude A, v_i is the annual activity rate for the ith seismogenic source for a threshold magnitude, and the function P yields the probability of the ground motion parameter a exceeding A for a given magnitude m at source-to-site distance r. Also considered is the standard deviation of the residuals (in log-normal distribution) associated with GMPE, denoted by σ. The corresponding probability density functions are represented by $f_m(m)$, $f_r(r)$, and $f_\sigma(\sigma)$.

When performing time-dependent PSHA, only major earthquakes are regarded as time-dependent, and smaller earthquakes are calculated as for the time-independent model. We plotted maps of PGA with 10% probability of exceedance in the next 50 years in the area for both the time-independent and time-dependent models (Figures 3 and 4).

Figure 3 shows that, with the time-independent model, the Jiaocheng and Hengshan faults have lower seismic hazard. From Figure 4, the Jiaocheng and Hengshan faults have much higher seismic hazard with the time-dependent model than with the time-independent model.

We calculated the ratio of PGAs with 10% probability of exceedance in the next 50 years in the area with the time-independent and time-dependent models. The distribution of the ratio is illustrated in Figure 5. Almost the whole area has higher seismic hazard with the time-dependent model than with the time-independent model. The seismic hazards of the Jiaocheng and Hengshan faults increase the most when the time-independent model is replaced by the time-dependent model.

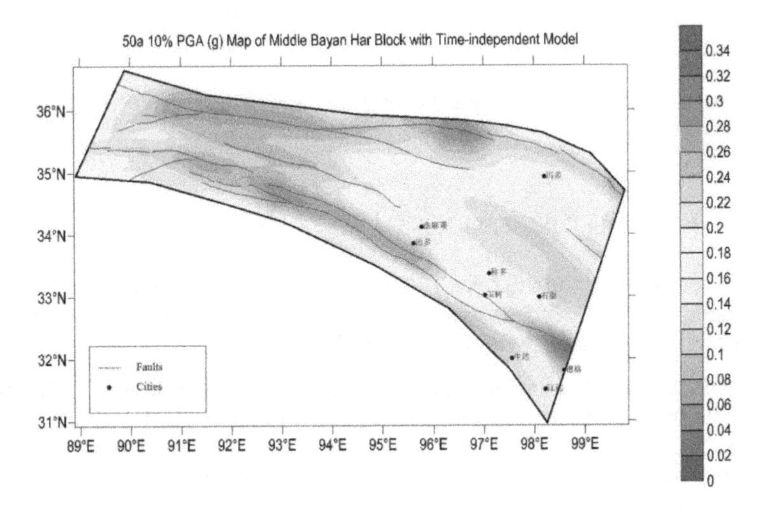

Figure 3. PGA with 10% probability of exceedance in the next 50 years in the area with the time-independent model.

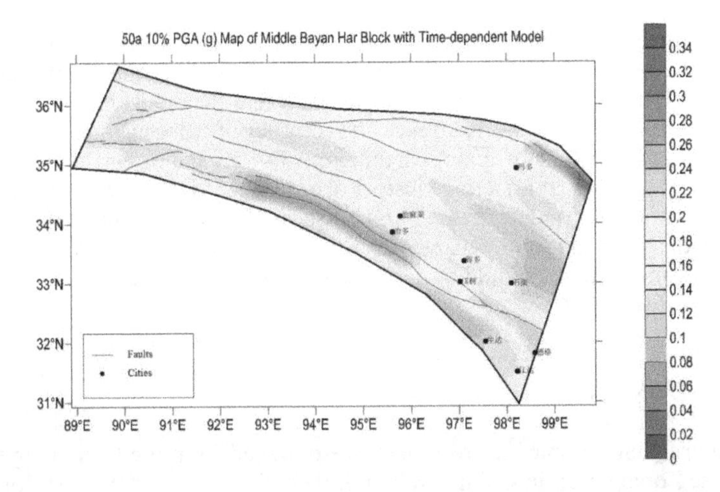

Figure 4. PGA with 10% probability of exceedance in the next 50 years in the area with the time-dependent model.

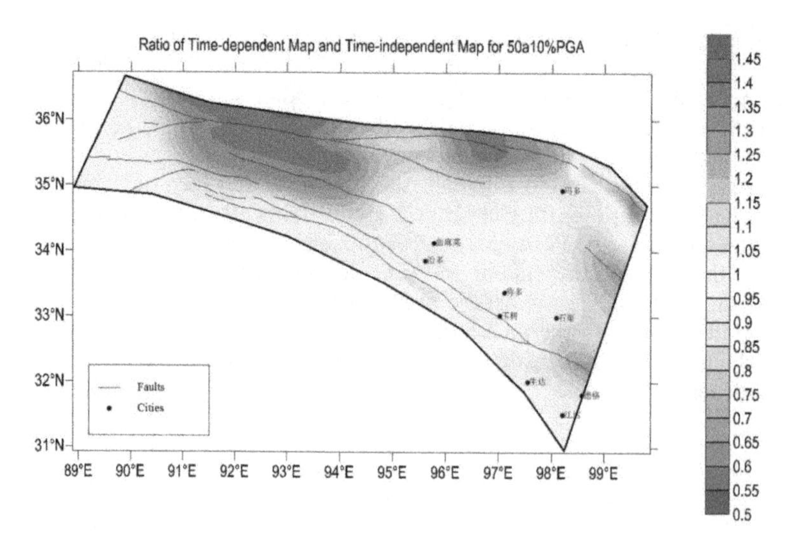

Figure 5. Ratio of PGAs with 10% probability of exceedance in the next 50 years in the area for the time-independent and time-dependent models.

5 CONCLUSION AND DISCUSSION

In this study, we calculated the PGA with 10% probability of exceedance in the next 50 years in the area using a time-dependent model, and compared the results with those of a time-independent model. Our main conclusions are as follows:

1. For the time-dependent seismicity model, the seismic hazard is higher on faults that are close to their expected event recurrence times, and is lower on the faults that have recently experienced a major event.
2. High-seismic-hazard areas in the next 50 years are the Maqumaqin Fault and the Wulanwula Fault.
3. If a time-independent model were used, areas with seismic hazards overestimated in the next 50 years would be the Kusai Fault, Yushu Fault and the Dengke Fault.
4. The Maqumaqin Fault is the area with the highest seismic hazard in the next 50 years with the new models.

The time-dependent seismic hazard model that we used is based on historical and paleoearthquake event data, and we suggest new conclusions about the seismic hazard of the area with this model. The uncertainty of the parameters in the model mainly result from the aperiodicity of major-earthquake recurrence, the accuracy of paleoearthquake dating and the complexity of the occurrence of major earthquakes. For the major-earthquake seismogenic structures that we selected, at least three historical or paleoearthquake records were available, and we used comprehensive data sources and paleoearthquake investigation methods; thus, the results of the study are quite reliable. In the future, more data should be collected and further theories and methods should be developed to assess the model uncertainty and verify the reliability of the results.

ACKNOWLEDGMENTS

The time-independent seismic hazard model was obtained from the Compiling Committee of the Seismic Zoning Map in China. We thank them for providing the area source model, seismicity model, GMPE model and site condition model data.

REFERENCES

Aki, K. (1968). Seismic Displacement near a fault. J Geophys Res 73(16):5359–5376.

Campbell, K.W., Bozorgnia. Y. (2008). NGA ground motion model for the geometric mean horizontal component of PGA, PGV, PGD and 5% damped linear elastic response spectra for periods ranging from 0.01 to 10 s. *Earthquake Spectra* 24(1): 139–171.

Cornell, C.A. (1968). Engineering seismic risk analysis. Bulletin of the Seismological Society of America 58(5): 1583–1606.

Du, Y., Huang, X.M., He, Z.T., et al. 2012. Paleoearthquake research on East Wulanwula Fault. *Journal of Chifeng University* (Natural Science Edition, in Chinese), 28(10): 131–132.

Esteva, L. (1970) Seismic risk and seismic design decisions. In Seismic Design for Nuclear Power Plants, ed. R. J. Hansen, 142–182. Cambridge, MA: Massachusetts Institute of Technology Press.

Fu, J.D. 2012. Paleoearthquake and large earhquake recurrence interval of Tazang Fault Luocha Segment of the east segment of the East Kunlun Fault. Beijing: Institute of Geology, China Earthquake Administration (in Chinese).

Gutenberg, B., Richter, C.F. (1944). Frequency of earthquakes in California. Bulletin of the Seismological Society of America 34(4): 185–188.

Hu, Y.X. (1999). *Seismic Safety Evaluation Technology Tutorials*. Beijing: Seismological Press (in Chinese).

Chen, P.S., Li, B.K., Bai, T.X. (1998). Prediction of peak horizontal acceleration in the light of tectonic ambient shear stress field. Chinese J. Geophys. (in Chinese) 41(4): 502–517.

Li, X.J. (2013). Adjustment of seismic ground motion parameters considering site effects in seismic zonation map. *Chinese Journal of Geotechnical Engineering* (in Chinese) 35(Supp. 2): 21–29.

McGuire, R.K. (1976). Fortran computer program for seismic risk analysis. USGS Open-File Report 76–67.

Pan H, Gao MT, Xie FR (2013) The earthquake activity model and seismicity parameters in the new seismic hazard map of China. Technology for Earthquake Disaster Prevention (in Chinese) 8(1): 11–23.

Schnabel, P.B., Seed, H.B. (1973). Accelerations in rock for earthquakes in the western United States, Bulletin of the Seismological Society of America 63(2): 501–516.

Wells, D.L., Coppersmith, K.J. (1994). New empirical relationships among magnitude, rupture length, rupture width, rupture area, and surface displacement. *Bull Seismol Soc Am* 84(4): 974–1002.

Xia, Y.S., Li, Z.M., Tu, H.W., et al. 2013. Ancient earthquake research of east Kunlun fault zone. Progress in Geophys. (in Chinese), 28(1): 146–154.

Yu, Y.X., Li, S.Y., Xiao, L. (2013). Development of ground motion attenuation relations for the new seismic hazard map of China. Technology for Earthquake Disaster Prevention (in Chinese) 8(1): 24–33.

Zhou, B.G., Chen, G.X., Gao, Z.W., et al. (2013) The technical highlights in identifying the potential seismic sources for the update of national seismic zoning map of China. Technology for Earthquake Disaster Prevention (in Chinese) 8(2): 113–124.

Risk Analysis and Management – Trends, Challenges and Emerging Issues – Bernatik, Huang & Salvi (Eds)
© 2017 Taylor & Francis Group, London, ISBN 978-1-138-03359-7

Exploring the reason for the dynamic problem of earthquake swarm in the Simao-Puer region, southwest Yunnan

Tieming Li & Yuzhu Bai
Key Laboratory of Active Fault and Volcano, Insititue of Geology, China Earthquake Administration, Beijing, China

Desheng Shao
Earthquake Administration of Yunnan Province, Kunming, China

Gang Chen
China University of Geoscience, Wuhan, P.R. China

ABSTRACT: There is a special structure combination and the concentration of a strong earthquake in the Simao-Puer block. The geological phenomena of motion and deformation caused by the motion of the active fault and the sub-block have attracted the attention of researchers in this field. Based on past research achievements, we constructed the 3-dimension finite element model of Simao-Puer region containing the active fault. Applying the velocity of block per year measured by GPS data and the researches on geology and geomorphology as the boundary conditions of the finite element model, we then computed the spatial distribution of deformation in the Simao-Puer region and simulated the deformation displacement along the NEE and NNW direction. Lastly, we analyzed the characteristic of the horizontal and vertical displacement field in the Simao-Puer region. The numerical result shows that extension motion of the southeastern margin of the Tibetan Plateau makes the movement of slump due to gravity. In the research area, the motion of fault with NEE strike draws the fault having NNW strike to move in a conjugate direction. The vertical deformation caused by the uplifting movement is consistent with the result of 60 years of larger regional standard measure and the deformation along the NNW direction distributed irregularly, which is possibly made by regional stress accumulation and readjustment after the earthquake. The deformation field along the NEE direction is larger in the west of Simao-Puer region than in the eastern part, especially the northwest part of the region corresponding to the 6.6 Jinggu earthquake that occurred recently. Lastly, we preliminarily analyzed the dynamic problems in the research area.

1 INTRODUCTION

Due to crisscross fault zones, complicated stress conditions and frequent moderately strong earthquakes (6.0~6.9), southwest Yunnan Simao-Puer seismic zone in China has become the focus of scientific research in geoscience. Previous research suggested that the tectonic movement, stress and strain, as well as the development and occurrence of strong earthquakes in southwest Yunnan is controlled by the north-east thrust of the India plate, especially the activities of structural knots in the east (Deng et al., 2003; Zhang et al., 2003; Xu et al, 2003; Li et al, 2003; Tapponnier et al., 1982; Paul et al., 2001; Clark and Royden, 2000; England and Houseman, 1989; Anne et al., 2006). In addition, the uplifting of the Tibetan Plateau and the lateral expansion of crustal material movement also contributes (Houseman and England, 1993; Engnd and Molnar, 1990; Replumaz and Tapponnier, 2003; Shen et al., 2001). On the west side of the Simao-Puer earthquake area is a large-scale seismic structural belt, the Longling-Lancang active fault zone (Guo et al., 2000), which has no surface penetration but good linear distribution of earthquakes. The clockwise rotation of Sichuan-Yunnan rhombic blocks constitutes the southeastward expansion of the plateau material, with most of the Yunnan region experiencing more than seven earthquakes. It can be seen that a circle-shaped seismic ring with a circle closed to Dali-Chuxiong-Tonghai-Honghe-Jiangcheng-Mengla-Lancang-Gengma-Longling-Baoshan is distributed in the study area (Li et al., 1980).

The Simao-Puer earthquake zone is located in the middle of the fault block and the moderate-strong earthquakes should be the result of strong activity of the fault within the fault block (Li et al., 1980). Due to the difference of the shielding effect and the absorption performance caused by the fault activity, the velocity and even the direction of movement of the fault block are unified and harmonized. There are obvious differences between the different secondary units divided by the fault zone and the study of kinematic and dynamic characteristics of local active fault blocks and fault zones is undoubtedly an important basis for understanding the location and dynamic conditions of strong earthquakes (Zhang et al., 2003). It can be seen that the study of strong earthquakes in the active fault zone with conjugate shear properties is very important for the occurrence of strong earthquakes in the subplots with a high degree of fractures and the interpretation and their relationship to regional crustal deformation and tectonic activity are the keys.

Based on previous studies, this paper started from the regional new (earthquake) tectonics, the active tectonics, and the present crustal deformation field according to multi-period GPS resurvey data from the most direct present crustal deformation Puer earthquake area. The authors analyzed the current crustal movement characteristics of the main faults and active blocks in the Simao-Puer earthquake area and analyzed the characteristics of the crustal movement of the Simao-Puer earthquake zone. The Simao-Puer earthquake group and its regional dynamic background are also discussed.

2 GPS DATA AND CALCULATION

The rapid development of modern space geodetic technology, especially GPS, makes it possible to obtain the observational data of the earth with unprecedented high spatial-temporal resolution. It becomes an important means to monitor the dynamic phenomenon of the earth. This study uses five GPS repetition measurement data of 1999, 2001, 2004, 2007, and 2011, and the data continuously for 30 days from the continuous observation station, which was determined by the site investigation in the Puer earthquake in 2007 to deduct from the observed velocity caused by the coseismic deformation speed to the area from seismic crustal movement and the speed of the coseismic effect field.

The GPS data was processed with the GAMIT/GLOBK software package in order to ensure the reliability of data processing. All the GPS original observation data in this study are processed by the same model, parameter, and method. Data from the global IGS station and the CMONOC base station during the same period were also processed. To improve the efficiency, data processing was divided into global, national, and regional levels, and they were combined through satellite orbit and the common GPS coordinates. Thus the global IGS station, CMONOC base station, and one-day solution of GPS point in the seismic region are obtained. The one-day solution provides the site coordinates, polar motion, orbit, and its variance-covariance matrix (Wang et al., 2008).

3 ESTABLISHING THE FINITE ELEMENT MODEL

To investigate the dynamic environment and evolution rule of such complex seismic tectonic zones with the distribution area and high temporal density, regional crustal deformation monitoring data at present, especially those laid in the use of inversion in the study area of

GPS data calculation is obviously far from enough to give the secondary fault block, single fault movement, and deformation accurately reliable results. The study of the deformation process and mechanism should be based not only on the basis of data of the surface deformation but also on deep tectonic characteristics. Therefore, this study attempts to establish a three-dimensional finite element model of the Simao-Puer block, which is clamped by the present tectonic framework of the chessboard through the Lancang River and the Honghe fault zone. The spatial distribution of crustal deformation in the Simao-Puer block is analyzed by the finite element method. The high-resolution analysis of the crustal deformation and the area of the vertical deformation field (perpendicular to the surface direction) are carried out in the direction perpendicular to the calculation area (NE) and along the calculation area (NNW), from the vertical to the calculation area (referred to as NEE) and along the strike calculation area (referred to as NNW), and two directions are of high-resolution horizontal crustal deformation field.

3.1 *Specific steps*

1. A three-dimensional finite element model of Simao-Puer area in Yunnan with 10 faults is established. The geometrical parameters of faults are shown in Table 1.
2. In order to simplify the finite element model, the layered geological structure model is not considered here and the computational medium is regarded as an isotropic medium without taking into account the topographical factors such as surface relief. When the fracture displacement of the Honghe and Lancangjiang faults is 4 mm/a, the boundary load of 3 MPa should be applied through the dislocation model. Therefore, compressive stresses of 4 MPa and 3 MPa are imposed on the western and northern boundaries of the model. In the south of the model, the displacements of the four GPS points near the south of the model are averaged, i.e., the southward displacement of 4 mm/a is applied to the south of the model. In the eastern boundary of the model, the same method takes displacement in the east direction, that is, the direction of application is east and the displacement constraint is 2 mm/a, as shown in Figure 1.

Table 1. Parameters of the active fault in the computational zone.

Name of the fault	Strike	Tendency	Angle of (°dip)	Nature of movement
Lancangjiang fault (F36)	N-N-W	S-W	80	Dextral strike-slip
HongheFault (F130)	N-N-W	N-E	80	Dextral strike-slip
Zhenyuan-Puer fault (F132)	N-N-W	N-E	70	Dextral strike-slip
Shangshi fault (F134)	N-N-W	N-E	70	Dextral strike-slip
Mohei-Qiaotou fault (F133)	N-N-W	N-E	70	Dextral strike-slip
Jinggu-Puwen fault (F135)	N- S	N-E	70	Dextral strike-slip
Sanlinchang-Siyongjie fault (F136)	N-E-E	S-W	70	Sinistral strike-slip
Zhendong-Mengxian fault (F137)	N-E-E	S-W	70	Sinistral strike-slip
Zhengwan-Laizidi fault (F138)	N-E-E	S-W	70	Sinistral strike-slip
Xiaomengyang-Xiangzhuang fault (F139)	N-E-E	S-W	70	Thrust

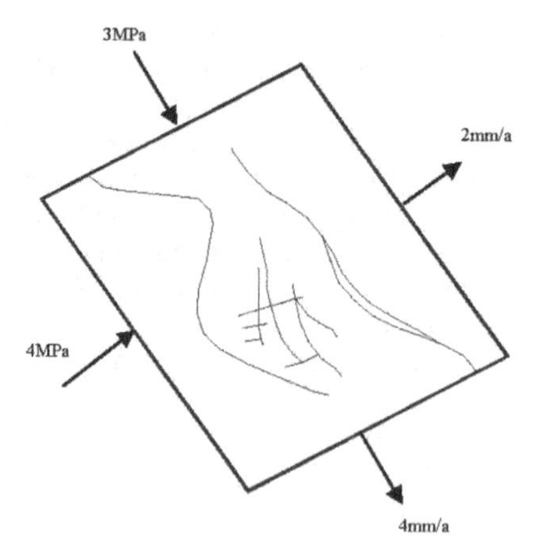

Figure 1. The boundary condition of computation.

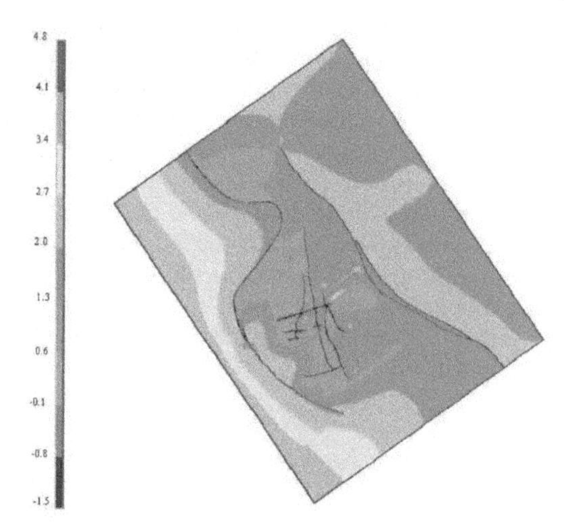

Figure 2. The deformation field along the direction normal to computational area (mm/a).

3.2 *Calculation results and analysis*

3.2.1 *North-East-East of the deformation field*

The Figure 2 shows that the deformation field has a large value on the Lancangjiang fault and the right side of the Red River fault. The deformation field of the Simao-Puer chessboard fault cutting block between the Lancangjiang fault and Honghe fault is much smaller than that of the Lancangjiang and the Honghe fault. In addition, we can see from the figure that the NE-trending faults with the exception of the Xiaomengyang-Xiangzhuang fault (F139) are characterized by left lateral strike slip with the left-hand rotation of the Sanlin-Siyongjie fault (F136) slip up to 1.1 mm/a. In the Simao-Puer block, the spatial distribution of the left-lateral rotation of the NE strike is also characterized by the fact that the larger value of the strike-slip component appears on the west side of the fault and forms the high-value region of the deformation field. From the analysis of the regional deformation field, we think that the phenomenon that the L-component is greater than the eastern part and the region where the maximum value appears should be the reflection of the Jinggu MS 6.6 earthquake.

3.2.2 *NNW strike deformation field*

The displacements of the north-northwest strike in the calculation area show the nature of dextral strike-slip. Figure 3 shows that the deformation field is different from that in the

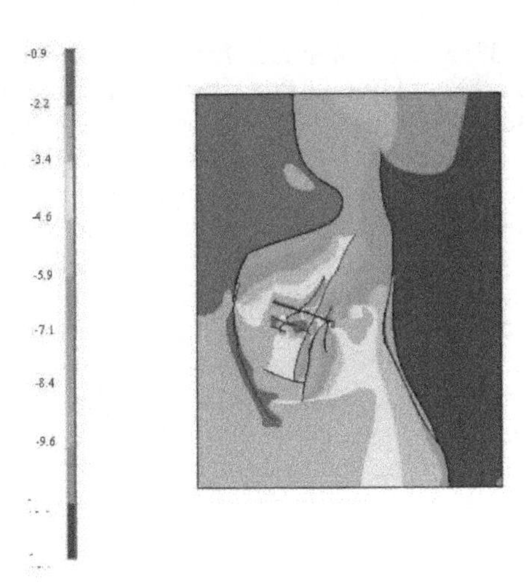

Figure 3. The deformation field along the direction of computational area strike (mm/a).

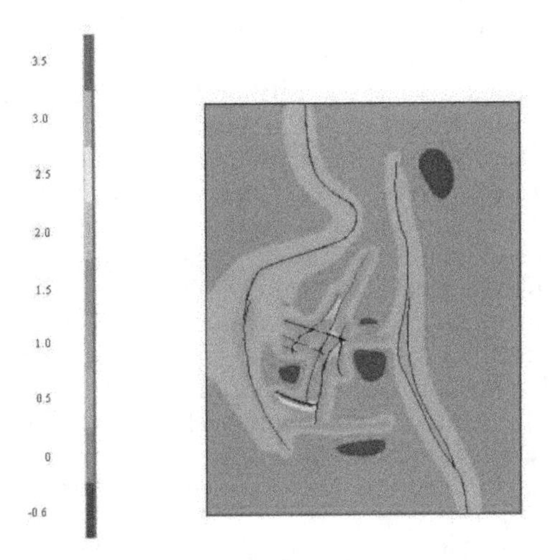

Figure 4. The deformation field along the direction normal to the ground surface (mm/a).

east-north direction and the deformation field caused by the north-west-west deformation field and the Lancangjiang fault (F36) and Honghe fault (F130) are numerically the maximum indicating that it is due to the dynamic environment of the southeast expansion of the Tibetan Plateau. The irregular and deformed images of the grid-like structure are the true reflections of the correlation activities of the two sets of conjugate fault zones. It can be seen that the anomalous image of crustal deformation is a complex tectonic framework. It is further proven that the existence of the relatively high displacement zone corresponds to the development of the Jinggu 6.6 earthquake. The north-north-west deformation field is more homogeneous in the north of the Zhenyuan-Puer fault (F132) and the south of the Xiaomengyang-Xiangzhuang fault (F139) outside the region where the active tectonics are obviously present.

Figure 4 shows the spatial distribution of the deformation in the vertical surface component caused by the fractured movement in which the deformation positive direction refers to the vertical upward direction. It can be seen from the figure that the uplifting deformation of the crust in the calculation area is mainly concentrated in the vicinity of all the fault exposures. The maximum value of deformation appears in the upper plate of the Xiaomengyang-

Xiangzhuang fracture (F139) reaching 4 mm/a. The main body of the uplifting deformation is concentrated in the Simao-Puer fault-cutting structure. The uplifting deformation ranging 1 mm/a ~ 3 mm/a is close to the regional level deformed field chaotic image echoes due to a number of verticals in the block on the vertical fracture of the destruction of each other cutting the integrity of the free surface. With leveling data from 1951 to 2011, Guo et al., (2013) calculated the vertical deformation field of the crust in Yunnan region. According to the vertical deformation rate in Yunnan from 1951 to 1980, 1980 to 1994, and 1994 to 2011, the south-central area went up and the north down in the first stage, the main area rose in the second stage, and the eastern and southern rose and the western descended in the third stage. Vertical deformation from 1951 to 2011 represents the region's vertical crustal movement during the past six decades. The descending rate in the northwest and north of Yunnan is 1 ~ 3 mm/a, while the rising rate in the south and middle east is 1 ~ 6 mm/a. This is the same as that in this paper and the uplifting deformation in Simao-Puer area is basically consistent with the calculated results.

4 DISCUSSION AND CONCLUSION

In conclusion, the crustal movement in the study area is bounded by the geological and geomorphological studies and the current multi-period GPS retest data, and the horizontal deformation field of the crust in the study area is given in detail by finite element analysis, accurate images, and regional vertical deformation field changes in the current characteristics. The east-east horizontal deformation field clearly shows that the displacement is greater in the west than in the east and the maximum appears in the vicinity of the NE-trending fault zone. The results of the horizontal deformation field and vertical deformation field in the northwest direction clearly reflect that the two sets of conjugate cross-fault zones are in mutual traction and co-movement under the regional dynamic background of the plateaus moving southward and a complex, scattered, irregular deformation field image. Further analysis suggests that the north-northwest oriented multi-point concentrated area is an earthquake-prone area and the obvious deformation field is the most important one superimposed on the post-earthquake adjustment process. In the study area, the vertical deformation field of the main body of the study area is continuously rising. It is inferred that the southeast to Diaobian (Luang Prabang) fault zone blocks the movement of some plateau materials in the southeast direction and the fault zone can be regarded as Simao South boundary of Puer Block.

The strain energy accumulated in the southern part of the Lancang and Honghe fault zones on the east and west sides of the east and the west is released by the conjugate activity of the network structure and the continuous occurrence of moderate earthquakes in the same dynamic environment. No serious earthquakes have been reported in the south of the deep fault zone. Of course, the seismicity of the southern section of the Lancang and Honghe fault zones and the origin of the Sipu earthquake belt are complex problems, and there is still difficulty in explaining the current data and degree of the study. It is hoped that this paper will lead to deep research of this problem.

The important problem in the study of Tibetan Plateau kinetics is that the amount of material consumed by plateau uplift cannot reach the order of mass produced by plateau shortening. So the redistribution of the remaining material is the key to maintaining the basic equilibrium during the uplift of the plateau. Based on the limited observation data, we propose the following simplified conceptual model: (1) the southeast flow of the Qinghai-Tibet Plateau caused by the post-tensioning action of Burma's arc (Ji et al., 2008), (2) Simao—the southeastward motion of the Puer area during earthquakes and (3) the clockwise rotation of the Sichuan-Yunnan rhombic blocks constitute the main body of the southeastward expansion of the plateau material and the frequent seismicity digestion, absorption, and regulation in the three regions. The southeastward movement of this plateau material makes the northeastward thrust of the Indian plate to be in a dynamic equilibrium with the Tibetan Plateau uplift and also the southeastward movement of plateaus similar to plate collision. This is the reason for the orogenic process in the southwestern part of China.

ACKNOWLEDGMENTS

The authors thank Prof. Wang Yan and Prof. Yang Xiaoping for useful discussion. They also thank Dr An Yanfen for drawing part of the illustrations.

REFERENCES

Anne, S., Christophe, V., Nicolas, C-R., et a1.2006. India and Sunda plates motion and deformation along their boundary in Myanmar determined by GPS [J]. *Journal of Geophysical Research*, 3:B05406.

Clark, M.K., Royden, I.H. 2000. Topographic ooze: Building the eastern margin of Tibet by lower crustalflow [J]. Geology(Boulder), 28(8): 7033–706.

Deng, Q., Zhang, P., Ran, Y., et a1. 2003. Basic characteristics of active tectonics of China [J]. Science in China (Ser D), 46(4): 356–372 (in Chinese).

England, P.C., Houseman, G.A., 1989. Extension during continental convergence with special reference to the Tibetan plateau [J]. *Journal of Geophysical Research*, 94(17): 561–597.

Engnd, P.C., Molnar, P., 1990. Right-lateral shear and rotation as the explarnation for strike-slip faulting in eastern Tibet [J]. Nature, 344: 140–142.

Guo, S., Xiang, H., Xu, X., et al.2000.Characteristics and deformation mechanism of the Longling-Lancang newly emerging fault zone in Quaternary in the southwest Yunnan [J]. Seismology and Geology, 22(3): 277–284 (in Chinese).

Guo, L., Ta, L., Chen, F., et al. 2013. Relation between crustal vertical motion and earthquake in Yunnan area [J]. *Journal of Geodesy and Geodynamics*, 33(1): 1–5 (in Chinses).

Houseman, G.A., England, P.C. 1993. Crustal thickening versus lateral expulsion in the Indian-Asian continental collision [J]. *Journal of Geophysical Research*, 98(B7):12233–12249.

Ji, S., Wang, Q., Sun, S., et al. 2008. Continental extrusion and seismicity in China [J]. *Acta Geologica Sinica*, 82(12): 1644–1667 (in Chinses).

Li, K., Zhao, W., Hou, X., et a1.1980. A discussion on the Intra-Block seismic geological features from the Simao-Puer earthquake [J]. *Journal of Seismological Research*, 3(1): 38–44 (in Chinese).

Li, T., Deng, Z., Lui, Y. 2003. Research on the crustal deformation data related to characteristics of strong earthquake (Ms≥ 6.0) distribution in the area of Chuandian (Sichuan, Yunnan), China [J]. Earthquake Research in China, 19(2): 132–147 (in Chinses).

Paul, J., Burgmann, R., Gaur, V.K, et a1. 2001. The motion and active deformation of India [J]. Geophysical Research Letters, 28(4): 647–650.

Replumaz, A., Tapponnier, P. 2003. Reconstruction of the deformed collision zone between India and Asia by backward motion of lithospheric blocks [J]. *Journal of Geophysical Research*, 108 (B6): 1–24.

Shen, F., Royden, L.H., Burchfiel, B. 2001. Large-scale crustal deformation of the Tibetan plateau [J]. *Journal of Geophysical Research*, 106: 6793–6816.

Tapponnier, P., Paltzer, G., Le, Dain., A.Y, et al. 1982. Propagating extrusion tectonics in Asia: New insights from simple experiments with plasticine [J]. Geology, 10: 611–616.

Wang, M., Shen, Z., Gan, W., et al. 2008. GPS monitoring of temporal deformation of the Xianshuihe fault [J]. Science in China (Series D: Earth Sciences), 51(9): 1259–1266 (in Chinses).

Xu, X., Wen, X., Zheng, R., et al. 2003. Pattern of latest tectonic motion and its dynamics for active blocks in Sichuan-Yunnan region, China [J]. Science in China (Ser D), 46 (Suppl): 210–226 (in Chinese).

Zhang, P., Deng, Q., Zhang, G., et a1. 2003. Active tectonic blocks and strong earthquakes in the continental China [J]. Science in China (SerD), 33 (Supp1): 12–20 (in Chinese).

Risk Analysis and Management – Trends, Challenges and Emerging Issues – Bernatik, Huang & Salvi (Eds)
© 2017 Taylor & Francis Group, London, ISBN 978-1-138-03359-7

Seismic risk analysis of the oil/gas pipeline system

Aiwen Liu, Jian Wu & Na Yang
Institute of Geophysics, CEA, Beijing, China

ABSTRACT: Long-distance oil/gas pipeline system connecting the oil/gas field with the users' place is usually associated with high seismic activity zones. The main factors of oil/gas pipelines damaged by strong earthquakes are summarized. In general, an oil/gas pipeline system includes pump stations, buried pipes, above-ground pipes, the pipe bridge crossing a river, and so on. At present, the seismic design of oil/gas pipelines is developed from the stress-based one to the strain-based one, and according to the new seismic design code for oil/gas pipeline of China, oil/gas pipelines are classified into two groups, with the exceeding probabilities of 5% and 10% in 50 years. The objective of the seismic design is to maintain the service function of the pipeline even after a strong earthquake. Two levels of earthquake resistance garrisons have been proposed, and the objective of the second level is to avoid oil/gas leakage after a possible strong earthquake. This performance-based pipeline design needs more parameters of the strong ground motion and the ground permanent deformation, requiring a higher request to the evaluation of seismic safety for the pipeline engineering sites. A uniform confidence seismic risk analysis approach should be adopted so that all components of the oil/gas pipeline system are designed consistent with their intended functions and degree of importance.

1 INTRODUCTION

Oil/gas pipelines are important lifeline facilities. In general they are designed and constructed as continuous steel pipelines that are usually buried in the ground for economic, safety, and environmental reasons. They spread over a large area and always encounter a range of seismic hazards and soil conditions, as shown in Figure 1. Many oil & gas pipelines in China run through areas of high seismic activity and hence face a considerable seismic risk.

Catastrophic oil/gas pipe failures have occurred in many areas, particularly with a large permanent ground displacement due to fault movement. Site investigations show that many gas supply pipelines were damaged by Chelungpu fault movement in the Jiji earthquake, as shown in Figure 2 (Takada, 2001).

In this paper, a new seismic design code of oil/gas pipeline was introduced first, and the method of evaluating the seismic hazard of the pipeline was discussed in detail.

2 SEISMIC DESIGN CODE OF THE OIL/GAS PIPELINE

At present, the oil/gas pipeline seismic design is developed from the stress-based one to the strain-based one, and two levels of earthquake resistance are proposed (JWWA, 1997, LIU, 2007). Pipelines running through active seismic zones should be designed in such a way that they remain functional even after being subjected to a high-intensity earthquake.

A newly revised seismic code GB50470 for oil & gas transmission pipelines in China will be issued in the near future. All oil/gas pipeline construction projects shall be designed to comply with the requirements for fortification against earthquakes and in conformity with this new code. This code is aimed at providing seismic design guidelines for the oil & gas transmission

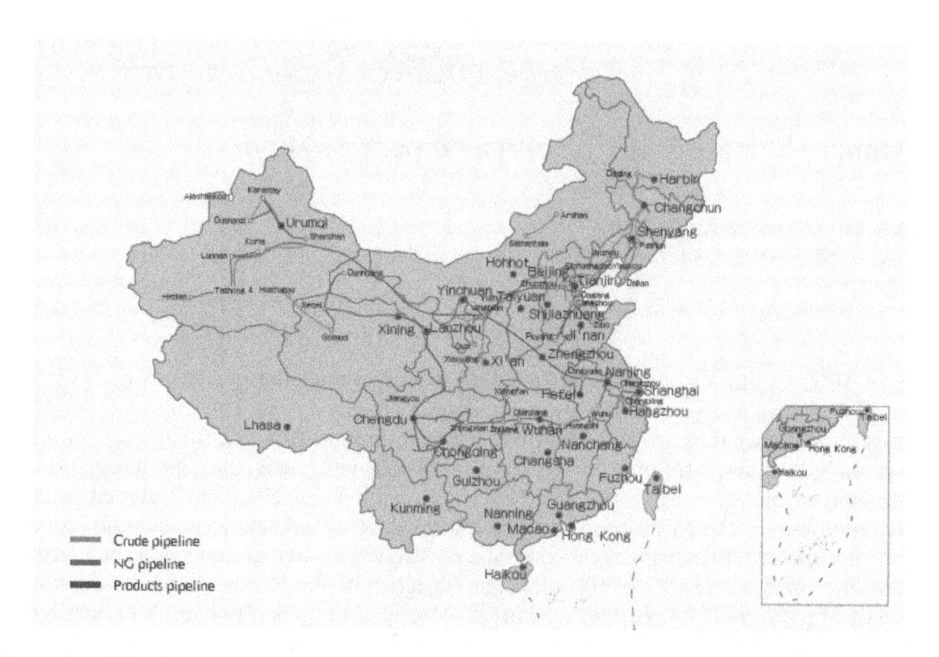

Figure 1. Oil and gas pipelines operated by China National Petroleum Corporation (CNPC, 2013).

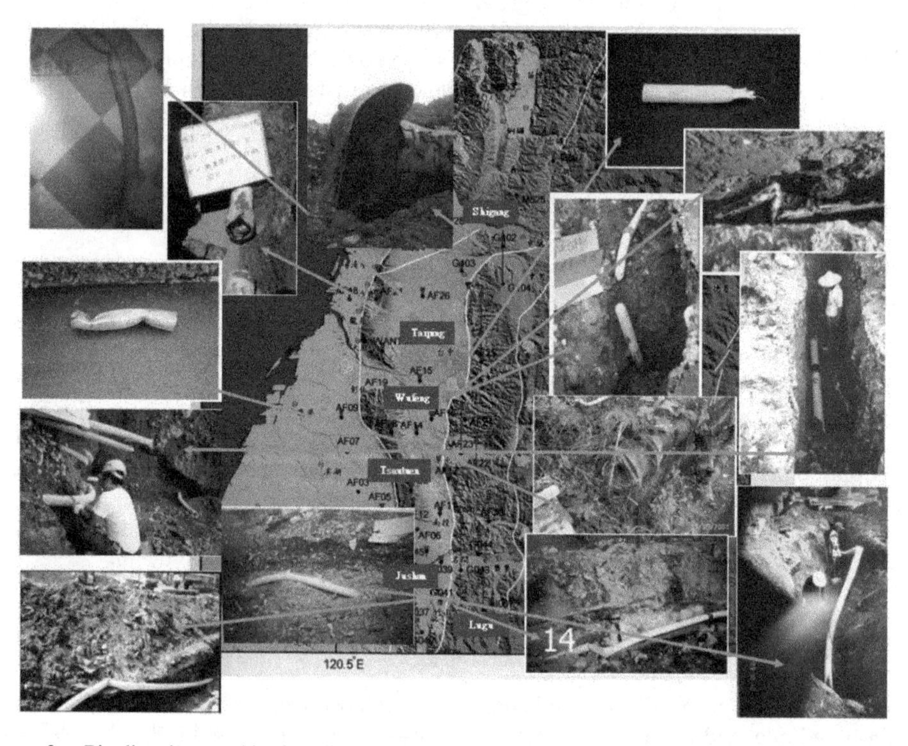

Figure 2. Pipeline damaged in the Jiji earthquake.

pipeline in mainland China. After functional design, a pipeline should be checked for all possible seismic hazards it may encounter, and pipeline safety should be checked for seismic loads simultaneously with the operating loads (pressure, temperature, initial bending, etc.). The seismic design is aimed at: (1) serviceability requirements: the pipeline system shall be designed and constructed in such a way as to be able to maintain its supply capability as much as

Table 1. Seismic performance requirements.

Ground motion level	Probability of exceedance in 50 years	Seismic performance requirements
GML1: general seismic motion	10% (475 years of return period)	Serviceability limit states: pipeline deformation shall be at a level that will allow normal operation without repair
GML2: very strong seismic motion due to earthquake	5% (975 years of return period)	Ultimate limit states: pipeline deformation should not result in leakage

possible, even with considerable local damage due to high-intensity earthquakes; (2) safety requirements: the location of the pipeline, size of the population that is exposed to the impact of pipeline rupture, and environmental damage due to the pipeline rupture shall be considered when establishing the level of acceptable risk during the design of the pipeline system.

The design level of seismic safety for an oil/gas pipeline in China depends on the importance of the pipeline and the consequences of its failure (consequence-based design). Oil/gas pipelines are classified as follows:

Class I: Oil and gas pipelines that are vital energy supply facilities are required to maintain their functionality during and following earthquakes. Failure of or damage to pipelines would cause an extensive loss of life or a major impact on the environment, for example, a pipeline crossing a main river in eastern China.

Class II: Oil and gas pipelines that pose a substantial hazard to human life and property in the event of failure, but their service can be interrupted for a short period until minor repairs are carried out.

The seismic hazard level applied during design depends on the class of pipe, adopting the probability of exceedance in 50 years, as shown in Table 1.

3 EVALUATION OF SEISMIC RISK

The scale of oil/gas pipelines can vary from few kilometers to thousands of kilometers. In contrast to a water pipeline network system, an oil/gas pipeline system is a spatially linearly distributed one, that is, formed by a series of segments, which are linearly distributed in space. When part of the oil/gas pipeline system is damaged by an earthquake, the pipe fails to transport oil/gas. Therefore, the pipeline system should be considered as a whole. In the PSHA models currently used for the seismic evaluation of important projects, microzoning and seismic zonation are carried out on a single site. Studies have shown that it is impossible to obtain total hazards for specific areas from the results of the hazard analysis of sites (Gao, 1993). Therefore, when we consider the total seismic hazard for an oil/gas pipeline system, seismic hazard analysis methods for sites alone are insufficient. In the decisions of seismic design and earthquake disaster prevention regarding an oil/gas pipeline system, not only the seismic hazard for segments of the system but also the total seismic hazard for the oil/gas system should be considered.

There are two methods for considering the total seismic hazard of oil/gas systems: (1) the Monte Carlo simulation method and (2) simplified methods. Considering an oil/gas pipeline with n pump stations, the ground motion seismic hazard has little effect on buried pipes, and the serious effects are those on the pump stations. The total seismic hazard for an oil/gas system due to ground motion, P, is:

$$P = 1 - \prod_{i=1}^{n} (1 - P_i) \tag{1}$$

where P_i is the exceedance probability of pump station i. When the exceedance probability of each pump station has the same value, Eq. (1) can be simplified to:

Table 2. Seismic hazards for oil & gas pipelines.

Hazard	Oil/gas pipeline system	Earthquake parameters	Obtained from
General shaking	Buried pipe	PGA, PGV, Tg	PSHA
	Pipe bridge, above-ground pipe, pump stations	PGA, PGV, Tg, Sa, a(t)	
Faulting	Buried pipe, above-ground pipe	Expected amount of fault displacement, crossing angle	DSHA or disaggregate PSHA
Liquefaction	Buried pipe, above-ground pipe, pipe bridge	PGA, magnitude L of pipeline exposed to PGD	Disaggregate PSHA
Differential settlement	Buried pipe, above-ground pipe, pipe bridge	PGA, magnitude L of pipeline exposed to PGD	Disaggregate PSHA

$$P = 1 - (1 - P_i)^n \tag{2}$$

Evaluation of seismic safety for pipeline engineering sites is carried out to identify the earthquake hazards to oil & gas pipelines, provide a general description of how the hazard affects pipelines, and define parameters needed to quantify the earthquake hazards for engineering design. Pipeline performance-based design requires more information on strong ground motion and permanent ground displacement and leads to a higher requirement for the evaluation of seismic safety for pipeline engineering sites.

Transient ground movement describes the shaking hazard by seismic waves and the amplifications due to surface and near-surface ground conditions and topography, including Peak Ground Acceleration (PGA), Peak Ground Velocity (PGV), response Spectrum (Sa), and ground motion history, a(t). Permanent Ground Displacement (PGD) describes the ground failures resulting from surface fault rupture, slope movements and landslides, liquefaction-induced lateral spreading and flow failure, and differential settlement. In general, an oil/gas pipeline system includes pump stations, buried pipe, above-ground pipe, and pipe bridges when crossing rivers. Table 2 summarizes the transient and permanent ground movement hazards considered during the design of pipelines.

3.1 *PSHA for transient ground movements*

Transient ground movements refer to the transient soil deformations (i.e., strain and curvature in the ground) due to seismic wave propagation and affect the pipeline over a large area with widely dispersed damage. Ground shaking does not have a serious effect on buried oil/gas pipelines under good condition, but the exceptions include points at the transition between very stiff and very soft soils, penetration into valve boxes, locations at or near pump stations, T-connections, pipe fittings, and valves. However, pipelines weakened by corrosion are more vulnerable to damage during ground shaking. The hazard due to ground shaking is often characterized in terms of PGV, rather than in terms of PGA.

The design basis of earthquake ground motion (in terms of PGA and PGV) should be estimated on the basis of site-specific pipeline seismic hazard analysis. Probabilistic Seismic Hazard Analysis (PSHA) has been widely used for seismic hazard evaluation. PSHA itself is also modified for the determination of the seismicity model, the regional attenuation law, and the analysis of uncertainties. In China, the following specific standards adequately deal with the seismic evaluation and design of oil and gas pipelines: Seismic Ground Motion Zonation Map of China, as shown in Figure 3 (CEA, 2001) and the Evaluation of Seismic Safety for Engineering Sites (CEA, 2005).

3.2 *Fault displacement hazard analysis*

Disaggregate PSHA has been proposed to estimate the PGD induced by liquefaction and landslide. The PSHA for fault displacement is developed within the framework of

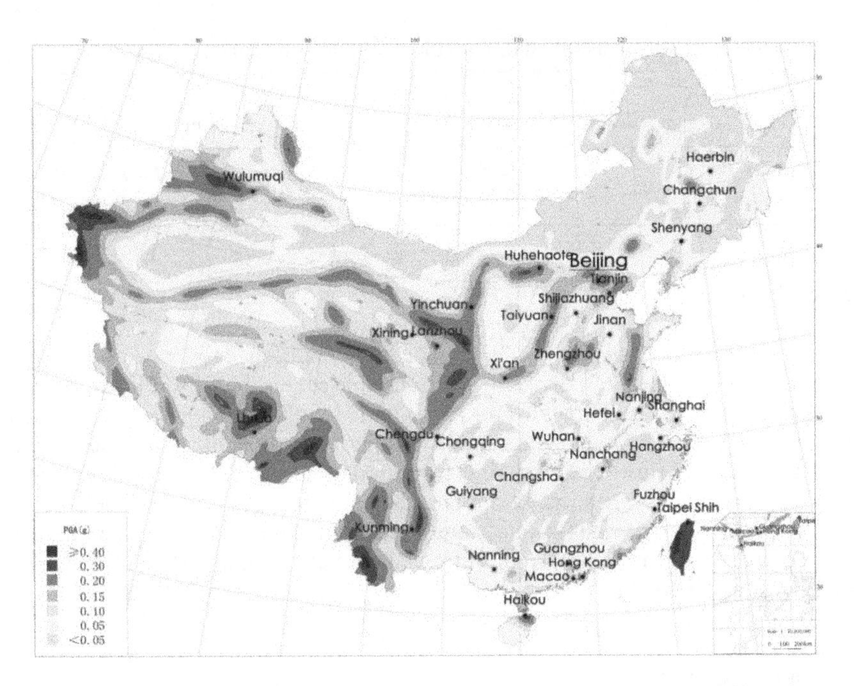

Figure 3. Seismic ground motion parameter zonation Map of China (10% probability of exceedance in 50 years, China Earthquake Administration, 2001).

probabilistic seismic hazard analysis (Youngs, 2003, Todorovska, 2007). The earthquake occurrences are also modeled as Poissonian sequences. In addition to earthquake occurrence rates, PSHA for fault displacement requires the probability distribution function of displacement along the fault line and probabilities that a rupture will affect the site. The distribution of displacement along the fault line is strongly dependent on the source mechanism of the earthquake, and the rupture of the site for pipelines crossing faults needs further research. Preliminary research results show that the hazard is small even for very small levels of displacement, in contrast to the ground shaking hazard, because only one fault contributes to the hazard and not every event on that fault necessarily affects the site.

Because PSHA for fault displacement hazard analysis needs further research, the PGD induced by fault movement is still primarily determined by DSHA in China. Fault displacement can be estimated using historical evidence, paleoseismic evidence, and/or slip rate calculations. Usually, the maximum fault displacement (D_{max}) is estimated by an empirical relationship with earthquake Magnitude (M) as follows:

$$\log D_{\max} = A \cdot M - B \tag{3}$$

where A and B are the regression coefficients.

At present, the maximum fault displacement determined by DSHA in China is for a 100-year period. This period is much shorter than the design return period (975 years), which would result in the design for a surface fault rupture corresponding to M less than the characteristic magnitude; thus, the maximum fault displacement could be exceeded. To consider the uncertainty in the magnitude of the earthquake as well as the uncertainty in the amount of displacement, given the occurrence of a particular magnitude earthquake, a more refined approach for DSHA is to define the design-basis fault displacements corresponding to the pipelines in Table 1. In other words, the design-basis fault displacement of Class I pipes is larger than that of Class III pipes, as shown in Table 3.

Table 3. Design-basis fault displacement for oil/gas pipelines.

Probability of exceedance in 50 years	Return period (years)	Design-basis fault displacement
10%	475	$\log D_{max} = A \cdot M - B$
5%	975	$\log D_{max} = A \cdot M - B + \sigma/2$

4 DISCUSSION

For the design of oil/gas pipelines in China, pipelines have been classified into two groups on the basis of the importance of the pipeline and the consequences of its failure, with probabilities of exceedance in 50 years of 5% and 10%. Considering that oil/gas pipeline systems are spatially linearly distributed, pipeline systems should be considered as a whole when analyzing seismic hazards. More attention should be paid to deciding the fault displacement used as a design basis for oil/gas pipelines. At present, the maximum fault displacement determined by DSHA needs to incorporate the uncertainties in the magnitude of the earthquake and the fault displacement, given the occurrence of an earthquake with a particular magnitude and the PSHA method need further study.

ACKNOWLEDGMENTS

This work was supported by the special research funds from the Public Institute of China, IGP, CEA (DQJB15C01), and the national scientific and technological support project MST (2006BAC13B02-0106).

REFERENCES

American Lifelines Alliance (ALA), (2005). *Design Guidelines for Seismic Resistant Water Pipeline Installations*, FEMA and National Institute of Building Sciences. www.americanlifelinesalliance.org.

China Earthquake Administration, (2001). *Seismic Ground Motion Parameter Zonation Map of China.* Beijing: China Standard Press(in Chinese).

China Earthquake Administration, (2005). *Evaluation of Seismic Safety for Engineering Sites*, Beijing: China Standard Press(in Chinese).

Davis Craig, A. (2005). *Uniform Confidence Hazard Approach for the Seismic Design of Pipelines*, 4th Japan-US Workshop on Seismic Measures for Water Supply, JWWA. and AWWARF, Kobe, Japan.

Gao, M.T. (1993). *Seismic hazard analysis for a spatial linearly distributed system*, ACTA Seismologica Sinica. 6(3), 713–719.

JWWA, (1997). *Seismic design and construction guidelines for water supply facilities*, Japan Water Works Association.

Liu, A., Feng, Q.M. (2007). *The new seismic code for oil & gas transmission pipelines of China*, Fifth China-Japan-US Trilateral Symposium on Lifeline Earthquake Engineering, Haikou, China.

Takada, S., Liu, A., Katagiri, S. Ban-jwu Wu and Walter W. Chen. 'Numerical Simulation On The Behavior Of Polyethylene Pipeline Under Fault Movement In Ji-Ji Earthquake', MEMORIES OF CONSTRUCTION ENGINEERING RESEARCH INSTITUTE, Vol. 43-B (papers), Page 1–12, Nov. 2001.

Todorovska, M.I., Trifunac, M.D., Lee, V.W, (2007). *Shaking hazard compatible methodology for probabilistic assessment of permanent ground displacement across earthquake faults, Soil Dynamics and Earthquake Engineering*, 27(6), 586–597.

Wells, D., Coppersmith, K. (1994). *New Emperical Relationships among Magnitude, Rupture Length, Rupture Width, Rupture Area, and Surface Displacement*, Bull. Seismol. Soc. Am., 84, 974–1002.

Youngs, R.R., Walter, Arabasz, M.J. et a1, (2003). *A methodology for probabilistic fault displacement hazard analysis* (PFDHA), Earthquake Spectra, 19(1), 191–219.

Risk Analysis and Management – Trends, Challenges and Emerging Issues – Bernatik, Huang & Salvi (Eds)
© 2017 Taylor & Francis Group, London, ISBN 978-1-138-03359-7

Entropy based streamflow uncertainty analysis using JSS estimator

Dengfeng Liu & Dong Wang
Key Laboratory of Surficial Geochemistry, MOE, Department of Hydrosciences, Nanjing University, Nanjing, P.R. China
State Key Laboratory of Pollution Control and Resource Reuse, School of Earth Sciences and Engineering, Nanjing, P.R. China

ABSTRACT: Entropy theory has been increasingly applied in hydrology in the descriptive and inferential ways. In this study, an innovative entropy estimators, the James-Stein-type Shrinkage (JSS) estimator, considering small-sample condition widespread in hydrological practice, are introduced. Multi-scale Moving Entropy-based Hydrological Analyses (MM-EHA) are employed to indicate the changing patterns of uncertainty of streamflow data collected from the Yangtze River, China. Some interesting results are revealed.

Keywords: Entropy; James-Stein-type shrinkage estimator; entropy weight; eutrophication; water environment

1 INTRODUCTION

The concept of entropy originated in physics which can be considered as a measure of the degree of uncertainty associated with a system. Shannon (1948) developed the theory of entropy and defined a metric that characterizes and quantifies the degree of uncertainty (or lack of information). In hydrology and water resources, Amorocho and Espildora (1973) used entropy to characterize the variability inherent in a hydrological process and evaluate the uncertainty of predictions made by a given hydrological model. Since then, many entropy-based studies have been carried out in the environmental or hydrological field (Singh, 1997; 2000; 2011). From the perspective of statistics, entropy-based applications are generally of two main types: (I) "Descriptive study", where one uses entropy (or any other variant of "entropy," e.g. approximate entropy, sample entropy, etc.) to assess the uncertainty of a system, which might be a hydrological process, model or parametric structure, etc. (II) "Inferential study," where one uses the principle of maximum entropy (Jaynes, 1957) or minimum relative entropy (K-L divergence) (Kullback and Leibler, 1951) to derive probabilistic or physical properties of a system and determine the model structure or parameters of a given model, etc.

The common strategy used to compute entropy is intuitive, to some extent, that entails first estimation of the Probability Density Function (PDF) of the data, and then computation of entropy from the definition of entropy usually in a continuous mode, that is, in a "distribution-based" way. Gong et al. (2014) discussed four important issues including "zero effect," "optimal bin width," "measurement effect," and "skewness effect" in hydrological applications that can bias the computation of entropy if not properly handled, and presented an approach which was verified to be accurate and robust with simulated and observed hydrological data.

However, there is a practical problem that should not be ignored in hydrological practice. That is, basins in many parts of the world are poorly gauged or may even be ungauged. Under this condition, hydrological data-based modelling might be highly uncertain. The IAHS (International Association of Hydrological Sciences) Decade on Predictions in Ungauged Basins (PUB), has been launched with the aim to formulate and implement appropriate programs to advance the capacity to make predictions in ungauged basins (Sivapalan et al. 2003). In hydrological entropy-based studies, accurate computation of entropy is likewise a big challenge for most observed hydrologic series whose length is small, especially when a distribution-based framework is employed. Different from the distribution-based entropy computation, another approach to obtaining entropy is the "cell probability-based," according to the definition of entropy in a discrete mode. In this approach, entropy is obtained with the estimation of cell probability which extricates entropy computation from considerable computation-expense and uncertainty of numerical integral computation in the distribution-based approach.

In this study, we introduce an"cell probability-based"entropy estimators, the James-Stein-type Shrinkage (JSS) estimator. In addition, entropy-based applications are presented, with Multi-scale Moving Entropy-based Hydrological Analyses (MM-EHA) to reveal the uncertainty of hydrological systems from the perspective of entropy. In MM-EHA, entropy of the hydrological data in moving windows with different time lengths are computed by the JSS method, with observed streamflow data from Yichang and Hankou on the Yangtze River, China.

The next section summaries entropy estimators, including the ML estimator and the JSS estimator. Section 3 presents the application of MM-EHA in the Yangtze River Basin. Conclusions are summarized in Section 4, with prospects for further research.

2 ENTROPY ESTIMATING APPROACH

2.1 *Entropy definition*

To define the Shannon entropy of a system, consider a categorical random variable which indicates the statement of the system with size p and associated statement probabilities $\theta_1, \ldots, \theta_p$ with $\sum_{k=1}^{p} \theta_k = 1$. The Shannon entropy in the discrete mode is given as:

$$H = -\sum_{k=1}^{p} \theta_k \ln(\theta_k) \tag{1}$$

In hydrological practice, for a given hydrological system (for example, streamflow series), the underlying cell probability θ_k corresponds to each statement and entropy H needs to be estimated from observed cell counts y_k.

2.2 *Maximum Likelihood estimate (ML)*

The widely used estimator of entropy is the Maximum Likelihood (ML) estimator:

$$\hat{H}^{ML} = -\sum_{k=1}^{p} \hat{\theta}_k^{ML} \ln\left(\hat{\theta}_k^{ML}\right) \tag{2}$$

constructed by plugging the ML frequency estimates:

$$\hat{\theta}_k^{ML} = \frac{y_k}{n} \tag{3}$$

into Equation (1), with $n = \sum_{k=1}^{p} y_k$ being the total number of counts.

2.3 *James-Stein Shrinkage estimate (JSS)*

The James-Stein Shrinkage estimator (James and Stein, 1961) employs shrinkage at the level of cell frequencies, and has been verified to be highly effective, both in terms of statistical accuracy and computational complexity (Hausser and Strimmer, 2009). The estimator is based on averaging two very different models: a high-dimensional model with low bias and high variance, and a lower dimensional model with larger bias but smaller variance. In this way, cell frequencies can be determined by the relative weights of the two models:

$$\hat{\theta}_k^{\prime Shrink} = \lambda t_k + (1-\lambda)\hat{\theta}_k^{ML} \tag{4}$$

where $\lambda \in [0, 1]$ is the shrinkage intensity that takes on a value between 0 (no shrinkage) and 1 (full shrinkage), and t_k is the shrinkage target. A convenient choice of t_k is the uniform distribution $t_k = \frac{1}{p}$, which is the maximum entropy target. The shrinkage intensity can be obtained as:

$$\hat{\lambda} = \frac{\sum_{k=1}^{p} \hat{V}ar\left(\hat{\theta}_k^{ML}\right)}{\sum_{k=1}^{p}\left(t_k - \hat{\theta}_k^{ML}\right)^2} \tag{5}$$

The resulting plugin shrinkage entropy estimate then is:

$$\hat{H}^{Shrink} = -\sum_{k=1}^{p} \hat{\theta}_k^{Shrink} \log\left(\hat{\theta}_k^{Shrink}\right) \tag{6}$$

3 MULTI-SCALE MOVING ENTROPY-BASED HYDROLOGICAL ANALYSES (MM-EHA)

Streamflow systems, including precipitation, evaporation, runoff, etc., are complex systems, especially with the background of significantly changing climate during recent decades (Allen and Ingram, 2002). It is therefore important to identify the uncertainty and its changing pattern in an accurate and effective way. However, the reality is that in many areas of hydrological applications, the uncertainty is largely ignored (Grayman, 2005). As a probability measure of disorder, entropy is an ideal identification of the degree of uncertainty of hydrological systems. Here, Multi-scale Moving Entropy-based Hydrological Analyses (MM-EHA) were conducted to indicate changing patterns of uncertainty or disorder of the hydrological series. In this analysis, "multi-scale" means entropy is obtained under time windows of different lengths which can indicate disorder of hydrological series from different time scales or different resolution ratios; and "moving" means when calculating entropy, time windows are moving continuously along the original series so that changes of disorder or uncertainty can be illustrated continuously. Here the JSS estimator was employed to estimate entropy.

3.1 *Data sources*

A representative basins in China were selected for the application of MM-EHA. Entropies were obtained by the JSS method, with monthly streamflow data collected from Yichang and Hankou in the Yangtze River Basin for a period from 1950/1 to 2012/12. The Yangtze River is the longest river in China and the third longest river in the world with a length of 6280 km. The total area of the Yangtze River basin is 1,800,000 km^2. Yichang station is at the boundary of upstream and midstream of the Yangtze River. Meanwhile, Yichang is also an important control station for streamflow out the Three Gorges reservoir. Hankou station is located at the junction of the main river and its largest tributary Han River and is an important control station in the midstream of the Yangtze River. Since the midstream suffers the highest flood risk for the complex channel condition, data collected from Yichang and Hankou are important references for flood control in the midstream.

3.2 *Summary of the results*

Table 1 lists statistical characteristics of streamflow data from Yichang and Hankou on the Yangtze River. From Table 1, it can be seen that streamflow of Hankou tends to be more significantly variant than that of Yichang, with higher Standard Deviation (SD) and range. In addition, streamflows of both stations are positively-skewed, indicating significantly more extreme (high) streamflow events.

In order to analyze the uncertainty of the two stations, entropies under moving windows with lengths of 400 month, 300 month, 200 month and 100 month were calculated by the JSS estimator, as illustrated in Figure 1 and Figure 2. From Figure 1 and Figure 2, LOESS smoothers (Cleveland et al., 1992) show similar trends of falling first then rising of the two stations. This means that uncertainty levels of streamflows of both stations are relatively high and then fall, and then rise again. From Figure 1, a significant jump can be found when it comes to 1998 in all windows, and a rapid fall is marked with blue (except in the window with $L = 100$ with two rapid falls). Analysis shows that the jump is directly related to the appearance of historical maximum streamflow during 1998, and the fall (before the jump) is related to the disappearance of second maximum streamflow during 1954 (another fall after 1998 in the window with $L = 100$ results from the disappearance of maximum streamflow during 1998). Since extreme events (both maximum and minimum) can greatly increase the potential "statements" of the system, the corresponding entropy will jump significantly indicating a rapid increase in the degree of uncertainty, given the same precision, or cell widths. In this way, the appearance of extreme values in a moving window usually results in a jump of entropy, and the disappearance of extreme values can result in a rapid fall of entropy. The same conclusions can be drawn from Figure 2.

Table 1. Statistical characteristics of monthly streamflow of Yichang and Hankou on the Yangtze River.

Sites	Mean	SD	Cv	Range	Relative range	Skewness	Kurtosis
Yichang	13584.46	9771.68	0.72	49110	3.62	0.94	0.07
Hankou	22375.44	12551.29	0.56	61880	2.77	0.65	–0.31

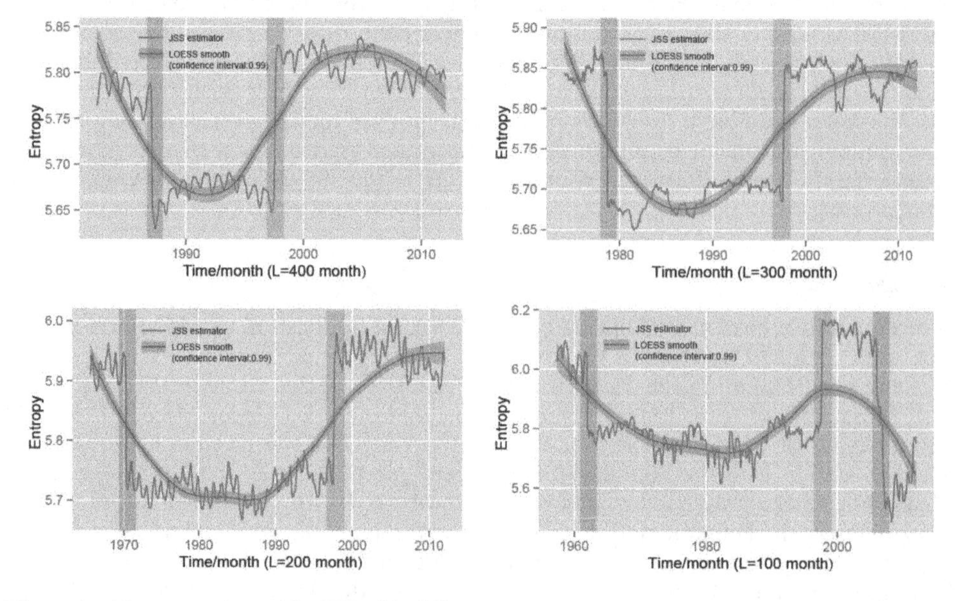

Figure 1. Entropy estimated by JSS with different moving window length (400 month, 300 month, 200 month, 100 month): monthly streamflow of Yichang, the Yangtze River.

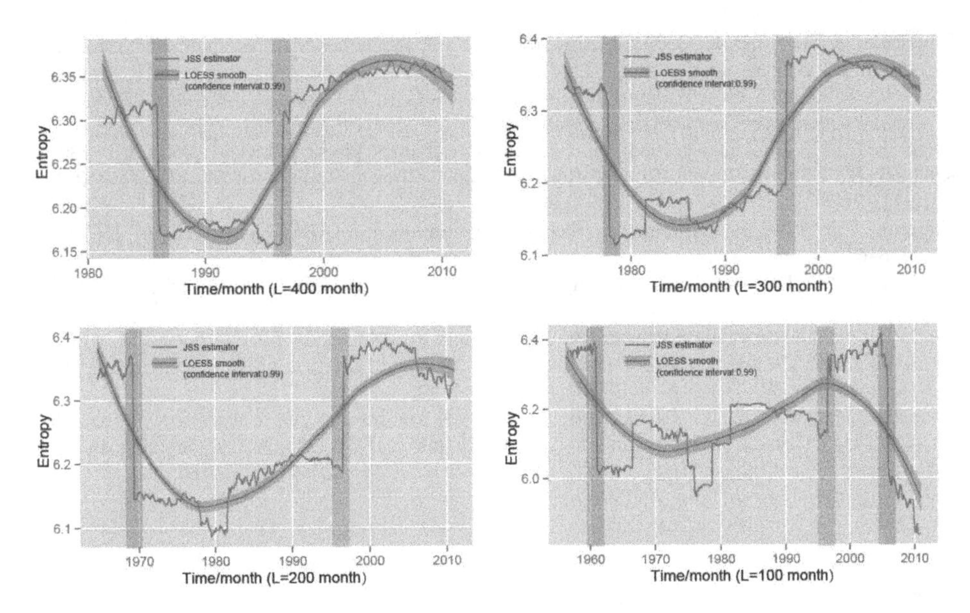

Figure 2. Entropy estimated by JSS with different moving window length (400 month, 300 month, 200 month, 100 month): monthly streamflow of Hankou, the Yangtze River.

4 CONCLUSION

This study focuses on entropy calculation under small-sample condition and introduces an innovative JSS entropy estimators developed in statistics. The JSS estimator has been verified to be a top-performer in simulation scenarios. For hydrological applications, MM-EHA based on the JSS estimator indicates that a falling first then rising trend of uncertainty of streamflow in the Yangtze River. Entropy is a powerful comprehensive statistic indicating uncertainty or disorder of a system and further studies are needed on computation techniques considering more practical conditions, on interpretations of entropy associated with hydrologic systems, and to make the entropy theory more adapted to hydrological applications.

ACKNOWLEDGEMENTS

This study was supported by the National Natural Science Fund of China (Nos. 41571017, 51190091, and 41071018), Program for New Century Excellent Talents in University (NCET-12-0262), China Doctoral Program of Higher Education (20120091110026), Qing Lan Project, the Skeleton Young Teachers Program and Excellent Disciplines Leaders in Midlife-Youth Program of Nanjing University.

REFERENCES

Allen, M.R., & Ingram, W.J. 2002. Constraints on future changes in climate and the hydrologic cycle. Nature, 419(6903), 224–232.

Amorocho, J., & Espildora, B. 1973. Entropy in the assessment of uncertainty in hydrologic systems and models. Water Resources Research, 9(6), 1511–1522.

Gong, W., Yang, D., Gupta, H.V., & Nearing, G. 2014. Estimating Information Entropy for Hydrological Data: One Dimensional Case. Water Resources Research.

Grayman, W.M. 2005. Incorporating uncertainty and variability in engineering analysis. *Journal of Water Resources Planning and Management*, 131(3), 158–160.

Hausser, J., & Strimmer, K. 2009. Entropy inference and the James-Stein estimator, with application to nonlinear gene association networks. *The Journal of Machine Learning Research*, 10, 1469–1484.
James, W., & Stein, C. 1961. Estimation with quadratic loss. In *Proceedings of the fourth Berkeley symposium on mathematical statistics and probability* (Vol. 1, No. 1961, pp. 361–379).
Jaynes, E.T. 1957. Information theory and statistical mechanics. Physical Review, 106(4), 620–630.
Kullback, S., & Leibler, R.A. 1951. On information and sufficiency. Annals of Mathematical Statistics, 22(22), 79–86.
Shannon, C.E. 1948. A mathemetical theory of communication, Bell Syst. Tech. J., 27(379–423), 623–656.
Singh, V.P. 1997. The use of entropy in hydrology and water resources. Hydrological Processes, 11(6), 587–626.
Singh, V.P. 1998. Entropy-based parameter estimation in hydrology (Vol. 30). Springer.
Singh, V.P. 2000. The entropy theory as a tool for modelling and decision-making in environmental and water resources. WATER SA-PRETORIA-, 26(1), 1–12.
Sivapalan, M., Takeuchi, K., Franks, S.W., Gupta, V.K., Karambiri, H., Lakshmi, V., ... & Zehe, E. 2003. IAHS Decade on Predictions in Ungauged Basins (PUB), 2003–2012: Shaping an exciting future for the hydrological sciences. *Hydrological Sciences Journal*, 48(6), 857–880.

Risk Analysis and Management – Trends, Challenges and Emerging Issues – Bernatik, Huang & Salvi (Eds)
 ISBN 978-1-138-03359-7

Water shortage risk assessment in the Beijing–Tianjin–Hebei region

Dengwei Liu
Development Research Center, the Ministry of Water Resources, P.R. China

Yuehong Zhang
Bureau of Development and Planning, the Chinese Academy of Sciences, China

ABSTRACT: The Beijing–Tianjin–Hebei Region (BTHR) is located in North China, where the average water resource amount per capita is only 15% that of the whole country and less than 4% that of the world. With the continuous increase of population, the higher demand for water use by industry and agriculture, and the impact of global warming, the water shortage situation has increasingly deteriorated. Water crisis in this region now has become a big concern for government and the public. For ensuring water safety of the country and promoting regional sustainable social and economic development, it must be a necessary strategic choice to manage the risks effectively and transform the water safety management from a "postdisaster response" mode to a "risk-prevention" mode. In this paper, risk assessment on water shortage from a multiple spatial scale using the GIS technology is conducted in BTHR, on the basis of the water resource risk assessment index. Studies have shown that the water shortage risk index is 7.5 in BTHR, indicating the high-risk water shortage state of this region. In order to maintain social sustainability and ecosystem equilibrium, it is necessary to strictly control the region's population and economic scale and to divert water from other areas, especially in the median-level hydrological years. And in order to reduce water shortage risks, it is necessary to strengthen water resources management, implement risk management strategies, establish emergency water diversion plans, and implement strategic water reserves plans.

Keywords: Water Shortage Risk, GIS, Beijing–Tianjin–Hebei Region (BTHR)

1 INTRODUCTION

Water shortage is affected by two major factors: supply and demand. When water supply cannot meet the demand, water shortage will occur. The study of water shortage risk includes risk identification, risk assessment, and risk decision. Here, we establish a system to identify and evaluate water shortage risks in BTHR and the impacts of South-to-North Water Diversion Project to the Haihe River Basin in BTHR. When water supply cannot meet demand and water supply system undergoes risks, it is very important for the function of the system to recover in order to guarantee water security. Furthermore, water shortage is subject to rainfall, runoff, population, economy, and other various random factors. Because of all these diversified factors, complicated interrelations, and their ambiguous nature, there are many uncertainties in water supply and demand.

In this paper, we analyze and assess water shortage risks in BTHR, carry out preliminary exploration on theories about water shortage risks, and then provide some measures to reduce them. The study will also provide an integrated water resources management policy, which will help boost the social and economic development of this region.

2 OVERVIEW OF THE REGION

Guaranteeing a high level of water supply in BTHR is necessary because BTHR is a core area around capital of China, which is related to sustainable development, people's living needs, and social stability in North China and even the whole country at large. Water crisis will cause huge economic losses and impose an adverse political influence. Thus, it is very important to strengthen the research on the risk of water shortage, analyze the conditions, trends, intensity, and spatial distribution of the risks, and provide effective data, information support, and effective solution, which will contribute to alleviating water shortage hazards, lessening the bottlenecks for economic sustainability, and strengthening water risk management in BTHR.

Water availability in BTHR has reduced from 280–290 $\times 10^8$ m^3 in the late 1950s to 140 $\times 10^8$–150 $\times 10^8$ m^3 in the early 21st century, far below the multiyear average level of 254.2 $\times 10^8$ m^3. This water reduction is indisputable. Although water in BTHR increased in the 1990s compared to that in the 1980s, it showed a significant decline with fluctuations around 2000, and the average of the past 5 years is only half of the multiyear average level. As water level is continuously decreasing in this region, water supply–demand contradictions have been exacerbated and water-carrying capacity has been lowered. Water consumption for domestic purposes in BTHR is growing continuously, whereas that in the industrial and agricultural sectors is declining surprisingly, and comprehensive water consumption per capita is steadily reducing as well. The reduction of agricultural water consumption and the promotion of water-saving projects will stabilize the region's total water consumption in the short term; however, the current situation of water shortage is still not fundamentally improved.

3 QUANTITATIVE ASSESSMENT ON WATER SHORTAGE RISKS

3.1 *Research methods*

On the basis of the water supply–demand balance theory, the water shortage risk index model is developed, which can reflect the water shortage risk degree and be used to further analyze the BTHR's development and exploitation prospects. The two key factors of the index are water supply and demand, where water supply is the sum of local water and diverted water, and the water demand reflects the region's rainfall, population, and socioeconomic development level. The equation is:

$$C = p \cdot K\sqrt{P \cdot G} / (W_1 + W_2) \tag{1}$$

Table 1. Water shortage risk index classification.

Level	Value C	Water shortage risk assessment	Response measures
I	>10	Very high risk	Control the population and economic scale, and water diversion is needed even in the wet years
II	5–10	High risk	Control the population and economic scale, and water diversion is needed in the median years or above
III	2–5	Sub-high risk	Control the population, and water diversion is needed in the dry years or above
IV	1–2	Moderate risk	Water diversion is needed in the extra-dry years
V	<1	Low risk	A balance between supply and demand can be achieved through proper deployment
VI	0	Risk-free or no human activity area	There is an adequate supply of water without water shortage risks

where C is the water shortage risk index; p is the drought probability; P is the population; G is GDP (in RMB100 million); W_1 is the local water gross (water supply factor 1, in 100 million m^3); W_2 is the diverted water gross (water supply factor 2, in 100 million m^3); and K is the rainfall-related factor, which is expressed as:

$$K = \begin{cases} 1.0 & R \leq 200 \\ 1.0 - 0.1(R-200)/200 & 200 < R \leq 400 \\ 0.9 - 0.2(R-400)/400 & 400 < R \leq 800 \\ 0.7 - 0.2(R-800)/800 & 800 < R \leq 1600 \\ 0.5 & R > 1600 \end{cases} \tag{2}$$

where R is the rainfall (mm). As it can be seen from the above equation, the greater the total amount of water resources in an area, the smaller the water shortage risk index; the larger the population and economic scale, the larger the water shortage risk index. According to this law, the water shortage risk levels can be specifically classified quantitatively (see Table 1).

3.2 *Multilevel analysis on water shortage risks*

3.2.1 *Municipal-level water shortage risks*

Studies have shown that the water shortage risk index in BTHR is 7.5, indicating the high-risk water shortage state of this region. In order to maintain social sustainability and ecosystem equilibrium, it is necessary to strictly control the region's population and economic scale and to divert water from the region, where water resource is abundant. Among them, the water shortage risk index is above 10 in Beijing, Shijiazhuang, and Changzhou, and it even peaks at 16 in Tianjin. In future, there will be more difficulties to develop water resources, and the water shortage situation will be more serious. Therefore, it is necessary to take more stringent measures to control the region's population and economic scale and to divert water from other regions even in the wet years. The corresponding values in Baoding, Langfang, and Tangshan are close to average level in BTHR, with a very high degree of development of water resources and a high-risk state of water shortage. The values in Zhangjiakou and Qinhuangdao are 3.5 and 3.7, respectively, with a moderate development and utilization, a high development potential, and a sub-high-risk state of water shortage. However, the index in Chengde is 1.5, with a low degree of development and utilization of water resources, a very high development potential, and a sub-moderate-risk state of water shortage (see Table 2).

After the implementation of South-to-North Water Diversion Project, the water shortage risks of intake areas were generally reduced to a lower level. In Beijing, Tianjin, Shijiazhuang, and Changzhou, the risk index reduced from very high risk to high risk, and in Baoding and Tangshan, the risk index dropped from high risk to sub-high risk. Although the implementation of South-to-North Water Diversion Project could somehow reduce the probability of shortage risks, water shortage risk level in the whole region is generally not reduced, which means the project plays a minor role in alleviating water shortage risks in whole BTHR. A fundamental way to reduce water shortage risks is to implement more stringent water-saving measures and to strictly control the population and economic scale.

3.2.2 *County-level water shortage risks*

From the county level, we can accurately understand water shortage risks of all counties in BTHR. As we can see from Figure 1, those differences that cannot be expressed at the regional level can be shown very clearly at the county level. For example, the water shortage risk index in Baoding is high at the municipal level, but at the county level, the index in Fuping County (subordinated to Baoding) is very low.

Overall, the water shortage risk degree in Zhangjiakou is not high, but there are huge differences inside it: only Chicheng County is at a very low risk, and Chongli County is at a low risk, whereas other six counties need water diversion, indicating the water resource state in Zhangjiakou is not optimistic. As for Chengde, where water shortage risk degree is the lowest

Table 2. Water shortage risk index in BTHR.

Area	Water resources (50%)	Total amount of water resources after diversion	Population (10,000 persons)	GDP (RMB100 million)	Index value	Risk assessment	Index value after diversion	Risk assessment after diversion
Chengde	37.6	37.6	342	235	1.5	Moderate-risk area	1.57	Moderate-risk area
Zhangjiakou	21.8	21.80	450	320	3.5	Sub-high-risk area	3.90	Sub-high-risk area
Qinhuangdao	16.46	16.46	273	387	3.7	Sub-high-risk area	3.81	Sub-high-risk area
Baoding	40.92	49.67	1077	925	5.0	High-risk area	4.20	Sub-high-risk area
Tangshan	26.26	38.51	706	1295	7.0	High-risk area	4.92	Sub-high-risk area
Langfang	11.9	16.91	387	529	7.8	High-risk area	5.48	High-risk area
Changzhou	12.64	18.59	680	629	10.3	Very high-risk area	7.22	High-risk area
Shijiazhuang	22.4	35.44	911	1378	10.4	Very high-risk area	6.78	High-risk area
Beijing	36.29	51.29	1456	3663	11.8	Very high-risk area	8.99	High-risk area
Tianjin	18.16	32.16	1011	2448	16.7	Very high-risk area	9.84	High-risk area
Whole region	244.43	318.43	7293/11809	11809/56623	7.5	High-risk area	5.7	High-risk area

Figure 1. County-level water shortage risk index.

at the municipal level, the development potential of water resources is very high, and the overall water shortage risk degree is very low; however, there are still large differences in it. Among the eight counties in Chengde, two—Chengde and Pingquan—have high water shortage risk levels, and the city-governed district of Chengde has a very high risk of water shortage.

Through the above research, we can further define key fields and areas that require enhancing water risk management. An analysis on the water risk index shows that the water risk index in BTHR is 7.5 in the median years, indicating a high risk level of water shortage. However, because of uneven distribution of water resources, the water shortage risk index of some areas even reaches a very high level, in which the water diversion from other areas is needed to meet water resources demand.

4 RECOMMENDATIONS

4.1 *Strengthen water supply and demand management to reduce water shortage risks*

The water shortage risks in the region are largely due to the imbalance of its water demand and supply, so the risk management at the technical level should include two aspects: water demand and water supply. The core of water demand management is to curb the excessive expansion of water demand and to promote the sustainable utilization of water resources, while the best way to strengthen water supply management is to increase water supply by water transfer project and make real-time adjustments for water use proportions. After the implementation of the first and second phases of the east route and the first phase of the middle route of South-to-North Water Diversion Project, water risk index in BTHR has dropped to a certain extent. Also, we cannot ignore water supply management. The balance of supply and demand in the water resource system in BTHR includes multiple water diversion proportions: first, the diversion proportions of Panjiakou Reservoir and Daheiting Reservoir for Tianjin and Hebei; second, the diversion proportions of Luanhe–Tianjin Water Diversion Project for Tianjin's Sihe Plain and Diandong Qingnan Plain. Considering the imbalance of the region's water demand and inflow, a fixed water diversion proportion may increase the regional water shortage risks to some extent. Thus, the real-time adjustments of diversion proportions among computing units may reduce the region's overall water shortage risks.

4.2 *Implement a variety of risk strategies to reduce losses caused by risk events*

As far as the water resource system in BTHR is concerned, the impact of water shortage risks can be effectively reduced through the evasion of risks, the transfer of shortage risks, and the implementation of pre-loss control strategies. The water shortage risks in the counties where water resource is abundant are undoubtedly engineering-oriented or quality-oriented, and for such water resource systems, the water shortage risks can completely be avoided through building of regional water conservancy projects and water treatment facilities.

For water shortage type that is resource-oriented in the plain areas, it is a big challenge to get rid of water shortage risks. Some measures must be taken to transfer risks, such as regional water diversion, trading of water rights, and insurance coverage of water shortage risks. Interbasin water diversion projects can share water shortage risks between diversion-in areas and diversion-out areas in dry years. Opening of the water right trading market enables the water demand sector to transfer some water shortage risks to the water supply sector. Insurance coverage of water shortage risks can transfer water shortage risks to all the policyholders of insurance companies and effectively diversify the risks. The economic losses caused by water shortage are related to not only the size of risks but also the regional socioeconomic scale. Thus, the pre-loss control of water shortage risks in BTHR should also take into account the socioeconomic scale matching its water resources carrying capacity.

4.3 *Establish the risk fund for water shortage*

The water security system project is an effective but expensive way of preventing water shortage risks. As it is not only a social welfare but also one of the basic industries in the national economy, the construction costs should be shared by all direct stakeholders, including companies, operators, and families, besides the part paid by the national finance. Considering it is still not mature to launch the water shortage risk insurance, establishing the water resource shortage risk fund may be a feasible way. The Risk Fund for Water Shortage will provide necessary replenishment of funds to the country's construction of water resource projects and contribute to the rolling development of river basin and the improvement of water use level in BTHR, which may benefit both the sustainable utilization of water resources and the regional socioeconomic sustainability. The risk fund may be studied seriously about its stakeholder range, levy pattern, fund use, the expected effect, and so on, and then be experimented in some pilot regions.

4.4 *Implement strategic reserves of water resources and establish emergency water diversion plans*

On the basis of the above study, we propose the emergence supply plan to cope with the water shortage crises in BTHR as follows. First, we implement a strict river-wide water quota distribution system for interprovincial rivers in dry and ultra-dry years, establish a unified water-scheduling system, and take emergency measures for interbasin or cross-regional temporary water diversion. Studies have shown that there is a low probability of extremely severe water shortage simultaneously occurring in different areas in BTHR. Thus, it is feasible and necessary to establish a sound and unified diversion system of water resources for a unified allocation and scheduling of water resources during drought periods. Second, we determine water supply priorities and quantities on the basis of the importance of water use sectors in water shortage period, such as set period and limit supply for urban and rural domestic water use; limit and halt production in plants with large water consumption and low socioeconomic impact; ensure the threshold irrigation water in agriculture according to the seasonal changes. Third, we tap the potential of water supply project and reinforce the storage management to increase water supply during water shortage periods. Water shortage management is to increase water supply as much as possible on the premise of ensuring basic water storage requirements for the next year. The water supply mode of water projects and reservoirs should be flexible, which is based on the development of drought, water shortage state, and water storage needs of key sectors. Fourth, we implement strategic water reserve plans in high risk areas of water shortage by groundwater regulation and storage. In dry periods, appropriate use of deep groundwater on the premise of not causing serious environmental or geological problems is the best way to alleviate urgent water shortage risk. And in wet years, the deep groundwater should be replenished.

5 CONCLUSION

At present, studies on water shortage risks are still very preliminary and need to go further in terms of techniques, methods, and theories. On the one hand, the water shortage risk index in the assessment still needs to be further improved and refined. Some parameters, such as water use efficiency, water supply frequency, and rainfall probability, and the index's application value and scope should be considered in the future study. On the other hand, quality-induced water shortage should be supplemented to further improve and optimize the water shortage risk assessment system.

REFERENCES

Benqing Ruan, Yuping Han, Hao Wang and Renfei Jiang, Comprehensive Fuzzy Assessment on Water Shortage Assessment [J]. *Journal of Hydraulic Engineering*, 2005, (08).

Cuina Xie, Shiyuan Xu, Jun Wang, Yaolong Liu and Zhenlou Chen, Establishment of Comprehensive Risk Assessment System and Model for Urban Water Resources [J], *Environmental Science and Management*, 2008, (05).

Guoxi Zeng and Yuansheng Pei, Research on Water Shortage Risk Control Model of River Basins, Water Resources & Hydropower of Northeast China, 2009, (03): 24–27, 71.

Li Zhang, Bo Zhang and Wei Zhang, Research on Risk Analysis Process of Water Resource System [J], *Journal of Anhui Agricultural Sciences*, 2009, (02): 738, 809.

Lin Qiu and Zhiliang Wang, Research on Theories, Methods and Application of Multi-Attribute Decision Making and Risk Analysis of Water Resource Management [M], Zhengzhou: The Yellow River Water Conservancy Press, 2007.

Mingcong Huang, Jiancang Xie, Benqing Ruan and Yamei Wang, Water Shortage Risk Assessment Model Based on Support for Vector Machines and Its Application [J]. *Journal of Hydraulic Engineering*, 2007, (03).

Ping Feng, Water Risk Management of Water Supply Systems in Dry Periods [J], *Journal of Natural Resources*, 1998, (02).

Qiyong Yang and Hui Yin, Assessment on Water Risks during Agricultural Drought in Hunan Province [J], *Journal of Water Resources Research*, 2007, (04).

Yuping Han and Benqing Ruan, Assessment on Water Shortage Risks Based on Geographic Information System [J], *Journal of Irrigation and Drainage*, 2008, (01).

Zhang S F, Jia S F. 2003. Water balance and water security in the Haihe RiverBasin. *Journal of Natural Resources*, 18(5):684–691. (in Chinese)

Zhang X, Xia J, et al. 2005. Water security of drought period and its risk assessment. *Journal of Hydraulic Engineering*, 36(9): 1138–1142. (in Chinese)

Risk Analysis and Management – Trends, Challenges and Emerging Issues – Bernatik, Huang & Salvi (Eds)
 ISBN 978-1-138-03359-7

Analysis and mitigation of project policy-related risks involving regional construction-purposed sea use in China

Shufen Liu, Wei Xu & Xin Teng
National Ocean Technology Center, Tianjin, China

ABSTRACT: China has established management policies regarding regional construction-purposed sea use and its planning (usually over 50 hectares), as well as projects therein. Based on policy analysis and the development trend of such projects, this paper analyzes the main risk factors influencing such sea-use construction projects and proposes how to mitigate their policy risks.

1 INTRODUCTION

Sea reclamation is a traditional way of utilizing marine space resources. Many coastal countries such as the Netherlands, South Korea and Japan follow this tradition. China is a country with more sea reclamation activities because along with the development of China's economy, the land price is increasing day by day. Due to the lower royalty of coastal sea area and its convenient traffic, more and more people turn to the choice of sea reclamation. According to statistics from the administrative department of China, since 2002, the sea reclamation area has been increasing by about 10,000 ha in China every year. Sea reclamation has become an important part of ocean development and utilization, and its management has become an important duty of ocean management departments in China. For better management of large-scale sea reclamation and land-filling, China has introduced the management system of RCSU.

2 RCSU DEFINITION

RCSU policy is aimed at large-scale sea reclamation and land-filling, which was introduced in 2006. According to the definition by the China Marine Administrative Department, the RCSU system refers to management policies regarding regional construction-purposed sea use and its planning (usually over 50 hectares), as well as projects therein. The RCSU management system is designed for whole planning and rational distribution of such projects so as to ensure scientific development and effective utilization of marine resources. At the same time, it helps to solve the contradiction between feasible single project planning and infeasible regional planning.

Local governments, through the RCSU system, can achieve better planning, layout, and management of RCSU projects, and the RCSU project investors can get more preferential policies and infrastructure facilities. At the same time, added to the overall sea use demonstration and the environmental impact assessment, this RCSU policy is extremely advantageous for marine environment protection.

Due to these advantages, RCSU has been quickly recognized by local governments and project investors under rapid development. According to statistics, by the end of November 2015, China has implemented 83 cases of RCSU planning involving more than 120,000 ha of sea area.

3 RCSU PROJECT-RELATED POLICIES

RCSU project-related policies include two categories: specific policies on ocean management and national macro guidances.

3.1 *Specific policies on ocean management*

From the aspect of the management process, the implementation of RCSU consists of the following links: draw-up of RCSU overall planning → demonstration of sea use and environmental impact assessment → approval of RCSU overall planning → RCSU project application for the right to use sea area → sea reclamation and land-filling → project implementation and operation. Regarding each link of the RCSU implementation above, there have been specific policies for the purpose of regulation and management.

1. Regulate the draw-up of RCSU overall planning. The State Oceanic Administration (SOA) of PRC issued the Specifications on Draw-up of Overall Planning of Regional Construction-purposed Sea Use (GHGZ [2011] No. 105) that established clear requirements on overall planning of RCSU regarding text content, format, maps, and other data. The Opinions on Improving Graphic Design of Sea Reclamation and Land-filling Projects (GHGZ [2008] No. 37) also put forward the instructional requirements on the graphic design of RCSU sea reclamation and land-filling.
2. Feasibility demonstration of sea use. Sea use demonstration and marine environmental impact assessment systems regulate marine development regarding suitability and marine environment risks. Due to the large scale of RCSU and the complexity of sea area development, the SOA issued the Opinions on Strengthening Management of Regional Construction-purposed Sea Use (GHF [2006] No. 14) and the Opinions on Regulating Environmental Impact Assessment in Planning of Regional Construction-purposed Sea Use (GHF [2011] No. 45), which put forward the special requirements on these two systems, i.e., the overall demonstration and environmental impact assessment are required. This solved the contradiction between feasible single project planning and infeasible regional planning and, to some extent, reduced the RCSU risks. In addition, the single RCSU project is required for sea use demonstration and marine environmental impact assessment separately.
3. RCSU overall planning must be reviewed by the SOA. As RCSU involves a large sea area and multiple sea use projects, the China government carries out strict management. According to the Opinions on Strengthening Management of Regional Construction-purposed Sea Use (GHF [2006] No. 14), RCSU overall planning must be reviewed and approved by the SOA, only after which the single project therein can apply for sea use, construction, and operation.
4. The single RCSU project needs to acquire the right to use sea area and the sea reclamation and land-filling is subject to approval. According to the provisions of law of the PRC on Administration of the Use of Sea Areas, the sea area is owned by the state, the State Council shall exercise ownership over the sea on behalf of the state, and the development and utilization of specific ocean resources must be subject to acquiring the right to use sea area. The use of sea area must comply with legal provisions. As an important content of sea reclamation approval, the scope of reclamation must comply with the scope of sea area approved for use.
5. China adopted an annual plan management policy on sea reclamation and land fillings. The annual-plan management policy is designed to control the scale and speed. The sea reclamation annual plan needs to be reviewed by the National People's Congress and the SOA bases on the annual plan approved to assign the annual quotations to the coastal provinces. When applying for the right to use sea areas, those people needing sea reclamation must first acquire the quotation of sea reclamation. The policy implementation achieves effective control of the scale and speed of sea reclamation and land-filling but, at the same time, brings the inevitable impact on RCSU development.

Among the above-specified ocean management policies, the (4) is based on the Law on Administration of the Use of Sea Areas, applicable to all marine development activities including RCSU projects.

3.2 *Macro guidance policies*

Macro policy here refers to the macro guiding policies issued by the China central government. These policies are often affected by the national economic development situation and overall social consciousness with distinct characteristics of a particular period, but usually with important influence on the country's allocation of resources. From 2006 to 2016, the two main RCSU-related macro guidance policies are investment expansion and economic stimulus plan in 2008, and ecological civilization construction concept put forward in 2012.

1. To cope with the international financial crisis in 2008, China introduced the investment expansion and economic stimulus plan. The marine department issued the Notice of Providing Service Assurance for Expanding Domestic Demands to Promote Steady and Rapid Economic Development which involved the RCSU content. RCSU is featured with scale investments and multiple investment projects. RCSU can effectively drive the local investment and economic development. Therefore the policy has greatly stimulated the development of RCSU.
2. Construction of ecological civilization. In 2012, under the concept of "construction of ecological civilization", the SOA put forward the requirements of "ecological use of sea", followed by a series of management policies for reasonable development and sustainable utilization of sea area resources, and promoting and maintaining the ecological balance of marine ecosystems. For carrying out the policy requirement, each RCSU management link is added with the requirements of ecological use of the sea equipped with corresponding measures. This policy has greatly increased the cost of investment in the RCSU project.

4 RCSU PROJECT DEVELOPMENT TREND AND POLICY CORRELATION ANALYSIS

4.1 *Annual changes and development analysis of RCSU projects*

Since 2006 when China started to implement the RCSU system, until the end of 2015, China adopted 82 cases of RCSU planning with a total area of about 119,900 ha, as shown in Figure 1.

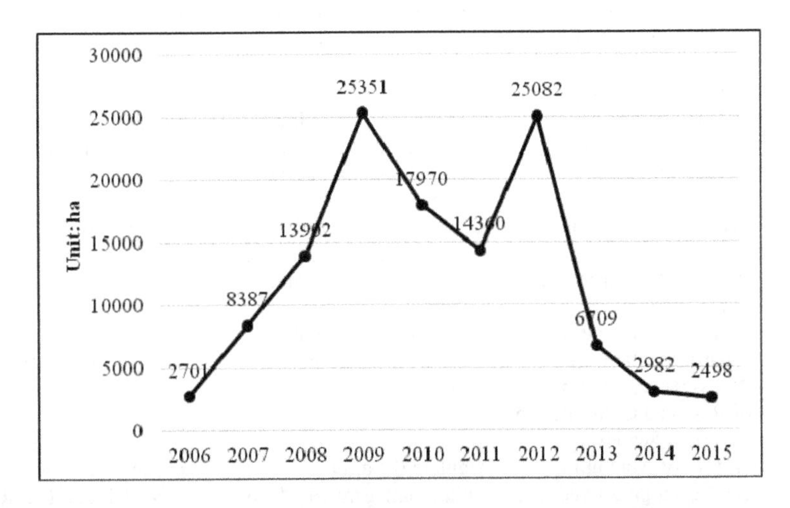

Figure 1. Annual changes of RCSU reclamation and land-filling areas in China.

We can see from Figure 1 that the areas present dramatic annual changes. The figure of 2009 (highest) is roughly ten times of that of 2015 (lowest). Such dramatic annual change means RCSU projects are faced with a huge risk. According to SOA statistics, from 2006 to 2015, China's marine economy development was smooth. There have been no big natural disasters. Therefore, the major cause of dramatic changes has nothing to do with marine economic development or natural disaster risks.

4.2 *Correlation analysis between specific ocean management policies and RCSU project development*

In the chronological order, RCSU-related specific management policies are listed in Table 1.

From Table 1, we can see that a specific ocean management policy did not bring large fluctuations to the RCSU areas. A relatively obvious influence happened during 2009–2012 during which the annual plan management policy was established and implemented. The main purpose of annual plan management is to regulate, manage and appropriately control RCSU development speed, and the management effect is proven to be good and the management goal achieved.

Table 1. Specific ocean management policies involving the RCSU project.

Time	Policy	Main content or RCSU content	Remarks
2006.04	Opinions on Strengthening Management of Regional Construction-purposed Sea Use	RCSU overall demonstration; RCSU overall planning must be reviewed by SOA.	RCSU system first presented and established; Issued by SOA.
2006.10	Regulations on Administrating the Right to Use Sea Areas	Single sea-use project to acquire the right to use the sea; Alternation of the right to use the sea.	A supplementary policy of the Law on Administration of the Use of Sea Areas; issued by SOA.
2007.06	Measures of Managing the Completion Approval of Sea Use by Land-filling Project	Regulate the completion approval of land-filling projects.	Issued by SOA.
2008.01	Opinions on Improving Graphic Design of Sea Reclamation and Land-filling Projects	Regulate the graphic design and overall layout of land-filling projects.	Issued by SOA.
2009.11	Notices of Enhancing the Management of Sea Reclamation and Land-filling Plans	Put forward the annual plan management of sea reclamation and land-filling.	Jointly issued by the NDRC and the SOA
2010.12	Notices of Strengthening the Management of Sea Reclamation and Land-filling	Strengthen annual plan management and regulate the sea reclamation land management.	Jointly issued by the MLR and the SOA
2011.02	Specifications on Draw-up of Overall Planning of Regional Construction-purposed Sea Use	Requirements on RCSU overall planning regarding text content, format and charts.	Issued by the SOA and amended in 2016
2011.09	Opinions on Regulating Environmental Impact Assessment in Planning of Regional Construction-purposed Sea Use	RCSU environmental impact assessment required.	Issued by the SOA and amended in 2016
2011.12	Measures of Managing Sea Reclamation and Land-filling Plans	Regulate the annual plan management of sea reclamation and land-filling.	Jointly issued by the NDRC and the SOA

Table 2. RCSU-related macroscopic guidance policies.

Time	Macro policy	Requirements
2008.12	Notices of Providing Service Assurance for Expanding Domestic Demands to Promote Steady and Rapid Economic Development	Strengthen the management of specific-purposed sea use planning and sea-use projects.
2012.11	The 18th National Congress of the Communist Party of China	Required to promote the construction of ecological civilization
2015.4	Opinions about Accelerating the Construction of Ecological Civilization by the CPC Central Committee and the State Council	Required for optimization of national spatial development, promotion of resource conservation, intensification of protection of natural ecosystem and environment, etc.
2015.9	Overall Reform Scheme of Ecological Civilization System by the CPC Central Committee and the State Council	Required to consummate the resources cap management and comprehensive conservation system, improve the ecological compensation system, and establish and improve the environmental governance system.

4.3 *Correlation analysis between macro guiding policies and RCSU project development*

In the chronological order, RCSU-related macro guidance policies are listed in the table given below.

From Table 1, we can see that the macro guiding policy has an obvious correlation with RCSU inter-annual changes. RCSU volatility peaks are in 2009 and 2012, which presents apparent fluctuation coming from macro policies issued.

During 2006–2009, there was a rapid growth of RCSU, especially between 2008 and 2009. The rapid development during this period arose from two reasons: first, the RCSU system established in 2006 conforms to the sea use demand and management requirements; second, in 2008, the investment expansion and economic stimulus plan and the Notices of Providing Service Assurance for Expanding Domestic Demands to Promote Steady and Rapid Economic Development played a significant role in the development of RCSU. The investment expansion and economic stimulus plan were to cope with the international financial crisis as a temporary event. Therefore the effect of RCSU development promotion only lasted for a limited period of time.

Since 2012, the RCSU area has presented a trend of sharp decline and this has an inseparable relationship with new requirements on the ecological use of the sea. For example, the original version of Specifications on Draw-up of Overall Planning of Regional Construction-purposed Sea Use was amended and added to such requirements as: "design the ecological corridor system, arrange a certain proportion of space for construction of artificial wetland and water system", and "ecological restoration" and so on. These requirements raised the threshold of RCSU, greatly increased the cost of investment, and reduced the intention of RCSU investment.

5 CONCLUSION

RCSU projects are faced with social risks, natural environment risks, project operation risks, and so on, in which the social risks include the factor of management policy. From the above analysis, we can see that RCSU and project risks therein mainly come from the macro guidance policies.

The specific ocean management policy involves RCSU and projects therein only regarding the specific operation, instead of the adjustment of industrial policy and resource allocation or the expected preferential policies and infrastructure facilities; in addition, the specific ocean management policy must have a basis from the host law, so its risk of uncertainty is lower.

Macro guidance policy often involves the adjustment of national economic overall development direction, industrial structure, and resource allocation, which has a direct impact on national and local RCSU preferential policies and then affects the investment cost, so it is one of the main risk factors of RCSU projects. In addition, the macro guidance policy often represents the management idea, generally not needing concrete law authorization, therefore, the macro guidance policy has much uncertainty and is hard to predict according to the existing laws and regulations.

According to the theory of comprehensive risk management, the RCSU project to mitigate policy risks must first put the focus on policies. The certainty policy risks mainly come from the ocean management policy and the uncertainty policy risks mainly come from the macro guidance policy. So RCSU project investors shall pay more attention to the macro guidance policy. RCSU project investment decisions must be based on understanding national macro guidance policies, analyzing their impact (positive and/or negative), and narrowing down investment cost differences. In the absence of official policy documents, RCSU project investors can turn to China central government's official media for important speeches, meeting material, and other files, so as to understand and analyze the direction of national macro guidance policies.

REFERENCES

Gao, J. 2015. Study of Management Optimization of Sea Reclamation and Land-filling in China [J], Marine Economy, 2015. 5 (3): 55–62.

Guo, X., Hu, D. 2010. Study on Necessity of Implementing Strategic Environmental Evaluation in Planning of Regional Construction-purposed Sea Use [J], Ocean Development and Management, 2010. 27 (9): 8–10.

Huang, Ch. 2005. Trapezoid Framework of Integrated Risk Management [J], *Journal of Natural Disasters*, 2005. 14 (6): 8 to 14.

Huang, Ch. 2011. Internet of Intelligence in Risk Analysis for Online Services [J], *Journal of Risk Analysis and Crisis Response*, 2011. 1 (2).

Kong, Y. 2016. SOA Issued the Measures for the Management of Regional Construction-purposed Sea Use (Trial)—Regional Construction-purposed Sea Use Shall not surpass the "ecological red line" [J], Ocean and Fishery, 2016. (3): 46–46.

Liu, S., Xu, W., Yue, Q. 2012. Introduction of Graphic Design in Regional Construction-purposed Sea Use [J], Ocean Development and Management, 2012. 29 (7): 22–24.

Liu, S., Zhang, J., Xu, W. 2016. Development of and Risk Analysis on Regional Construction-purposed Sea Use [C], Risk Analysis and Crisis Response in Big Data Era, 2016.

Wang, S., Zhang, Z. 2015. Problems in Sea Reclamation and Construction of Ecological Civilization System [J]. China Ocean Sociology Studies, 2015 (00).

Yao, X., Huang, F., Chen, Q., Guan, B. Lin, J. 2013. Applied Practice of "Five Sea-use Principals" in Planning Regional Construction-purposed Sea Use—Taking Jinjiang Municipal Planning of Regional Construction-purposed Sea Use as Example [J], Ocean Development and Management, 2013. 30 (9): 22–27.

Yue, Q., Zhao, M., Liu, S., Xu, W. 2014. Sea Reclamation and Land-filling in China, and Related Potential Risk Analysis, INFORMATION TECHNOLOGY FOR RISK ANALYSIS AND CRISIS RESPONSE [C], 2014.

Risk Analysis and Management – Trends, Challenges and Emerging Issues – Bernatik, Huang & Salvi (Eds)
© 2017 Taylor & Francis Group, London, ISBN 978-1-138-03359-7

Risk analysis and management of countermeasures on sugarcane meteorological disasters under the global climate anomaly

Xiaojun Pan
Economics School, Southwest University for Nationalities, Chengdu, China.
State Key Laboratory of Earth Surface Processes and Resources Ecology, Beijing Normal University, Beijing, China
Ministry of Civil Affairs and Ministry of Education, Academy of Disaster Reduction and Emergency Management, Beijing, China

Chongfu Huang
State Key Laboratory of Earth Surface Processes and Resources Ecology, Beijing Normal University, Beijing, China
Ministry of Civil Affairs and Ministry of Education, Academy of Disaster Reduction and Emergency Management, Beijing, China

Chengyi Pu & Heru Wang
Economics School, Southwest University for Nationalities, Chengdu, China

ABSTRACT: Agriculture is the backbone of Chinese economy, and stable agriculture leads to a steady growth of any country. However, agriculture is a process of intertexture of natural reproduction and economic reproduction, and it is a weak industry of high risk. Because of the global climate change, the frequently extreme agricultural meteorological disaster results in shaping unstable food production, being volatile grate prices, and the food security crisis becoming intensified. On the basis of the data of temperature, rains, and sugarcane yield obtained from 1998 to 2015 in Sichuan Province, through climate mutation diagnosis, meteorological yield measurement, and cobweb theory analysis of sugarcane meteorological disaster risk, risk bonds for sugar meteorological disaster are worked out, revealing a risk sequential chain structure: climate change→extreme agricultural meteorological disasters→crop yield changes→grain market price fluctuations→economic and social development risks. Strengthening meteorological risk analysis can intensify the effect of disaster prevention and mitigation. Giving play to the coordination of government can smooth market price fluctuation. Expediting the constructions of finance and legal institution can reduce social risk and negative effects caused by climate change.

Keywords: climate change; meteorological disaster; meteorological yield; cat bonds

1 INTRODUCTION

Agriculture lays the foundation of Chinese economy, and the stable development of agriculture is one of the prerequisites of social stability. Agriculture is a comprehensive process of natural reproduction and economic reproduction to maximize the production under specific agrarian, geographical, and meteorological circumstances. The main reason is that, which differs by agricultural region, the complexity and the systematic nature of risk occurrence are the technical obstacles that lead to an accurate measurement of agricultural risk identification, frequency of disaster, range of disaster, and degree of loss. In recent years, a great deal of extreme weather disasters was caused frequently on account of climate change worldwide, while increasing loss of agriculture, instability of agricultural output, and large fluctuations in grain prices.

The United Nations Environment Program has pointed out that over 850 million people in the world do not have sufficient food and clothing. Climate change will highlight the food security crisis, and the survival of human security issues has become more severe and urgent and even endangers social stability and global economic sustainability development. There are many types of meteorological disasters in China, and the occurrence of disasters is the first in terms of frequency and intensity. The impact of climate change on agricultural production is particularly evident. According to statistics, in 2013, all kinds of natural disasters affected 31.35 million hectares of crops and 3.84 million hectares of land and caused direct economic losses of 580.84 billion Yuan. Therefore, under the trend of global climate anomaly, it is not only possible to measure the loss of agricultural production and economy and society caused by extreme climate change; however, this information can be used by government and related departments to deal with agro-meteorological disasters, providing scientific disaster prevention and mitigation measures and new theoretical and practical reference for resolving international climate cooperation.

2 LITERATURE REVIEW

With the development of modern science and technology, crop production has accelerated. However, in the face of the frequent occurrence of extreme meteorological disasters, agriculture is still the most fragile industry. In the late 1980s, risk analysis was rapidly developed and widely used in agriculture, environment, engineering technology, and other fields. Petak and Atkisson focused on the expected accident, natural disaster risk assessment and public policy in the analysis of the natural disaster risk and its management policy in the United States, but less focused on the content of the agricultural meteorological disasters. Moreover, some scholars argued that in the next 100 years, extreme weather events will lead the net profit of agricultural production in the United States to decrease by about 5% (Reilly, etc., 2003) Precisely, the research on agricultural meteorological disasters in other countries is focused more on the economic field (agricultural insurance).

Domestic research about agricultural meteorological disaster risk, dating from the late 1990s (Du, Li 1997, Tan 1998), started on the disaster risk analysis method (Wang 2005) soon afterward combining with the disaster index (Li Chun, etc., 2011), weather index insurance, and agricultural catastrophe insurance (Zhang, Bao 2012). The occurrence of agricultural meteorological disasters is closely related to the changes of temperature, precipitation, and other meteorological parameters (Han, Li 2013) (Tuo, Zhu 2010) to assess the risk of agricultural meteorology. However, agricultural meteorological disasters seriously restrict the development of the national economy (Chen, Zhang, Xue 2010) because of the underdeveloped infrastructure for agricultural production, the activities of which being susceptible to natural havoc and strong dependence on the environment (Kun 2012).

In summary, although the technical methods used by scholars worldwide studying meteorological disasters and crop yield vary, with more emphasis laid on the relationship between meteorological factors and crop yield, improvements are still needed in the capacity of the quantitative assessment of risk and the mechanism of disaster and problems to tackle over limited management of countermeasures and the imbalance, where studies on grain crops are frequent compared to those on economic crops. On this basis, this paper takes Sichuan Province's 1998–2015 sugarcane temperature, precipitation, sugarcane yield, planting area, sugar price data, analysis of the food price volatility caused by the risk of economic and social relations, climate change and meteorological disasters of meteorological disasters affecting crop yield, and crop yield of sugarcane, exploring climate change disaster sequential chain risk and transfer characteristics of generation and development, and put forward the corresponding countermeasures of disaster reduction.

3 SUGARCANE UNDER CLIMATE CHANGE BY METEOROLOGICAL DISASTERS AND THEIR RELATIONSHIPS

3.1 *Sugarcane planting and the statistical description of risk*

Sichuan basin in the middle and upper reaches of the Yangtze River Valley has more sugarcane-growing areas, with a high frequency of meteorological disasters such as drought, floods, and rain. Sugarcane is a typical economic crop, and the low temperature and rainy weather seriously restrict its growth and development. The sown area and the actual yield of sugarcane in Sichuan during 1998–2015 showed a decreasing trend (Figure 1). The average annual sown area of sugarcane during 1998–2003 (30633.33 hectares) decline in 2004–2015 to 22377.78 hectares. Furthermore, the actual average annual yield of sugarcane decreased from 164,400 tons to 108,000 tons. Sugarcane production per unit area also showed a certain degree of declination (Figure 2); in 1999, the highest yield of sugarcane was 55.15 tons/ha and the lowest in 2015 was 41.45 tons/ha. Because of the frequent price fluctuation of agricultural product market sucrose due to natural disasters, sugarcane yield decreased from 9.72% to 8.30%, thereby significantly affecting the farmers.

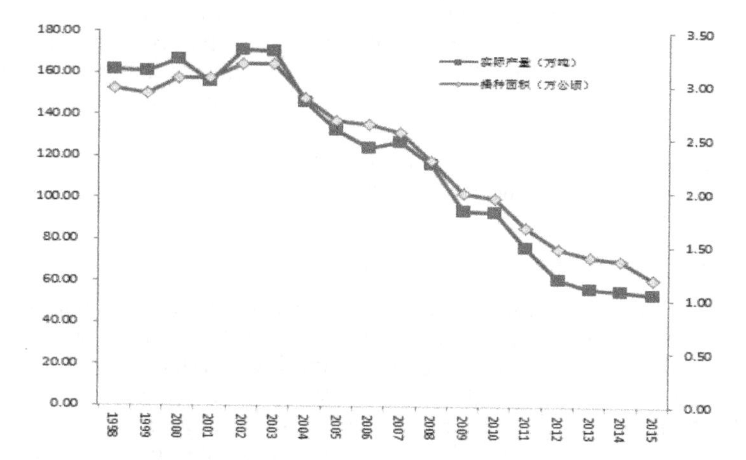

Figure 1. Sichuan Province's sugarcane sowing area and actual production (1998–2015).

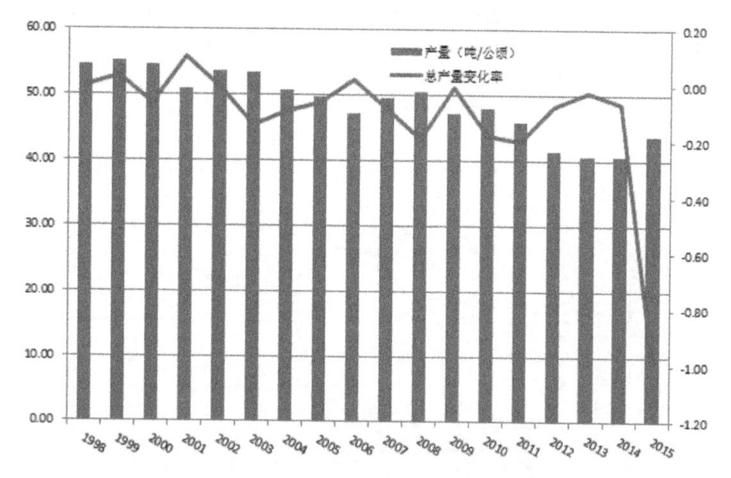

Figure 2. Sichuan Province's sugarcane yield and yield rate (1998–2015).
Source: Statistical yearbook (1999–2016) in Sichuan Province.

3.2 *Determination of sugarcane meteorological disasters*

The National Climate Center released 160 meteorological observation stations in Sichuan Province, including Chengdu, Ganzi, Daxian, Kangding, Mianyang, Huili, Nanchong, Neijiang, Xichang, and Ya'an and 11 stations in Yibin during 1998–2012, considered precipitation and temperature as measures of the meteorological data of Sichuan index, and analyzed the relationship between climate change and meteorological disasters.

Sugarcane meteorological disasters were determined by the sliding t-test of abrupt climate change. Sliding t-test by examining the difference between the two groups of sample average is significant to test the mutation. Its basic idea is that the climate sequence is divided into two subsequences; if the mean difference two scripts sequence exceeds a certain level of significance, then it generates a mean change sequence.

In the analysis of the relationship between climate change and meteorological disasters in Sichuan Province, if meteorological disasters are mainly affected by temperature and precipitation within a period, it is considered that the annual meteorological disaster occurred. A 2-year sliding t-test was conducted. At the significant level of $\alpha = 0.1$, the degree of freedom of T distribution was 13, and the Upper Limit (*UL*) and the Lower Limit (*LL*) of the T values were 1.35 and −1.35, respectively. Sliding t-test in Sichuan Province during 1998–2012 for annual precipitation and temperature change (Figure 3) shows that during 1998–2010, the meteorological disasters climate change by 4 points, the precipitation mutation point changes by 2 points, and the temperature abrupt climate changes by 3 points for an average period of 2.9 years. It is obvious that an abnormal climate change is closely related to the occurrence of meteorological disasters. In 2002–2003, the precipitation and temperature are abrupt; moreover, compared to that in 1998–2005, the frequency of abrupt change of climate in Sichuan Province has decreased in recent years.

3.3 *Determination of the influence factors of sugarcane yield*

To find the reason for the constant decrease of sugarcane yield in Sichuan Province and explain the influence factors relative to elastic changes in sugarcane yield (including meteorological factors), multivariate linear regression models were established between sugarcane yield (OUT) and sucrose Price (PRICE), Sown Area (SA), precipitation (RAIN), and Temperature (TEM). First, a nondimensional method is used to deal with the data of sugarcane yield, sucrose price, sown area, precipitation, temperature, and so on. The index data can be used to eliminate the influence of variable unit.

ADF test was used to test the stability of each variable, the optimal lag period was determined according to the SIC criteria, and the critical value was determined by the critical value of Mackinnon. The results show that RAIN and PRICE are stable at 10% level of significance; OUT, SA, and TEM are not stable; and the first-order difference sequences OUT, PRICE, SA, RAIN, and TEM are all stable (Table 1).

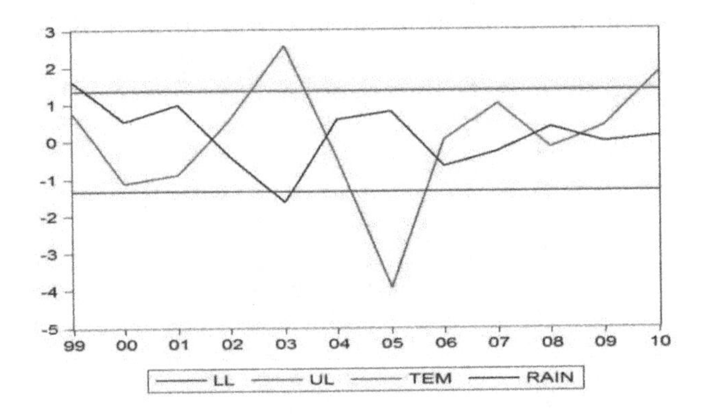

Figure 3. Climate mutation of sliding t test.

Table 1. Unit root test.

Variable	The ADF statistics	Inspection form(c, t, l)	1% critical value	5% critical value	10% critical value	Conclusion
OUT	0.959376	(1, 0, 0)	–4.057910	–3.119910	–2.701103	Nonstationary
SA	1.522124*	(1, 0, 0)	–4.057910	–3.119910	–2.701103	Nonstationary
RAIN	–3.715104	(1, 0, 0)	–4.057910	–3.119910	–2.701103	Stationary
TEM	–3.700705	(1, 0, 0)	–4.057910	–4.057910	–4.057910	Nonstationary
PRICE	–3.700705	(1, 0, 0)	–4.057910	–3.119910	–2.701103	Stationary
Δ OUT	–2.706743***	(1, 0, 0)	–4.121990	–3.144920	–2.713751	Stationary
Δ SA	–3.517366*	(1, 1, 1)	–4.886426	–3.828975	–3.362984	Stationary
Δ RAIN	–5.067174***	(1, 0, 0)	–4.121990	–3.144920	–2.713751	Stationary
Δ TEM	–3.336120**	(1, 0, 0)	–4.200056	–3.175352	–2.728985	Stationary
Δ PRICE	–3.336120**	(1, 0, 0)	–4.200056	–3.175352	–2.728985	Stationary

Note: c, t, and l are, respectively, the constant term, the trend, and the intercept term.

Table 2. EG test of OUT, SA, PRICE, RAIN, TEM, cointegration relationship.

The model form	Lag truncation parameter	EG statistics	1% critical value	Consequence
Intercept term, a trend term	0	-3.604339	-2.5658	Have a cointegration relationship

$$\Delta OUT = -3348561.7 + 10.66SA + 55.19RAIN + 58208.2TEM + 133490.07PRICE(-1)$$

$$\overline{R^2} = 0.97 \quad F = 315.09 \quad DW = 2.01$$

From the viewpoint of the estimated equation, there is a long-term stable equilibrium relationship between sugarcane yield and sown area, sugar price, precipitation, and temperature in Sichuan Province during 1998–2012. Among them, the biggest impact on sugarcane production is the first year of the market changes in the price of sucrose, 1% point increase in the average price of sugar will lead to a 133,490% point increase or decrease in the yield of sugarcane. The second is the temperature of sugarcane growth environment; sugarcane is a tropical economic crop, and each 1% point change in the average temperature can lead to a change of sugarcane yield by 58,208% points, whereas the precipitation and planting area of sugarcane production is the effect of strength and temperature, which is small when compared with the price of sugar.

3.4 *Sugarcane meteorological yield is calculated*

There are many factors that influence the production of sugarcane in addition to the meteorological factors posterity sugarcane yield components into:

$$y = y_t + y_m + y_e$$

where y is the sugarcane yield per unit area, y_t is the trend yield, and y_e is the meteorological yield for the stochastic production.

The trend of yield refers to the meteorological factors under normal conditions due to the impact of irrigation, fertilization, variety improvement, and cultivation measures of the yield. Stochastic production is the part of the output that is affected by random factors, such as economic and social conditions. Meteorological production is the part of the output due to the fluctuation of meteorological factors, which is the main natural factor causing the fluctuation of production, because the weather conditions are good and bad and hence

Table 3. Different ways to determine the trend of sugarcane yield and meteorological yield (unit: t/ha).

		Conic method		5-times moving average method		H-P filter		Exponential smoothing	
Year	Yield per unit area	Trend yield	Climate yield	Trend yield	Climate yield	Trend yield	Climate yield	Trend yield	Climate yield
1998	54.55405	54.61066	–0.0566	51.34564	3.20842	55.21214	–0.65809	54.55405	0.00000
1999	55.15068	54.24393	0.90676	51.80162	3.34907	54.55692	0.59376	53.87084	1.27984
2000	54.47059	53.81529	0.6553	53.0794	1.39119	53.89512	0.57547	53.95553	0.51506
2001	50.87255	53.32475	–2.4522	53.46706	–2.59451	53.22609	–2.35354	53.58135	–2.7088
2002	53.67085	52.7723	0.89855	53.08466	0.58619	52.55495	1.1159	51.27285	2.398
2003	53.45768	52.15793	1.29975	53.74374	–0.28606	51.86328	1.5944	52.02843	1.42925
2004	50.72917	51.48167	–0.7525	53.52447	–2.7953	51.1438	–0.41463	52.20276	–1.47359
2005	49.77154	50.74349	–0.97195	52.64017	–2.86863	50.4052	–0.63366	50.63539	–0.86385
2006	47.20076	49.9434	–2.74265	51.70036	–4.4996	49.65202	–2.45126	49.43386	–2.2331
2007	49.60156	49.08141	0.52015	50.966	–1.36443	48.88244	0.71912	47.41078	2.19078
2008	50.55652	48.15751	2.39901	50.15214	0.40438	48.07015	2.48637	48.04203	2.51449
2009	47.19598	47.1717	0.02428	49.57191	–2.37593	47.19602	–0.00004	48.86751	–1.67153
2010	47.89744	46.12398	1.77345	48.86527	–0.96784	46.2658	1.63164	47.18137	0.71607
2011	45.91018	45.01436	0.89582	48.49045	–2.58027	45.28522	0.62496	46.92779	–1.01761
2012	41.44595	43.84283	–2.39688	48.23234	–6.78639	44.27634	–2.83039	45.63401	–4.18806

the meteorological yield is also positive and negative. Therefore, Shibo Fang (2011) used the conic method, the two-curve method, the 5-times moving average method, the H-P filtering method, and the exponential smoothing method to simulate the trend of sugarcane yield and the yield of sugarcane from meteorological (Table 3).

By the regression equation, the meteorological yield is positive, and precipitation and temperature have a positive impact on the yield of sugarcane by the exponential smoothing method. Compared with the two-curve method and the H-P filter method, the 5-time moving average method has the highest correlation with the meteorological factors, which can reflect the influence of climatic factors on the yield of sugarcane. Moreover, compared with the precipitation, the temperature and the yield of sugarcane are higher.

3.5 *Effect of sugarcane yield on price fluctuation*

Sugarcane is the main sugar crop in China; however, its production and supply are affected by market demand changes. Obviously, its production decision behavior, planting area, and sucrose supply are fully consistent with the assumptions of the cobweb theory. In this paper, the price of sugar in Sichuan Province was selected as a measure of the price index of the supply and demand situation of sugarcane market. According to the method proposed by Shuai Zhang (2013), sugarcane yield and temporarily stored for other use were added into the model, and the cobweb theory model, which relates sugarcane yield and the prices of sucrose, is established as follows:

$$\begin{cases} D_t = 2117975 - 208.5437P_t \\ S_t = 2177532 - 226.5551P_{\text{预}} \\ P_{\text{预}} = 0.742751P_t + 0.257249P_{t-1} \\ D_t = S_t \end{cases}$$

where D_t is the demand, S_t is the supply, P_t is the current price, P_{t-1} is the previous price, that is, the expected price, which consists of current price and previous price. In the expected

Table 4. Correlation comparison between meteorological factors and meteorological yield obtained by different methods.

Methods	Precipitation	Temperature
Conic method	0.2036	–0.0891
5-times moving average method	0.2399	–0.1213
H-P Filter	0.1214	–0.079
Exponential smoothing	0.1854	0.1067

price, the weight of the current price is 0.74 and that of the early price is 0.26. The correlation coefficient is:

$$|(1-0.742751)\times(-226.5551)| = 58.2811 > |0.742751\times(-226.5551)+208.5437| = 40.2698$$

Obviously, the cobweb model is diverging over time; if it only relies on the spontaneous adjustment of the market, the real price of sugar will be bigger and bigger of amplitude around the equilibrium price (3307 Yuan/ton) with ups and downs, and finally, an infinite deviation will occur in the equilibrium price, which has a great impact on the market. Therefore, because of the price, the cobweb model between sugarcane yield and sugar is discrete; the spontaneous regulation of market has certain blindness, hysteresis, and price fluctuations, which will cause serious damage to sugarcane farmers and related sugar mills, affect residents' consumption, interfere with the national macroeconomic regulation and control policy, and may threaten social stability and economic development.

Because sugar is a necessity of life, its demand preferences within a certain period of time tend to be stable. On the contrary, sugarcane supply is higher, and the influence of the external factors of sugar on sugarcane farmers is significant. The crop income of sugarcane farmers is not stable due to continued volatility, and it fluctuates in a vicious cycle. Because of the weak risk ability, farmers tend to pursue short-term profit; when the price risk is big, market price is below the equilibrium price of sugar. On the one hand, it will reduce planting area. On the other hand, it reduces spending on land so that the sugarcane yield declines (Figure 2) and decreases total sugarcane production (Figure 1). When the market price is higher than the equilibrium price, the remainder will be blind to expand the production scale, even if overproduction results, which results in the nonsale or reduced sale of agricultural products. And, when the sugarcane planting income decreases, sugarcane-driven benefit will adjust the structure of agricultural production and crop varieties; the larger the land market value, the higher the alternative economic crops.

In addition, the price risk is the enterprise in the direct production of the most significant risk. When the market's supply and demand change lead prices to continuously fluctuate, sugarhouse will not adjust to the expected scale of production in accordance with the requirements to defense, and it can only be produced according to the provisions of the contract price, passively bearing the loss of all the market risk, often can increase the enterprise production cost, and reduce profits, ultimately through credit leverage, leading to bad debt risks for the banks. And the risk of prices to consumers inevitably lean, thereby reducing consumers' real purchasing power and lowering their living standards. Price risk affects production changes that increase the national budget allocation and the difficulty of the macroeconomic policy. A greater risk of price will also fluctuate the economic growth and even bring social harmony and unstable factors. In larger sugarcane areas, the use of insurance in financial insurance market and financial derivatives will disperse proactively the whole industry chain risk from sugarcane production to sugar sales and make the output of the disaster losses to a minimum.

4 DESIGN OF METEOROLOGICAL RISK BOND FOR SUGARCANE

4.1 *Sugarcane meteorological bond interest rate risk*

Usually, the interest rate is the primary factor affecting the bond price changes. Bond prices are decreasing market interest rates; the higher the market interest rate, the lower the price of the bond. Therefore, for an effective evaluation of the price of bonds, in this paper, using the risk-free bond yields, as shown in Table 5, we estimate China's 3-year bond market's interest rates.

Table 5 shows that the yield to maturity on the 1-year treasury i_1 is = 1.42%, which is equal to the bond market interest rates in the first year. Because bond prices should be equal to the same period, the market prices of the same yield bones, assuming the market price of 100. According to the basic principle of bond pricing, the price of the bond is equal to the present value of future earnings; in the second year of market, interest rate i_2 shall be calculated as:

$$100 = \frac{1.75}{1+i_1} + \frac{1.75+100}{(1+i_1)(1+i_2)}, \text{ get } i_2 = 2.76\%$$

In the same way:

$$100 = \frac{2.08}{1+i_1} + \frac{2.08}{(1+i_1)(1+i_2)} + \frac{2.08+100}{(1+i_1)(1+i_2)(1+i_3)}, \text{ get } i_3 = 2.76\%$$

The 3-year bonds market interest rates are: $i_1 = 1.42\%$, $i_2 = 2.09\%$, and $i_3 = 2.76\%$.

4.2 *Sugarcane weather risk bond structure*

4.2.1 *Sugarcane weather risk bond structure*

Bonds is the interest-bearing debt and coupon payment is certain. Debt to repay debt and its guarantee to repay the amount due are sure. In this paper, we assume that for 3-year catastrophe bonds, the coupon rate is r, $Nr = 8$, and bond value for $N = 100$.

8 (year 1), 8 (year 2), 100+8 (year 3); time axis: 0, 1, 2, 3

Thus, according to China's 3-year bond interest rate structure, we obtain the cash flow of the bond as follows.

Therefore, a security bond offering price for the first year, by $i_1 = 1.42\%$, $i_2 = 2.09\%$, $i_3 = 2.76\%$:

$$p_{(1)} = \frac{Nr}{1+i_1} + \frac{Nr}{(1+i_1)(1+i_2)} + \frac{Nr+N}{(1+i_1)(1+i_2)(1+i_3)}, \text{ get } p_{(1)} = 117.12$$

In the second year of book value:

$$p_{(2)} = \frac{Nr}{(1+i_2)} + \frac{Nr+N}{(1+i_2)(1+i_3)}, \text{ get } p_{(2)} = 110.78$$

In the third year of book value:

$$p_{(3)} = \frac{Nr+N}{(1+i_3)}, \text{ get } p_{(3)} = 105.1$$

Table 5. Cane weather risk bond interest rate forecasts.

Bond number	Year to maturity	Yield to maturity
010506	1	1.42
010570	2	1.75
015008	3	2.08

4.2.2 *Sugarcane weather risk bond structure in different assumptions*

Bonds is the interest-bearing bonds and coupon payment depend on the loss of disasters beyond the trigger level. When the loss exceeds the index, remaining coupon will not be paid. Bond to guarantee to repay the debt and to repay the amount due is certain.

If T is the duration of the bonds, $I(t)$ is sugarcane weather risk losses caused at time t, K is the loss of the trigger level, r is the coupon rate, coupon for Nr, bond value for N, then $P(t)$ is the amount paid for moment t.

If $T = 3$, $K = 1.5$ million Yuan, $Nr = 8$ Yuan, $N = 100$ Yuan, to trigger conditions of the agreement, the probability of coupon payments is:

$$P = P\,(\text{x}150) = 0.918,\ q = 1 - P = 0.082$$

The following is a discussion:

a. *Only has the interest risk; the principal will pay in the period of the last time.*

If a catastrophe occurred and reached the trigger condition in [0, T] period, only the final principal is paid.

If no hail occurred, its probability is $p = 0.918$, and the investors can obtain 100 Yuan at the end of the principal in time T and get the 8 Yuan coupon to receive at the end of each year. The above bond structure is expressed mathematically as follows:

When $t = 0$, $I(t) = 0$

When

$$0<t<T,\ P(t)=\begin{cases} N_r & I(t)<K \\ 0 & I(t)\geq K \end{cases}$$

When

$$t=T,\ P(t)=\begin{cases} N_r+N & I(t)<K \\ N & I(t)\geq K \end{cases}$$

Have the bond prices

$$p_{(t)}=\begin{cases} p_{(t)}=\dfrac{p_{(t+1)}+Nr}{1+i_1}\times p+0\times q & \\ & t=1,2 \\ p_{(3)}=\dfrac{1.08}{1+i_3}\times p+100\times q & \end{cases}$$

Can solve catastrophe bonds issue price for the first year $p_0 = 98.97$.

The book value of bonds at the beginning of the second and third years are, respectively:

$$p_{(2)} = 101.33 \text{ and } p_{(3)} = 104.68$$

If we ignore the time value of the short-term bond coupon, each 100 Yuan bonds offset the risk of lines as follows: (24 + 16 + 8) / 3 = 116.

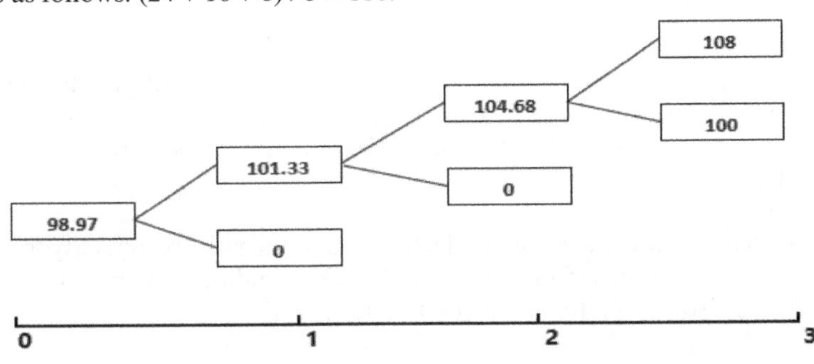

b. All interest and principal have risks

If catastrophe does not occur in the period [0, *T*] under the condition, the principal and interest will be paid. That is, in the absence of catastrophe, its probability is $p = 0.918$, investors can obtain 100 Yuan at the end of time t, and the principal of the interest income is 8 Yuan.

The above bond structure is expressed mathematically as follows:

When $t = 0$, $I(t) = 0$,

When

$$0 < t < T, P(t) = \begin{cases} N_r & I(t) < K \\ 0 & I(t) \geq K \end{cases}$$

When

$$t = T, P(t) = \begin{cases} N_r + N & I(t) < K \\ N \times 50\% & I(t) \geq K \end{cases}$$

Have the bond prices

$$p_{(t)} = \begin{cases} p_{(t)} = \dfrac{p_{(t+1)} Nr}{1 + i_1} \times p + 0 \times q \\ p_{(3)} = \dfrac{1.08}{1 + i_3} \times p + 50\% \times 100 \times q \end{cases} \quad t = 1, 2$$

The catastrophe bond issuance price for the first year is p0 = 95.63.

The book value of bonds at the beginning of the second and third years are, respectively:

$$p_{(2)} = 97.65 \text{ and } p_{(3)} = 100.58$$

Each bond risk limit for: 1/3 [(168 + 24)] + 0.5 × 100 = 66.

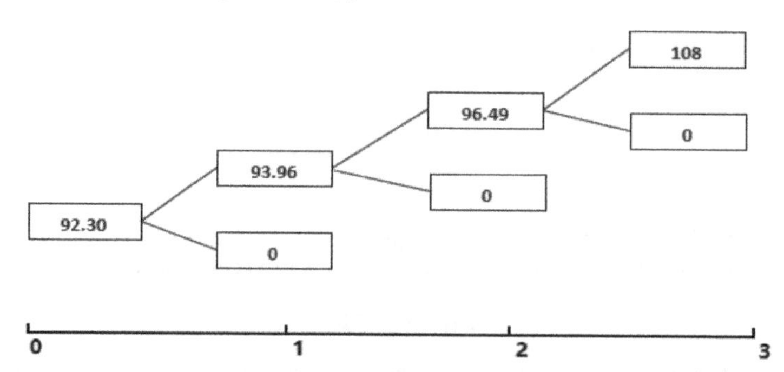

4.2.3 *Sugarcane weather risk bonds operation design*

a. Operation agreement

Considering the parties' bearing ability, public bond of 3–6% and private equity funds of 2–3% are advisable. In the absence of the conventional risk of accidents, the investors' bond premium is paid. If it is risky, investors will discount bonds. Losses are not only the coupon but also a part or all the principal.

b. Bond strengths

In a closed system to shift risk, adverse selection and moral risk are low, easy to lock the risk, and the effect is better than self-preservation. If policy conditions permit, after the improvement, bonds can also be sold to public, for a better effect.

5 CONCLUSION AND REVELATION

5.1 *Research conclusion*

Through Sichuan 1998–2012 data of temperature, precipitation, sowing area, cane yield, sucrose price, and so on, we use the sudden change diagnosis of climate, the factor of yield effects, the estimates of meteorological yield, and the cobweb theory of improved market supply function to analyze sugarcane meteorological disaster risk and reveal the climate change of Sichuan and the occurrence of sugarcane meteorological disaster, and that meteorological factors have a close relationship with sugarcane yield. And the decline in the total sugarcane production in Sichuan Province is mainly affected by the price of sugar, and temperature, the yield per unit sown area, and a relatively low precipitation of the sugarcane area. In addition, we found that the cobweb model of sugarcane in Sichuan market is divergent. Obviously, when we only rely on the regulatory role of the market, a spontaneous equilibrium cannot reach the supply–demand balance of the sugarcane market, which may cause serious damage to sugarcane farmers and sugar companies and disturb a nation policy of macroeconomic readjustment and control, endangering the social stability and economic development.

Second, according to climate change and meteorological disasters of sugarcane, meteorological factors on the relationship between the yield of sugarcane, the use of insurance and futures, and other financial derivatives in the financial insurance market actively dispersed from the sugarcane sucrose production to sales of the whole industry chain risk, which minimizes the yield loss due to disaster.

In addition, we found that climate change leads to extreme meteorological disasters, which caused some impact on crop yield, further bringing price impact of the grain market and causing risk to the stability and development of the economy and society. This verifies the sequential chain structure of "agricultural meteorological disaster risk and development of agrometeorological disasters: "climate change → happening of extreme agrometeorological disasters → crop yield changes → fluctuations of grain market price → economic and social development risk," which helps to accurately grasp the futures market hedging function to achieve the transfer price risk.

5.2 *Countermeasures and suggestions*

To conclude, we should strengthen the analysis of agricultural meteorological disaster risk and evaluation technology research, establish meteorological disaster monitoring and warning mechanism, enhance disaster prediction ability, and strengthen disaster prevention to loss reduction effect. At the same time, according to the reasonable layout of sugarcane planting season, we should create shelter forest to withstand drought and hot gas disasters and carry out reasonable irrigation and fertilization to optimize the soil structure, improve the antidisaster and the ability of sugarcane planting, and strengthen the disaster prevention and loss reduction effect.

In addition, to maximize the autonomous organizations' role of grassroots community and to widen the knowledge of disaster prevention and mitigation for farmers, we provide disaster relief measures and key grab farmers resettlement transfer scheme, which minimizes the yield loss due to disaster. For years, farmers are active through the formation of cooperatives, improve the agricultural service system in the market under the guidance of sugarcane planting scale, optimize the structure of sugarcane and sugarcane planting area, and gradually form the best and make full use of the insurance and futures and other financial derivatives in the financial insurance market, being proactive to disperse the risk from production to sales of sugarcane in the whole industry chain.

ACKNOWLEDGMENTS

This paper was supported by the Fundamental Research Funds for the Central Universities (2014SZYTD01), the Training and Funding Project for Provincial Academic Leaders

(Sichuan, No. [2012]280), the National Social Science Fund (12XJL012), the Social Science Planning of the Ministry of Education (11YJA850016), the China Postdoctoral Fund (2011M501409, 2013T60851), and the National Natural Science Foundation of China (41671502).

REFERENCES

Chun Li et al. Research and Application on risk assessment technology of facility agricultural meteorological disaster [C]. The twenty-eighth annual meeting of China Meteorological society, 2011.

Chunyi Wang, Shili Wang, Zhiguo Huo, Jianping Guo, Weijun Li. Advances in research on monitoring and early warning and assessment technology of main agro meteorological disasters in China in the past 10 years [J]. *Journal of Meteorology*, 2005, 63(5): 659–671.

Guozhu Tuo, Junsheng Zhu. Comparison and selection of agricultural insurance catastrophe risk dispersion system [J]. Insurance Research, 2010, 9: 47–53.

Heng Zhang, Wen Bao. Construction of agricultural meteorological disaster insurance and agricultural disaster prevention and reduction [J]. Agricultural modernization research, 2012, 33(2): 166–169, 248.

Huailiang Chen, Hongwei Zhang, Changyin Xue, Extreme weather events in China and agricultural meteorological services [J], Meteorology and Environmental Science, 2010, 33(3): 67–77.

Hui Han, Yaohui Li. Trend analysis of climate change and its agricultural meteorological disasters: a case study of Gansu Province [J]. *Journal of Glaciology and Geocryology*, 2013, 35(4): 999–1006.

Kun Cao, Zhenzhi Bai, Ministry of Civil Affairs: all kinds of natural disasters caused a total of 1851 people died in the country in 2013 [W], 2014.1.04, http://politics.people.com.cn/n/2014/0104/c1001–24023 367.html.

Kun Jin. Dynamic monitoring system of agro meteorological disasters [C]. Abstracts of the seventh annual conference of the National Geographic society, 2012.

Peng Du, Shikui Li, Agricultural meteorological disaster risk assessment model and its application [J], Journal of Meteorology, 1997, 55(1): 95–102.

Reilly J, Tubiello F et al. US Agriculture and Climate Change [J]. Claim Change, 2003, 57: 43–69.

Rycroft W, Regens L, Dietz T. Incorporating risk assessment and benefit-cost analysis in environmental management [J]. Risk Analysis, 1988, 8(3): 415–420.

Shibo Fang. Discussion on the method of separating trend yield and climatic yield [J]. *Journal of natural disasters,* 2011, 20(6): 13–18.

Shuai Zhang. With the improvement of the cobweb model [J]. Modern property (a), 2013, 12(7): 10–12.

Vincent T, Covello, Kazuhiko Kawamura et al. Cooperation versus confrontation: A comparison of approaches to environmental risk management in Japan and the United States [J]. Risk Analysis, 1988, 8(2): 247–260.

Zongkun Tan. Meteorological disaster risk assessment and disaster reduction countermeasure analysis [J]. Guangxi Meteorology, 1998, 19(1): 38–41.

Risk Analysis and Management – Trends, Challenges and Emerging Issues – Bernatik, Huang & Salvi (Eds)
 ISBN 978-1-138-03359-7

Variations and trends in climatic extremes in China in the period 1959–2010

Jun Shi & Haizhen Mu
Shanghai Climate Center, Shanghai Meteorological Bureau, Shanghai, China

Linli Cui
Shanghai Satellite Remote-Sensing and Application Centre, Shanghai Meteorological Bureau, Shanghai, China

Kangmin Wen
School of Environmental Studies, China University of Geosciences, Wuhan, China

ABSTRACT: The temporal and spatial variations of the daily maximum temperature (Tmax), the daily minimum temperature (Tmin), the daily maximum precipitation (Pmax), and the daily maximum wind speed (WSmax) in China were examined using the high-quality data provided by the National Meteorological Information Center, China Meteorology Administration. The results indicated that the annual Tmin tended to increase during 1959–2010, and the annual WSmax inclined to decrease during 1972–2010. The trends of annual Tmax and Pmax were not statistically significant in China. Spatially, the annual Tmax increased in the northeastern China, the northern parts of northern China, and most parts of western China; and the annual Tmin increased in most parts of China during 1959–2010. The trend of annual Pmax was not statistically significant in most parts of China. The annual WSmax decreased in most parts of northeastern China, northern China, eastern China, Qinghai, Yunnan, Hainan, and some areas of Xinjiang, Tibet, and Sichuan during 1972–2010. With the global climate change, more strategies and measures of mitigation and/or adaptation of climatic extremes are necessary for the Chinese government and the publics in the future.

Keywords: climatic extremes; characteristics; trends; climate change; vulnerabilities; China

1 INTRODUCTION

The globally averaged combined land and ocean surface temperature data, as calculated by a linear trend, show a warming of 0.85°C during the period 1880–2012 (IPCC 2013). Many studies based on the observations or simulations confirm that the changes in climatic extremes are greater than those in mean climate (Katz and Brown 1992; Groisman et al. 2005; Jiang et al. 2009), and the changes in climatic extremes are likely to have larger impacts on agriculture, ecology, infrastructure, and human activities (Kunkel et al. 1999; Fu et al. 2010; IPCC 2012). Thus, the analysis of long-term characteristics of climatic extremes is indispensable for the adaptation and mitigation strategies in the context of sustainable development (de Vyver 2012).

China is a region with complex topography, where a strong monsoon system operates. Xu et al. (2011) analyzed the variations of temperature and the precipitation extremes under the rapid warming over China in two periods: 1960–1989 and 1990–2007. Chen and Sun (2014) investigated the changes in temperature- and precipitation-based extreme indices using CMIP5 simulations of a warming of 1, 2, and 3°C in China. The objectives of this study were: (1) to investigate the temporal variations and trends of the annual averages of four climatic elements: the daily maximum temperature (Tmax), the daily minimum temperature (Tmin), the daily maximum precipitation (Pmax), and the daily maximum wind speed (WSmax) in China, and (2) to evaluate the spatial trends of the four climatic elements in the whole of China.

2 DATA AND METHOD

2.1 *Data*

The daily surface climate data of maximum temperature, minimum temperature, precipitation, and maximum wind speed were provided by the National Meteorological Information Center (NMIC) of China Meteorological Administration (CMA) (http://www.nmic.gov.cn). On the basis of a combination of criteria related to the spatial scatter of the series and data length, completeness, and quality, 604 among 756 available stations, with relatively complete data series, were used. Stations with continuous missing data of more than 30 days were filled in by their neighboring stations through the simple linear regression method (Zhang et al. 2008), and stations with missing data of less than 30 days were filled in by the average value during 1981–2010 at the same station.

2.2 *Method*

2.2.1 *Construction of a series for climatic extremes*

On the basis of the daily surface climate data, the extremes of four climatic elements were calculated for individual stations as well as for China as a whole. For individual station, one extreme value was selected in each year according to the characteristics of each climatic element, and finally four series of annual extreme values were constructed. In each station, the series for annual Tmax, Tmin, and Pmax values all began in 1959 and ended in 2010, but the series for the annual WSmax began in 1972 and ended in 2010, because before 1972, the WSmax data contained more gaps and missing values. For China as a whole, the time series were calculated first as simple arithmetic averages on the basis of the series of annual extreme values over all stations in each province, and then the mean annual series of climatic extremes in China were calculated as area-weighted averages according to the area of each province.

2.2.2 *Trend analysis and significant test*

On the basis of the annual series of climatic extremes from individual stations or whole China, the linear trends in the time series of Tmax, Tmin, Pmax, and WSmax were calculated with the method of ordinary least-squares regression, which was widely applied in extreme temperature and precipitation studies (Rahimzadeh et al. 2009; Kruger and Sekele 2013; de Lima et al. 2013). Annual trends in the values of climatic extremes were tested for statistical significance at the 0.05 confidence level using a two-tailed t-test (Skansi et al. 2013; Wang et al. 2013a). The spatial variations of the trends in the values of four climatic extremes were determined using the Kriging interpolation technique, to show the linear regression coefficient of each station on the timescale of a decade.

3 RESULTS

3.1 *Temporal variations in the values of climatic extremes in China*

During 1959–2010, the mean annual daily maximum temperature (Tmax) in China increased at a rate of 0.2°C per decade, and the long-term trend was statistically significant (Figure 1a). The mean annual Tmax increased slowly during 1959–1980, and in the 1990s (1991–2000) it

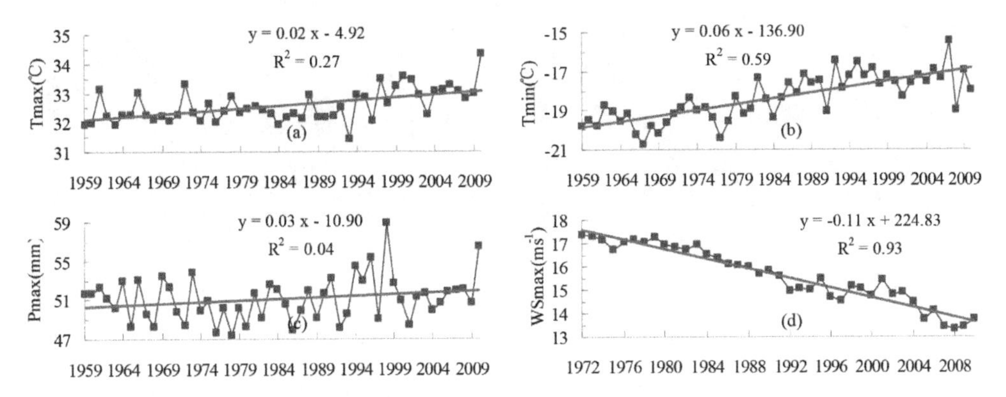

Figure 1. Temporal variations of mean annual (a) daily maximum temperature, (b) daily minimum temperature, (c) daily maximum precipitation, and (d) daily maximum wind speed in China (The blue lines are the annual extreme values, and the red lines are the linear trends, the same as below).

increased rapidly. In 1993, the mean annual Tmax was the lowest, with the value of 31.4°C, but in 2010, it was the highest, with the value of 34.3°C. The highest Tmax in China was recorded on August 4, 2008 at Turban station, Xinjiang Province, with the value of 47.8°C.

Mean annual daily minimum temperature (Tmin) increased at a rate of 0.6°C per decade in the past 52 years, and the long-term trend was also statistically significant (Figure 1b). Mean annual Tmin increased continuously during 1959–1995 and then decreased slightly. During 2000–2010, it changed little. The mean annual Tmin was the lowest in 1967 (−20.7°C) and the highest in 2007 (−15.5°C). The lowest Tmin (−52.3°C) was recorded on February 13, 1969 at Mohe Station, Heilongjiang Province.

The long-term trend of the mean annual daily maximum precipitation (Pmax) was not statistically significant in China during 1959–2010 (Figure 1c). The mean annual Pmax was the highest in 1998 (58.8 mm) and the lowest in 1978 (47.4 mm). The highest daily precipitation (633.8 mm) was recorded on September 6, 1995 at Xisha Station, Hainan Province.

During 1972–2010, mean annual daily maximum wind speed (WSmax) decreased at a rate of 1.1 ms^{-1} per decade, and the long-term trend was also statistically significant (Figure 1d). The mean annual WSmax decreased continuously in the past 39 years, and it was the lowest (13.3 ms^{-1}) in 2008. The highest WSmax was 140 ms^{-1} in China, which occurred on January 27, 1975 at Xisha Station, Hainan Province.

3.2 *Spatial trends in the values of climatic extremes in China*

During 1959–2010, the annual Tmax increased at a rate of 0–0.5°C per decade in most parts of China, but the trend was statistically significant only in southeastern Xinjiang, southern and eastern Qinghai, southern Gansu, central Sichuan, northern and eastern Heilongjiang, most areas of Tibet and Inner Mongolia, and the Yangtze River Delta and the Pearl River Delta (Figure 2a). In the Huang–Huai–Hai plain, including Henan, northeastern Hubei, northern Anhui, northern Jiangsu, western and central Shandong and southern Hebei, and some small areas in other provinces, annual Tmax decreased at a rate of 0–0.5°C per decade, but the trend was only significant in central Henan and western Shandong.

The annual Tmin increased in the whole China during 1959–2010 except for several stations, and the rate of increase was 0–0.9°C per decade in most parts of China (Figure 2b). In some areas of Xinjiang, Qinghai, Tibet, northeastern China, Hebei, and Shanxi, the annual Tmin increased at a rate of 0.9–2.0°C per decade. The increasing trend of the annual Tmin was statistically significant in most parts of China except in western Inner Mongolia, Gansu, Shannxi, southern Qinghai, and western Liaoning.

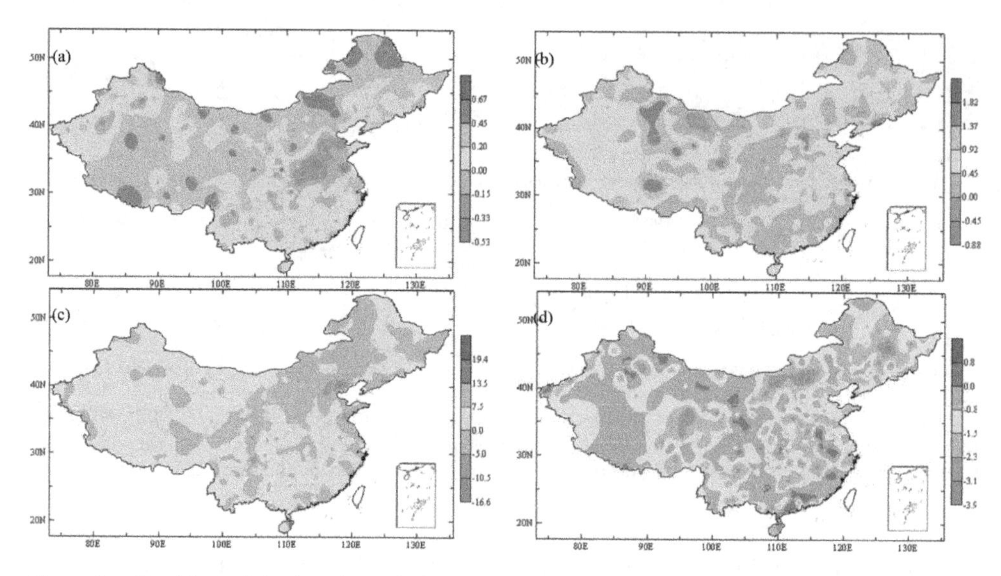

Figure 2. Spatial trends in the values of climatic extremes in China (a: daily maximum temperature; b: daily minimum temperature, c: daily maximum precipitation; d: daily maximum wind speed).

The annual Pmax varied from −5.0 to 7.5 mm per decade in almost the whole China during 1959–2010, and it was not statistically significant in most parts of China (Figure 2c). Only in Beijing, northern Tianjin, and central and southeastern Hebei, the annual Pmax decreased at a rate of 5.0–15.0 mm per decade, and the trend was statistically significant. In central and western Inner Mongolia, Shanxi, most of Hebei, northern Shannxi, and eastern Heilongjiang, the annual Pmax decreased at a rate of 0–5.0 mm per decade, but the trend was not statistically significant. In other regions, the annual Pmax increased mainly at a rate of 0–7.5 mm per decade, and the trend was also not statistically significant.

Except for some areas of Anhui, Fujian, Guangdong, Sichuan, Gansu, and Xinjiang, the annual WSmax decreased in the whole China during 1972–2010 (Figure 2d). In most areas of northeastern China, northern China, eastern China, Qinghai, Yunnan, Hainan, and some areas of Xinjiang, Tibet and Sichuan, the annual WSmax decreased at a rate of 0.8–2.3 ms^{-1} per decade, and the trend was statistically significant. The decreasing trend of the annual WSmax was not statistically significant in central Tibet, Gansu, western Inner Mongolia, southern Guizhou, western Hubei, Guangxi, Guangdong, Fujian, Anhui, and most areas of Xinjiang.

5 CONCLUSIONS

During 1959–2010, the mean annual Tmax and Tmin increased at rates of 0.2 and 0.6°C per decade, respectively, and the mean annual WSmax decreased at a rate of 1.1 ms^{-1} per decade in China during 1972–2010. The long-term trend of the mean annual Pmax was not statistically significant in China. Spatially, the annual Tmax increased in the northeastern China, the northern parts of northern China, and most parts of western China. The annual Tmin increased at a rate of 0–0.9°C per decade in most parts of China. The trend of annual Pmax was not also statistically significant in most parts of China, and it decreased significantly only in Beijing, northern Tianjin, and central and southeastern Hebei. The annual WSmax decreased in most parts of northeastern China, northern China, eastern China, Qinghai, Yunnan, Hainan, and some areas of Xinjiang, Tibet, and Sichuan.

ACKNOWLEDGMENTS

This work was supported by the National Natural Science Foundation of China (Nos. 41571044, 41401661, and 41001283) and the China Clean Development Mechanism (CDM) Fund Project (No. 2012043).

REFERENCES

Chen, H., Sun, J. (2014). Changes in climate extreme events in China associated with warming. Int J Climatol, doi: 10.1002/joc.4168.

de Lima, M.I.P., Santo, F.E., Ramos, A.M. et al (2013). Recent changes in daily precipitation and surface air temperature extremes in mainland Portugal, in the period 1941–2007. Atmos Res 127: 195–209.

de Vyver, H.V. (2012). Evolution of extreme temperatures in Belgium since the 1950s. Theor Appl Climatol 107: 113–129.

Fu, G.B., Viney, N.R., Charles, S.P. et al (2010). Long-term temporal variation of extreme rainfall events in Australia, 1910–2006. J Hydrometeorol 11: 951–966.

Groisman, P.Y., Knight, R.W., Easterling, D.R. et al (2005). Trends in intense precipitation in the climate record. J Climate 18: 1326–1350.

IPCC (2012). Managing the risks of extreme events and disasters to advance climate change adaptation. Cambridge University Press, Cambridge, UK, and New York, NY, USA, 582 pp.

IPCC (2013). Climate Change 2013: The Physical Science Basis. Cambridge University Press, Cambridge, United Kingdom and New York, NY, USA, 1535 pp.

Jiang, Z., Ding, Y., Cai, M. (2009). Monte Carlo experiments on the sensitivity of future extreme rainfall to climate warming. Acta Meteorologica Sinica 67(2): 272–279 (in Chinese).

Katz, R.W., Brown, B.G. (1992). Extreme events in a changing climate: variability is more important than averages. Climatic Change 21: 289–302.

Kruger, A.C., Sekele, S.S. (2013). Trends in extreme temperature indices in South Africa: 1962–2009. Int J Climatol 33(3): 661–676.

Kunkel, K.E., Pielke, R.A., Changnon, S.A. (1999). Temporal fluctuations in weather and climate extremes that cause.

Rahimzadeh, F., Asgari, A., Fattahi, E. (2009). Variability of extreme temperature and precipitation in Iran during.

Skansi, M.M., Brunet, M., Sigró, J. et al (2013). Warming and wetting signals emerging from analysis of changes in.

Wang, S., Zhang, M., Wang, B. et al (2013a). Recent changes in daily extremes of temperature and precipitation over the western Tibetan Plateau, 1973–2011. Quatern Int 313–314: 110–117.

Xu, X., Du, Y., Tang, J. et al (2011). Variations of temperature and precipitation extremes in recent two decades over China. Atmos Res 101(1–2): 143–154.

Zhang, Q., Xu, C.Y., Zhang, Z. et al (2008). Climate change or variability? The case of Yellow river as indicated by extreme maximum and minimum air temperature during 1960–2004. Theor Appl Climatol 93: 35–43.

Risk Analysis and Management – Trends, Challenges and Emerging Issues – Bernatik, Huang & Salvi (Eds)
© 2017 Taylor & Francis Group, London, ISBN 978-1-138-03359-7

Implementation of a web based map for the internet of intelligences to analyze regional disaster risks

Yanan Sun, Chongfu Huang & Qiuyu Wang
State Key Laboratory of Earth Surface Processes and Resources Ecology, Beijing Normal University, Beijing, China
Ministry of Civil Affairs and Ministry of Education, Academy of Disaster Reduction and Emergency Management, Beijing, China
Faculty of Geographical Science, Beijing Normal University, Beijing, China

ABSTRACT: To develop the Internet of intelligences for online analysis of the risks of natural disasters in a township region in China, in this paper, we discuss on the implementation of a Web base map of disaster risk. First, we collect the geographic information of the research region and establish the database. Then, we use WebGIS API for JavaScript combining GeoServer to achieve the WebGIS development and release the related map services. Finally, we develop and design the geographic information system for the Internet of intelligences.

1 INTRODUCTION

The Internet has changed every aspect of our life from our daily activities to professional activities, including the promotion of online analysis of natural disaster risk. Huang (2011) proposed the method of Internet of Intelligences (IOI) to mitigate the problems of traditional probabilistic risk analysis such as difficulty in confirmed its statistic rule, difficulty in acquiring data, and difficulty in guaranteeing the reliability of the analysis result.

Any IOI platform serving for natural disaster risk analysis must be supported by Web base maps, if the customers ask for risk maps. A base map is a digital line graph consisting of several layers, such as boundaries, hydrographs, roads and trails, and man-made features. A Web map is an online atlas that enables to store, recover, manage, and analyze spatial data. It allows a customer to view and download geospatial data.

To develop the Internet of intelligences for online analysis of natural disaster risks in a township region in China, in this paper, we study the implementation of a Web base map of disaster risk.

2 WEB MAPS FOR THE INTERNET OF INTELLIGENCES

2.1 *The internet of intelligences*

A network connecting agents by the Internet and integrating scattered wisdoms to be a great wisdom by embedded models is called an Internet of Intelligences (IOI). It can be formally defined as:

Definition 1 If A is a set of agents, N is a network used by A, and M is a model to process information provided by A, then the triple $\langle A, N, M \rangle$ is called an Internet of intelligences, which is denoted by Φ.

Obviously, an important issue to construct an IOI for online analysis of natural disaster risks is to have the same Web maps, which could be regarded as the geographical information provided by the agents. The model M in the IOI is a recognized technology that is mainly composed of data-handling tools.

2.2 *WebGIS*

Because of the strong ability of GIS to comprehensively process and analyze geospatial data, it has been gradually applied in the field of risk analysis and decision-making of natural disasters. The traditional Geographic Information System (GIS) focuses on their own system, and the barriers between different systems are quite distinct. They have different background, data model, and functional organization structure. A set of professional GIS system is very expensive and needs complex operations for development and deployment and a high threshold. WebGIS overcomes all these drawbacks and brings GIS to the hands of people. It reduces the need to create custom applications, provides a platform for integrating GIS with other business systems, and enables cross-organizational collaboration.

WebGIS is a combination of Internet and Geographic Information System (GIS), which is a recognized technology that is mainly composed of data-handling tools for storage, recovery, management, and analysis of spatial data. WebGIS is a kind of distributed information system. The simplest architecture of a WebGIS must have at least one client and one server; the client is a desktop application or a Web browser application that allows users to communicate with the server, and the server is a Web server application. It adopts HTTP protocol to transmit data. It not only has most functions of the traditional GIS software, but also provides interactive map, data, and interactive data query analysis through the Internet. This facilitates mapping, querying spatial database, and so on anywhere through the Internet (Yan, 2015). The development of WebGIS has gradually become an important way of GIS application, transforming it from professional to personal. It can adapt to natural disaster risk analysis, where data collection is difficult, has strong spatial characteristics and strong dynamic variability, and so on (Huang, 2012).

WebGIS brings analytics to spatial data in a way that was not possible before. Previously, spatial data had to be processed, modified, and extracted to answer a predetermined set of questions. Now, the data are transformed into Web maps or services that are mashed up with different layers into a Web GIS, which provides the flexibility to answer any possible question. WebGIS is a much more flexible and agile workflow with the development of Web technology, component technology, distributed systems, and so on. When the idea of Web server was put forward, users can find and get interesting data by spatial information Web service, which can be distributed on the network. And different service providers can provide data in different regions, topics, and quality (Liu and Zeng, 2016).

2.3 *Maps in WebGIS for the internet of intelligences*

Taking into account the powerful geographic information processing capability of WebGIS and the strong interactive Web server philosophy, if the WebGIS technology is applied to the Internet of intelligences system, theoretically, it can provide a visualized geographic interface for intellectual agents and realize the functions of collecting, storing, and visualizing the flexible geographic information in the Internet of intelligences. The Web mapping is a service by which consumers may choose what the map will show. WebGIS emphasizes that geodata processing aspects are more involved with design aspects, such as data acquisition and server software architecture, such as data storage and algorithms, than the end-user reports themselves (Kraak, Menno Jan, 2001).

The intelligent agent can add the geospatial information data for the interested risk events through the risk base map issued by the Internet of intelligences and provide clear and accurate geographical position information for the distribution of spatial elements, which reflects the interrelationship between the risk scene and the distribution of factors. It helps the agents to understand the regularity of the risk scene more clearly, clarify their flexible perception of the risk problem, and give the geospatial attributes to the flexible risk perception so as

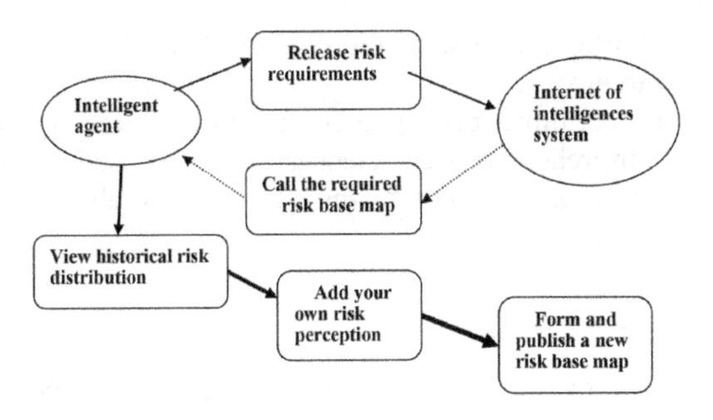

Figure 1. Diagram of the conceptual flow of the online risk analysis in the internet of intelligences with the networked risk base map function.

to optimize the risk analysis performance of the Internet of intelligences system. A specific concept flow is shown in Figure 1.

3 APPROACH TO IMPLEMENT THE WEB BASED MAP

The current Web development technology is quite mature, and WebGIS development has many successful examples and design models that are widely used in various industries. Realization of networked disaster risk base map for the online risk analysis in the Internet of intelligences system, combing with the advantages of GIS technology in spatial data processing and database technology, is a feasible idea from a technical point of view.

3.1 *Selection of development tools*

Commercial GIS software is expensive to use and maintain, for example, a full suite of ESRI ARCGIS software for both client and server costs about $100,000. Moreover, the data and operations of commercial GIS software cannot be fully converted and shared.

Considering the high cost and the need for the popularization of the Internet service, there is no barrier to the operation of open-source GIS, which is active in the opposite of commercial GIS. Open-source GIS has flexibility and low threshold, which greatly reduces the software cost and labor (Liu and Zeng, 2016).

There are many free open-source GIS software, such as large-scale desktop GIS with QIS and GRASS GIS; server software such as Geoserver and MapServer; and open-source GIS databases such as PostGIS.

GeoServer, as an open-source software (Yuan and Zheng, 2007), is free to use and can be modified, copied, and redistributed, and hence, it has become one of the major WebGIS solutions. GeoServer is an open-source mapping server that fully complies with the Open Geospatial Consortium (OGC) open standard and is compliant with the J2EE specification. It provides mapping services to the clients in the network. It can receive requests for Web Map Service (WMS) and Web Feature Service (WFS) and return data in multiple formats. The definition of the WMS/WFS specification in this process creates the possibility for a public map service. GeoServer can easily publish the map data, allowing the user to update, delete, and insert the feature data and hence it is easy to share the spatial geographic information among the users. It can be used to publish data as maps/images (WMS implementation) or to publish actual data directly (WFS implementation). It also provides the ability to modify, delete, and add new functionality (WFST implementation). GeoServer can also be configured to integrate third-party mapping services, such as Google Map and ArcGIS, and can also be used for pre-generated map slices. Because of its open-source nature, GeoServer is widely used in many projects. It can be combined with other new technologies, such as PostGIS, Shapefile, and other connections, access data, and publish them as a map. The integrated use of GeoServer, PostGIS, and Openlayer provides a solution for WebGIS framework.

With benefits such as economic, open source, freedom, simplicity, scalability, and other functions to achieve a comprehensive view, Geoserver can be used as a technical solution to publish networked disaster risk base map for the online risk analysis in the Internet of intelligences system.

3.2 *Geographic information database*

The database is a platform for a unified organization and a centralized management of system data. It is used to store the attribute data, spatial data, and metadata information of the system. It includes spatial database engine, relational database system, and map slice library. It is a unified organization of system data, which can provide standardized and efficient data services to the Internet of intelligences of online risk analysis system.

PostgreSQL is an object-oriented relational database (Zhang and Lin, 2015), which is powerful, shows good performance and good adaptability, and creates a precedent, in which object-oriented thinking was applied to the database technology. PostGIS is an extension to PostgreSQL, which can store and analyze spatial-type data. It has open-source features, is powerful, easy to use, and the default engine of GIS. PostGIS supports a wide range of spatial data types, including points, lines, polygons, and collections of object sets.

The Internet of intelligences of online risk analysis system can store and manage the spatial geographic information data by using PostgreSQL's spatial extension component PostGIS. The spatial data and the spatial index can be used to improve the efficiency of spatial data retrieval. The map slice library stores the pre-rendered map slice. When the browser sends the same map service (WMS) request, it fetches the map directly from the map slice library and avoids the map image rendering process, thus realizing fast system response and reducing the pressure of the map service module.

3.3 *Technical architecture design*

In order to realize the networked disaster risk base map of the intelligent online risk analysis system, the designed system adopts a four-layer B/S architecture (browser/server mode) structure (Shi and Zhang, 2009), which is divided into client browser, Web server, map server, and database server. Each layer has its division of labor as well as mutual interaction with others. The client browser is responsible for interacting with the user and displaying the data; the Web server is responsible for WWW services, including the release of HTML, the implementation of JSP pages and components, and database server communications; the map server is responsible for handling the base map request and the implementation of the risk analysis of the base map; and the database server is responsible for managing geospatial data.

In detail, the client using a common browser integrating JavaScript and Openlayers achieves visual interface. The widely used mainstream Web server is Tomcat, which is stable and reliable, and is the first choice of development of Web systems. Map server using the classic open source GeoServer is actually a Java software package that provides spatial data publishing as a layer through its WMS service. Publishing various spatial data information into a map provides base map service and data query for the Internet of intelligence's online risk analysis system. Database server layer uses spatial data files and relational database (PostgreSQL and PostGIS) to build a geographic information database to manage and maintain geographic data. The spatial database file is responsible for storing spatial feature data, and the relational database file is responsible for storing attribute data.

1. When the user carries out related operations through the browser, JavaScript calls XMLHttpRequest object to sends request to the Web server, which sends the user's operating information to the Web server.

2. According to different requests, Web server finds out the relevant attribute data from the postgreSQL database and passes the corresponding parameters to the map server—GeoServer. Then, as desired, GeoServer requests the PostGIS database to obtain spatial information data and processes them. When MapServer receives the request, it calls the appropriate method, accesses data (usually a picture or data layer like point, line and so on), and conveys the data obtained to the Web server.
3. Web server package contains the data obtained in the query as a JSON object or a Web page, JavaScript receives the returned results through the Ajax engine callback function in client and parses the data of spatial characteristics and attribute information. The former is displayed on the map as the form of WMS, and the latter is displayed in the form of a property window so that the user can see the results in the browser.

The specific technical architecture design diagram is shown in Figure 2.

3.4 *Realizing path design*

The design flow is divided into four parts: collecting and processing geographic information data of the study area, constructing geographic information database (Wang and Zhao, 2015), Geoserver publishing map service, and system development and design.

On the basis of a careful analysis and comparison of the data, we collect and process the geographical information data and form the risk base map of the study area. We need to determine the data suitable as the basic data of the map and analyze them, to provide basic and accurate geographical contours for the base map, such as latitude and longitude lines, rivers, and settlements. As for the expression degree of the geographical basis elements, it is necessary to clarify the content of a topic environment, help reading, and make it legible, and the subject matter should not be interfered (Xie, 1999); Simultaneously, it requires the base map just reflect some of the major basic elements as far as possible and provide sufficient space for the user to add new information elements later.

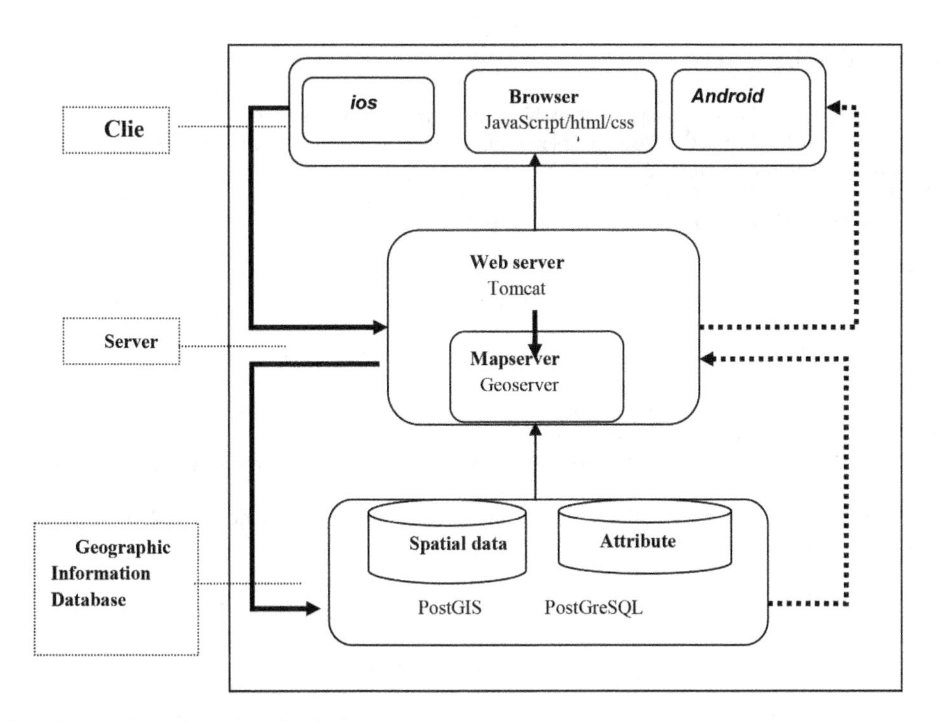

Figure 2. Technical architecture design.

For building a Geographic Information Database, we use PostgreSQL's pgAmin III graphical administration tool to connect to the database server, create a new database, import spatial data in the shapefile format, and set SRIDs (Spatial Reference Identifiers) for spatial data.

Geoserver publishes the map service. It is used to connect with the spatial database, publish the data to the layer, return to the browser, display on the interface (WMS service), use its WFS service, inquire the object attribute information, and display the results in the form of text symbols.

Finally, we configure the development environment and design the visual interactive page.

4 CONCLUSION

In this paper, in order to employ the Internet of intelligence to the online analysis of integrated risk in a small area, we mainly discuss the realization path of the networked disaster risk base map and choose the open-source Geoserver as the technical tool to solve the issue of releasing risk base map for IOI. We used PostGIS to construct the spatial geographic information database, design the technical framework of the four-layer B/S system, and implement the four-step path. The networked risk base map provides the geographic attribute information with good visibility for the analysis of the regional risk problem. At the same time, the corresponding base map layer can be called depending on different characteristics of the regional nature disasters, which reduces the interference of non-effective factors.

ACKNOWLEDGMENT

This study was supported by the National Natural Science Foundation of China (41671502).

REFERENCES

Huang, C.F. 2012. Natural disaster risk analysis and management. Beijing: Science Press.

Huang, C.F. 2014. Mitigation four problems of the risk analysis with the Internet of intelligences. *Journal of natural disasters* 23 (2): 1–7.

Kraak, Menno Jan. 2001. Settings and needs for web cartography. Kraak and Allan Brown (eds), Web Cartography, Francis and Taylor, New York, p. 3–4.

Liu, G., Zeng, J.W. 2016. 09.Web GIS principles and application development. *Tsinghua University Press*.

Shi, R.M., Zhang, H.M. 2009. Application Research of WebGIS Based on GeoServer. *China Science and Technology Information*, (24): 93–94.

Wang, H.Y., Zhao, R.Y. 2015. Digital Campus Map Design and Implementation Based on WebGIS. *Software Guide*, (11): 71–74.

Xie, L.Z. 1999. Geographic Base Map of Geo-thematic Maps. *Quaternary Research*, (03): 260–267.

Yan, S.L. 2015. Datashared WebGIS platform based of GeoServer.

Yuan, Y., Zheng, W.F., Wang, X.B. 2007. WebGIS development based on GeoServer. *Software Guide (05): 96–98.*

Zhang, D.J., Lin, Q.Y. 2015. GeoServer-based tourism information system design and implementation. *Journal of Sanming University, (06): 60–64.*

Risk Analysis and Management – Trends, Challenges and Emerging Issues – Bernatik, Huang & Salvi (Eds)
 ISBN 978-1-138-03359-7

Selection of a research area to construct a virtual case for integrated risk assessment of earthquake and flood

Qiuyu Wang, Chongfu Huang & Yannan Sun
State Key Laboratory of Earth Surface Processes and Resources Ecology, Beijing Normal University, Beijing, China
Ministry of Civil Affairs and Ministry of Education, Academy of Disaster Reduction and Emergency Management, Beijing, China
Faculty of Geographical Science, Beijing Normal University, Beijing, China

ABSTRACT: Reliability of a method used in integrated risk assessment comes from a large number of practical evidences. When we employ the information diffusion theory and methods for integrated risk assessment in a small administrative area, we do not have much data. It needs a virtual case construct to test if the method can reliably assess integrated risk with a small sample, meaning that the result is not far away as the one that results from a big sample. In this paper, we review the data situation on earthquake and flood in provinces of mainland China and eventually select Sichuan province as a research area to construct a virtual case. In the meantime, we analyzed natural geographical features, population, economics, transportation, post and telecommunications of the Sichuan province to demonstrate its superiority and rationality. In the virtual case, data of earthquake, flood and disaster, as much as we need, can be produced by a random number of generators whose parameters are defined with the data in Sichuan province.

1 INTRODUCTION

With the development of disaster risk study, more and more attention has been paid on multi-risk. Some theoretical works of multi-risk have been done and several methods for multi-risk assessment have also been developed. After clarifying the concepts of multi-hazards and multi-risk, the methods of multi-risk assessment which have been better known and used internationally or domestically are systematically summarized and the classification system of the methods has been discussed (Ming and Xu, 2015).

The simple sum of risks is not equal to the integrated risk caused by multi-hazards. In most cases, different natural disasters do not occur at the same time; however, in the longer term, there may be a large number of natural disasters (Huang, 2012). It is easy to improve the availability of independent multiple hazards of property loss assessment. Due to that, risk analysis is usually done with incomplete information, i.e., involving both random uncertainty and fuzzy uncertainty. There are some geographical units which lack the required sample information and so we are unable to carry out assessment work. In conditions of incomplete information, how to form a universal integrated risk assessment method for a given period is a key scientific problem that must be solved to improve the usability of multi-hazards comprehensive risk assessment.

Theoretically speaking, any natural disaster system which includes the subset of a society system is a physical system. We would use a group of differential equations to describe it. However, a natural disaster system is too complicated to construct available equations. Even if we have the equations; it is very difficult to get a solution because: (1) the solution must be existing; (2) the solution must be unique; (3) the solution must depend continuously on the data of the problem.

Then, we regard a natural disaster system as a stochastic system to reduce complexity and the probability theory is introduced to study the risks of natural disasters. Particularly, most of the stochastic systems evolving over time in the study period are considered as stationary Markov processes, i.e., the systems in the future are the same as the stochastic behavior in the past so that we can use observations of the sample record historical events that occurred to analyze the risks.

In fact, most of the disaster systems are in change and the corresponding stochastic processes do not satisfy the stationary Markov process hypothesis. If we collect observations of historical disasters across hundred years to assess disaster risks, the reliability of the assessments will be low. For an integrated risk caused by multi-hazards, we face a greater challenge to analyze it with a given sample. We are developing the information diffusion technique (Huang, 2002) to construct a joint probability distribution and a disaster function to calculate an integrated risk. In this study, small samples are studied and the models must be verified.

Until today, it is impossible to verify if a calculated risk is real as it depends on used models and available data. In most cases, what we can do is it to verify if the used models are reliable. The verifying of a model used in analyzing probability risks would be done by probability simulation, also called Monte Carlo simulation (Fishman, 1996).

This simulation technology allows us to use virtual disaster cases to study how to calculate disaster risk. In a virtual case, we can produce as many random numbers as we need. The random numbers are drawn from a population with given probability distribution. The task of verifying a model is to check whether we can use it to estimate the probability distribution much better than other models.

A virtual case for verifying a model used in the integrated risk assessment of earthquake and flood should be constructed with historical disaster data. In this paper, we review the data situation on earthquake and flood in provinces of mainland China and select a research area to collect the data we need. The data determines the probability distribution for the virtual case.

2 HISTORICAL EARTHQUAKE AND FLOOD IN MAINLAND CHINA

According to China Earthquake Administration (CEA) report, 743 tremors of magnitude 5 or higher have struck mainland China between 1984–2015, including 182 times in Xinjiang Uygur autonomous region, 132 times in Tibetan autonomous region, 124 times in Qinghai Province, 120 times in Yunnan Province and 95 times in Sichuan Province followed by the top five bit; earthquake occurred 653 times in these five provinces in total, accounting for 87.89 percent of the total earthquake, and the remaining provinces with less earthquake, basically less than 10 times. They are shown in Figure 1.

The earthquake occurred mainly in five areas: Plateau earthquake zone, north China earthquake zone, Xinjiang earthquake zone, Taiwan earthquake zone and the southeast coastal seismic zone, a total of 23 seismic belts (Shi, Peijun 2014). Mainland China has significant seismic activity featuring "higher frequency in west, lower in east". From 1984 to 2015, China's average annual number of disappearances and death is 4440 due to the earthquake, which accounts for 52.3 percent of the whole natural disaster number of deaths and missing. The average annual direct economic loss caused by the earthquake is 43.1 billion RMB, accounting for 17.1 percent of natural disasters in direct economic losses.

Flood disaster in China mainly concentrated in the Yangtze River, the Huaihe River and the southeast coast and other regions (Xu, Wei. 2014). Forty percent of the country's population, 35 percent of arable land and 60 percent of industrial and agricultural output value has long been under the threat of flood. Losses caused by flood on food production are less than those caused by drought hazards. The average annual losses of food due to floods caused 25 percent of the total losses.

From 1984 to 2015, the number of China's annual average dead and missing people caused by floods is 2338, accounting for 27.5 percent of all natural disasters; the direct economic loss is 80.128 billion RMB accounting for natural disasters, which is 41.7 percent of the total economic loss. Flood is a major hazard which caused direct economic losses.

Then the flood that occurred between the years 1984–2015 were analyzed and it was found that the largest annual direct economic losses were 9.44 billion RMB in Hunan Province, and following it was Sichuan Province where the annual direct economic losses reached 7.767 billion RMB.

Anhui, Jiangxi, Hubei and other provinces and regions have also a high incidence of floods and the annual direct economic losses have all reached 50 billion RMB. They are shown in Figure 2.

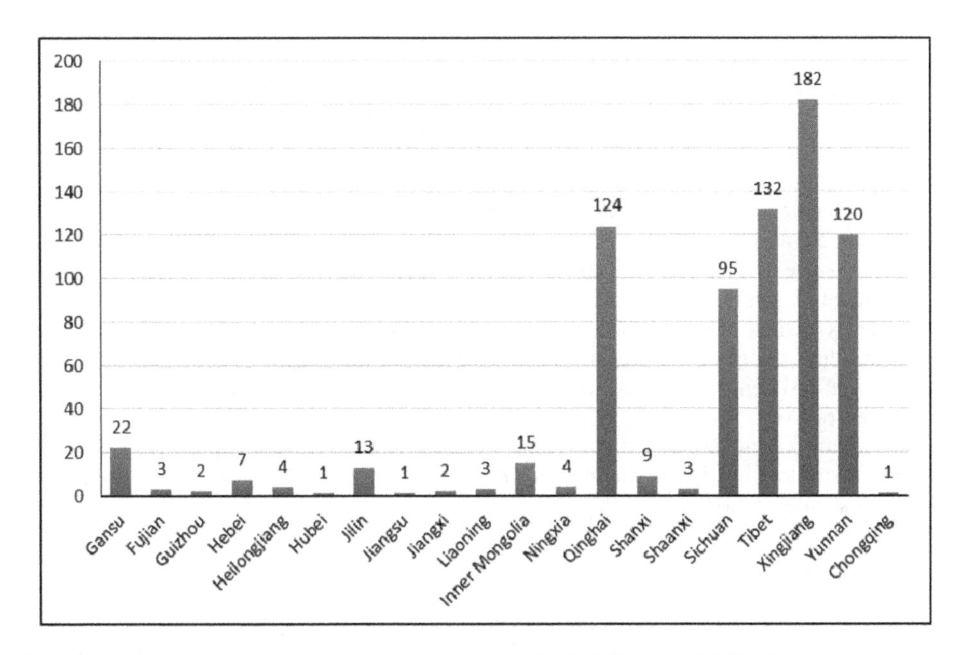

Figure 1. Frequency of earthquakes in provinces of mainland China, 1984–2015.

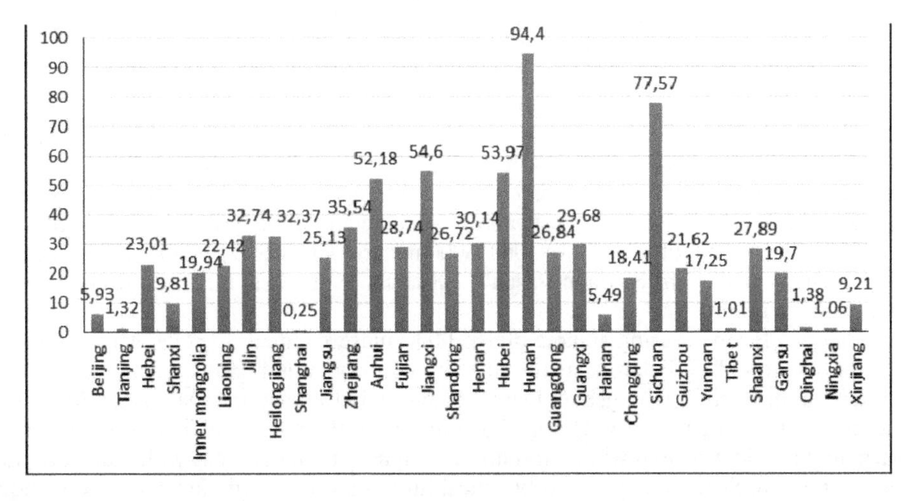

Figure 2. Annual average direct losses of floods in provinces of mainland China, 1984–2015 (100 million RMB).

The flood high-risk areas are Yangtze River Delta, Yangtze River and Gan River, Hubei and Hunan provinces, the Huaihe River, Haihe and Luanhe River, Pearl River, Liaohe River, the Northeast Plain, the Weihe Plain (Du, Xingxin. Wang, Ze.1997), and Sichuan Basin. The flood prevention capacity of small rivers, including some large river tributaries, is low.

3 SELECTED REGION FOR A VIRTUAL CASE

On the one hand, earthquakes in mainland China are mainly concentrated in the western regions, as we can see above, especially in Sichuan, Yunnan, Xinjiang, Qinghai, and Tibet, which are the most frequent areas of earthquake and geological disasters.

On the other hand, as we can see in Figure 2, floods caused the greatest losses in Sichuan, Hunan, Anhui, Jiangxi, Hubei and other places, the main reason being that most of these areas have large rivers or are flash flood prone areas, however, the flood control capacity of these areas is low and the economy is relatively developed.

Therefore, a comprehensive analysis shows that by studying the characteristics of multiple earthquakes and flood-prone regions in China, it is more appropriate to extract Sichuan Province, which has a relative frequency of earthquakes and flood disasters and is economically developed, as a virtual case area in our study. One of the more developed economies in the region to meet the requirements of multi-hazards research; what's more, in this region, we can obtain enough high quality and quantity of disaster data.

Sichuan is located in west China as an important junction with southwest China, northwest China and middle area, and a critical intersection and transport corridor connecting south China and central China, linking southwest China and northwest China as well as uniting central Asia, south Asia, and southeast Asia. Sichuan's governing area amounts to 486,000 square kilometers, ranking fifth in China. There are 21 cities (prefectures) and 183 counties (districts) in Sichuan. Sichuan is a province with abundant resources, large population and a strong economy.

3.1 *Natural geographical features*

Sichuan Province in southwest China is located in the upper reaches of the Yangtze River between east longitude 92° 21' to 108° 12' and between north latitude 26° 03' to 34° 19', measuring 1075 km from its east to west and over 900 km from north to south. Its neighbors are Chongqing to east, Yunnan and Guizhou to south, Tibet to west, Qinghai, Gansu and Shaanxi provinces to the north. Sichuan covers an area of 486,000 square kilometers, ranking fifth in the country after Xinjiang, Tibet, inner Mongolia and Qinghai.

The topography over the whole province differs enormously from east to west with complex and various terrains. The terrain of China's mainland forms a flight of three steps in terms of altitude. Sichuan lies between the first step, namely the Qinghai-Xizang Plateau and the second step, namely the middle-lower Yangtze Plain. It is high in the west while low in the east demonstrating a distinct disparity in altitude. Plateaus and mountainous regions can be seen in west Sichuan at altitudes above 4000 meters, while basins and hilly areas can be found in the east at altitudes above 1000–3000 meters. The whole province can be divided into three major parts: The Sichuan Basin, the Sichuan Northwest Plateau and the Sichuan Southwest Mountains.

The Sichuan Basin covering an area of 165,000 square kilometers is one of the four largest in the country. This basin is surrounded by mountainous regions having the Qinling Mountain in the north, Micangshan, Daba Mountain in the east, the Daloushan Mountain in the south, and Longmenshan, Qionglai Mountains in the northeast. The regional climate is warm and humid, warm in winter and hot in summer and most areas with annual rainfall of 900–1200 mm. Sichuan enjoys a subtropical humid monsoon climate and its vegetation is subtropical evergreen broadleaf forest. The pattern of agricultural use is double cropping in a year. The west of Sichuan Basin has the Chuanxi Plain which is the irrigation area of Dujiangyan. The lands in West Sichuan Plain are fertile and with high productivity. The middle of Sichuan Basin is the purple hill area, 400–800 meters above sea level, whose terrain is slightly tilted to the south. Minjiang River, Tuojiang River, Fujiang River and Jialing River run to the south through its north mountains and join the Yangtze River. Besides, on the east of Sichuan Basin is the east Sichuan Parallel Range Gorge region including Huaying Mountain, Tongluo Mountain, and Mingyue Mountain.

The northwest of Sichuan Province is the northwest Sichuan Plateau belonging to the southeast corner of the Tibetan Plateau with an average elevation of 3000–5000 meters and alpine climate covered with alpine meadow vegetation.

The southwest of Sichuan Province is the north section of Hengduan Mountains with high mountains and stiff valleys. Rivers flow through mountains alternatively. The rivers and mountains from east to west are the Minshan Mountain, Minjiang River, Qionglai Mountain, Dadu River, Daxue Mountain, Yalong River, Shualuli Mountain and Jinsha River. The climate and vegetation here are in a vertical distribution, mainly including boreal coniferous forests, temperate mixed broadleaf-conifer forests, northern subtropical mixed evergreen and deciduous forests, and middle subtropical mixed evergreen and broadleaf forests.

3.2 *Population*

According to the estimation of 2015 population change sample survey, the year 2015 saw 801,000 births, a birth rate of 9.9 per thousand, and 558,000 deaths, a death rate of 6.9 per thousand. The natural growth rate of the population was 3.0 per thousand. By the year end, the population of constant residents was 81.07 million, which grew by 308,000 persons over the previous year. Of this total, the urban population was 36.4 million and the rural population 44.67 million, an urbanization rate of 44.9 percent, up by 1.37 percentage points.

3.3 *Economics*

Authorized by the National Bureau of Statistics of the People's Republic of China, Sichuan province regional Gross Domestic Product (GDP) reached 3.01 trillion RMB in 2015, an increase of 7.9 percent over the previous year. Among them, the first industrial added value was 367.73 billion RMB, up by 3.7 percent; the second industry added value was 1.42 trillion RMB, up by 7.8 percent; the tertiary industry added value was 1.21 trillion RMB, an increase of 9.4 percent. The contribution rate of three types of industry on economic growth is 5.0 percent, 53.9 percent, and 41.1 percent, respectively. The per capita GDP is 36,836 RMB, an increase of 7.2 percent.

The non-public economy totaled 1.8266 trillion RMB, up by 8.3 percent over the previous year, as a share of 60.7 percent of GDP with the contribution to GDP growth of 63.8 percent. Among them, the primary industry reached 147.38 billion RMB, up by 3.3 percent; the secondary industry reached 1.063 trillion RMB, up by 8.2 percent; and the tertiary industry reached 615.44 billion RMB, an increase of 9.8 percent.

The annual Consumer Price Index (CPI) rose 1.5 percent over the previous year, the total level of food prices rose 2.9 percent, and housing prices rose 0.5 percent. The retail prices rose 0.2 percent and prices for agricultural supplies rose by 1.5 percent. Producer Prices Index (PPI) fell by 3.6 percent over the previous year, the means of production prices fell by 4.7 percent, and life material prices fell 0.2 percent. Industry Production Index (IPI) fell by 3.3 percent over the previous year.

3.4 *Transportation, post, and telecommunications*

The total freight traffic in highways, railways, civil aviation, waterways and others reached 250.01 billion tons, up by 6.3 percent over the previous year; the passenger flows were 162.38

Table 1. Traffic by all means of transportation and the growth rates in 2015 in Sichuan province.

Item	Unit	Volume	Increase over 2014 (%)
Total freight traffic	100 million ton	2500.1	6.3
Highways	100 million ton	1693.3	–10.5
Railways	100 million ton	606.3	0.3
Civil aviation	100 million ton	9.7	10.3
Waterways	100 million ton	190.8	23.7
Total passenger traffic	100 million person-kilometers	1623.8	6.0
Highways	100 million person-kilometers	632.8	0.4
Railways	100 million person-kilometers	271.8	0.1
Civil Aviation	100 million person-kilometers	716.6	14.1
Waterways	100 million person-kilometers	2.6	–1.8

The data is from Sichuan Almanac (2015 Volume).

billion person-kilometers, which grew by 6.0 percent. Railway operation mileage reached 4710 kilometers and highway to 6016 kilometers; container shipping of ports reached 233,000 TEUs as we can see in Table 1.

The turnover of post and telecommunication services totaled 128.11 billion RMB, up by 24.7 percent over the previous year. Of this total, post services were 13.86 billion RMB, up by 18.1 percent, and telecommunication services 114.25 billion RMB, up by 25.6 percent. In the year end, the total capacity of office switchboard stood at 7.547 million lines and that of telephone switchboard 156.91 million lines. By the end of 2015, the number of fixed-line telephone subscribers was 13.534 million.

The popularity rate of telephones reached 100 percent, of which 15.6 percent was that of fixed-line telephones and 84.4 percent mobile phones. The number of fixed Internet users reached 10.264 million and that of mobile Internet users 54.766 million. The length of fiber-optic cable totaled 1.615 million kilometers.

In general, the earthquake occurs frequently because of seismic zones with undulating terrain and broken rock mass in the Sichuan province. Besides, it is located in a large river basin with low flood control capacity. Sichuan's unique natural and socio-economic conditions determine that the region is deeply affected by natural disasters, especially earthquake and floods. So the study of integrated risk assessment in the Sichuan province has comprehensive significance to a degree.

4 DISCUSSION AND CONCLUSION

By comparing the data situation on earthquake and flood in provinces of mainland China and analyzing natural geographical features, population, economics, transportation, post, and telecommunications in the Sichuan Province, we select the Sichuan Province as a research area to construct a virtual case for integrated risk assessment. Next, we will continue to use a pseudo-random number generator to generate a large number of theoretical hazards and disaster samples by using the Monte Carlo method to construct a virtual case. From these large samples, the effectiveness of this method can be evaluated by integrated disasters in the research area. Finally, we will develop the information diffusion theory and methods to deal specifically with incomplete information issues.

ACKNOWLEDGMENTS

This project was supported by the National Natural Science Foundation of China (41671502).

REFERENCES

Du, Xingxin, Wang, Ze 1997. Comprehensive risk analysis and damage evaluation of flood and earthquake in weihe lower reaches. *Journal of Catastrophology* (2): 39–43.

Fishman, G.S. 1996. *Monte Carlo: Concepts, Algorithms, and Applications*. New York: Springer-Verlag.

Huang, C.F. 2002. Information diffusion techniques and small sample problem. *International Journal of Information Technology and Decision Making* 1(2): 229–249.

Huang, C.F. 2012. Risk Analysis and Management of Natural Disaster. Beijing: Science Press. (in Chinese)

Ming, X.D., Xu, W. 2015. An overview of the progress on multi-risk assessment. *Journal of Catastrophology*. 28(1): 126–132.

Shi, Peijun. 2014. World Atlas of Natural Disaster Risk. Beijing: BNUP and Springer.

Xu, Wei. 2014. Natural disasters integrated disaster risk prevention. Beijing: Science Press. (in Chinese)

Risk Analysis and Management – Trends, Challenges and Emerging Issues – Bernatik, Huang & Salvi (Eds)
 ISBN 978-1-138-03359-7

Ecological vulnerability assessment of Minjiang River in the Yangtze River Basin, China

Yuankun Wang & Dong Wang
State Key Laboratory of Pollution Control and Resource Reuse, Nanjing University, Nanjing, P.R. China
Department of Hydro-Sciences, School of Earth Sciences and Engineering, Nanjing, P.R. China

Dong Sheng
Hunan Water Resources and Hydropower Research Institute, Changsha, P.R. China

ABSTRACT: A new assessment model for river ecological vulnerability is developed based on the Variable Fuzzy Theory on the basis of river eco-environment characteristics. The model is applied to assess ecological vulnerability of Minjiang River in the Yangtze River Basin, China. Results show that the proposed model rationally determines ecological vulnerability level. We hope this study could provide a scientific reference for environment management in Yangtze River Basin.

Keywords: Ecological vulnerability; Variable Fuzzy Theory; assessment model; Minjiang River

1 INTRODUCTION

Eco-environmental vulnerability has become a central focus of the global change and sustainability research communities (Ford et al., 2006). It has also become a popular topic in the domain of environmental resource research, especially eco-environmental vulnerability assessment (Eakin and Luers, 2006). Understanding of ecological vulnerability helps to ascertain the key ecological characteristics of a study area.

Eco-environmental vulnerability systemic evaluation has been fast developed in recent years. Many studies have been conducted on eco-environmental vulnerability evaluation approaches, such as artificial neural network (Kia et al. 2012), fuzzy decision analysis (Navas et al. 2012), landscape ecology approach (Mortberg et al. 2007), Environmental Sensitivity Index (ESI) method (Mao et al. 2013), the P-S-R model (Wang et al. 2010), and the analytic hierarchy process (Donevska et al. 2012). Although each of these methods possesses individual merits, further improvements are required for objectivity and comprehensiveness. The systemic assessment for eco-environmnet is a fuzzy concept with multiple indicators and classes. A small increase/decrease in pollutant data near the boundary value results in a change in classification. Fuzzy theory introduced by Zadeh (1965) was designed to interpret the fuzziness and uncertainties of real situations. The fuzzy set theory has been proved effectively in solving problems of fuzzy boundaries and controlling the effect of monitoring errors on assessment results (Guleda et al., 2004). Chen (2006) developed variable fuzzy set theory for solving the problem of the assessment standard as with interval values in comparison with conventional fuzzy evaluation methods. Application of variable fuzzy set theory has been widely used in many fields (Wang et al., 2011; Li et al., 2012) Therefore it is a potential approach for improving assessment results accuracy. The objective of this study is to develop a new environmental vulnerability assessment method by using variable fuzzy theory.

Table 1. Eco-vulnerable assessment standard class (Yao et al., 2004).

Index	I	II	III
A_1 (kg/hm^2)	5250	4800	4500
A_2 (%)	30	60	80
A_3	0.069	0.251	0.667
A_4	1.3	2.0	3.0
A_5	0.5	1	2
A_6 (%)	0	20	40
A_7 (t/km^2)	300	600	1000
A_8 (%)	80	60	30
A_9 (goat unit/hm^2)	15	10	4
A_{10}	0.02	0.25	0.50
A_{11} (‰)	0	10	14
A_{12} (%)	0	20	40
A_{13} (%)	0	10	30
A_{14} (%)	0	5	15

2 RIVER ECOLOGICAL VULNERABILITY INDEX

In the present study 14 indices are selected to assess ecological vulnerability level include re land productivity (A_1), rate of disasters (A_2), ground waviness (A_3), Evaporation coefficient (A_4), dryness index (A_5),Water Loss Rate(A_6), soil erosion rate (A_7), The forest covering rate (A_8), the carrying capacity for live stock (A_9), ratio of cultivable land (A_{10}), natural growth rate of population (A_{11}), grassland degeneration rate (A_{12}), poverty rate (A_{13}), species disappearing rate (A_{14}). These indices have strong performance in reflecting the status of ecological vulnerability as reported in the literature (Yao et al. 2004; Wang et al., 2013). According to the vulnerability, every grade is granted a quantified value, respectively, which is as follows: slight vulnerability is I, medial vulnerability is II, and heavy vulnerability is III (shown in Table 1). Slight vulnerability means relatively stable ecosystem and anti-interference ability, rich soil, and relatively low altitude. Medial vulnerability means unstable ecosystem, poor anti-interference ability, deteriorated soil, and dominated by alpine shrub grassy marshland. Heavy vulnerability unstable ecosystem and poor anti-interference ability, deteriorated soil, and sparse vegetation dominated by extreme-coldness plants.

3 VARIABLE FUZZY EVALUATION MODEL

We use the variable fuzzy set theory developed by Chen (2006) to propose the ecological vulnerability assessment model. The implementation of the model consists of the following steps.

Let the number of sample sets of ecological vulnerability evaluation be denoted by n, we obtain:

$$X = \{x_1, x_2 \cdots, x_n\} \tag{1}$$

where x is the sample set of ecological vulnerability evaluation.

The index eigenvalue of the sample j can be expressed as:

$$x_j = \left(x_{1j}, x_{2j} \cdots, x_{mj}\right) \tag{2}$$

where m is the number of sample indices.

Then the sample set can be expressed as:

$$X = (x_{ij}) \tag{3}$$

where x_{ij} is the eigenvalue of the index i of the sample j; $i = 1, 2..., m$; $j = 1, 2,..., n$.

According to the definition of variable fuzzy set, let a fuzzy subset A in the domain be of interest. In continuum on the number axis, a number $\mu_A(u)$, $\forall\ u \in U$, is named as the relative membership degree of u to A with attraction. Consider X_0 as interval $[a, b]$ of attraction domain of V on the number axis, which means $0 < D_A(u) \leq 1$, then, X is a certain interval $[c, d]$ including X_0 ($X_0 \subset X$).

Step (1) On the basis of index standard matrix, the interval criterion matrix can be expressed as:

$$I_{ab} = ([a,b]_{ih}) \tag{4}$$

where $[a, b]$ denotes a (closed) interval, a could be greater than b.

According to the interval criterion matrix denoted by I_{ab}, variable interval matrix I_{cd} can be constructed as:

$$I_{cd} = ([c,d]_{ih}) \tag{5}$$

where $[c, d]$ also denotes a (closed) interval, c could be greater than d.

Step (2) Based on the physical meaning analysis of assessment index, the matrix M determining index class standard is expressed as:

$$M = (M_{ih}) \tag{6}$$

Step (3) The element ω_i in weight set W represents the weight coefficients of each factor. In this study, the entropy theory is used to obtain the weight of every factor.

Step (4) According to the definition of variable fuzzy set, relative difference degree $D_A(u)$ can be expressed as:

$$\begin{cases} D_A(u) = \left(\dfrac{x-a}{M-a}\right)^{\beta}, & x \in [a,M]; \\ D_A(u) = \left(\dfrac{x-a}{c-a}\right)^{\beta}, & x \in [c,a]; \end{cases} \tag{7}$$

$$\begin{cases} D_A(u) = \left(\dfrac{x-b}{M-b}\right)^{\beta}, & x \in [M,b]; \\ D_A(u) = \left(\dfrac{x-b}{d-b}\right)^{\beta}, & x \in [b,d]; \end{cases} \tag{8}$$

where case β is a non-negative number with a defined default value of $\beta = 1$.

$$\mu_A(u)_{ih} = [1 + D_A(u)]/2 \tag{9}$$

Step (5) Relative membership matrix can be denoted as $\mu_A(u)$ with dimensions of the assessment class index i and the class standard h as:

$$\mu_A(u) = (\mu_A(u)_{ih}) \tag{10}$$

The difference between $\mu_A(u)h$ and $\mu_A^c(u)_h$ of eigenvalue of the index i in the reference continuum to the two apices can be expressed as generalized weighted distance d_{gh} and d_{bh} as:

$$d_{gh} = \left\{ \sum_{i=1}^{m} \left[\omega_i \left(1 - \mu_A(u)_{ih}\right)\right]^p \right\}^{\frac{1}{p}} \tag{11}$$

$$d_{bh} = \left\{ \sum_{i=1}^{m} \left[\omega_i \left(1 - \mu_A^c(u)_{ih}\right)\right]^p \right\}^{\frac{1}{p}} \tag{12}$$

where p is a distance parameter. When $p = 1$, it is named the Hamming Distance. When $p = 2$, it is named the Euclidean distance.

Let $v_A(u)_h$ denote as the relative membership degree of u to A. let $v_A(u)_h$ and $v_A{}^c(u)_h$ be the weight factors, then the generalized weighted distance D_g and D_b can be written as:

$$D_g = v_A(u)_h \cdot d_{gh} \tag{13}$$

$$D_b = v_A^c(u)_h \cdot d_{bh} = \left(1 - v_A(u)_h\right) \cdot d_{bh} \tag{14}$$

Constructing objective function yields:

$$\min\left\{ F\left(v_A(u)_h\right)\right\} = \left(v_A(u)_h\right)^2 \cdot d_{gh}^a + \left(1 - v_A(u)\right)^2 \cdot d_{bh}^a \tag{15}$$

where a is the rule parameter of the model optimization, $a = 1$ is the least single method and $a = 2$ is the least square method.

By differentiating equation (19) with $v_A(u)_h$ and equating to zero, we obtain:

$$\frac{dF\left(v_A(u)_h\right)}{dv_A(u)_h} = 0 \tag{16}$$

Equation (20) becomes:

$$v_A(u)_h = \frac{1}{1 + \left(\dfrac{\sum_{i=1}^{m} \left(w_i \left(1 - \mu_A(u)_{ih}\right)\right)^p}{\sum_{i=1}^{m} \left(w_i \mu_A(u)_{ih}\right)^p} \right)^{\frac{a}{p}}} \tag{17}$$

This is the assessment model based on the variable fuzzy set theory.

Step (6) Form equation (21), one obtains:

$$U_C = \left(v_A(u)_h\right) \tag{18}$$

Then the normalization processing for the comprehensive relative membership vector $v_A(u)_h$ is implemented, one obtains $v_A^o(u)_h$:

$$U = \left(v_A^o(u)_h\right) \tag{19}$$

$$\sum_{h=1}^{c} v_A^o(u)_h = 1 \tag{20}$$

where

$$v_A^o(u)_h = \frac{v_A(u)_h}{\sum_{h=1}^{c} v_A(u)_h} \tag{21}$$

Step (7) By using ranking feature value method ecological vulnerability level can be obtained:

$$H = (1 \quad 2 \quad \cdots c) \cdot U^T \tag{22}$$

4 STUDY SITE

Minjiang River is a major tributary of the upper Yangtze River Basin, China. The upper reaches of the Miniiang River area include Wenchuan County, Li County, Mao County, Songpan and Heishui County, the total area is about 25426 km^2. It is a green-reservoir and eco-fence of the Chengdu Plain, and one of the water-resource areas of the Changjiang River. Located on the eastern edge of Qinghai-Tibet Plain, the topography of the area is characterized by the complex distribution of hills and valleys, and ranging in elevation from 700 to 6260 m with average elevation difference above 1000 m (Li et al., 2006). This region is a typical fragile ecological system based on dry valley. The eco-environmental fragility displays in many respects, such as soil erosions increases, water and soil loss seriously and ecological function reduces. Eco-environmental fragility is caused by the nature fragility and human activities.

5 RESULTS

We apply the proposed model in the present study to assess the ecological vulnerability status of the Minjiang River step by step. Assessment results are shown in Table 2. As shown in Table 2, ranking feature values are relatively stable by varying the model and its parameters. The mean value of four patterns can be taken as the final ranking feature values. By the method proposed here, the ecological vulnerability of the case is 2.42, which is between medium level and heavy level, and close to grade 2. The results of ecological vulnerability obtained by the utilization of the variable fuzzy set assessment model are compared with those by use of other models (Table 3). From Table 2 and Table 3 it can be seen that we take the mean value 2.42 as the assessment results. Compared with the use of other methods, the variable fuzzy set assessment results are therefore more reliable.

Table 2. Values of ecological vulnerability status of the Minjiang River.

Ranking feature value				
P=1, a=1	P=1, a=2	P=2, a=1	P=1, a=2	Mean value
2.32	2. 21	2.45	2.50	2.42

Table 3. Comparison of ecological vulnerability values by using several methods.

Method	FCA	PPM	ANN	SPA	VFT
Result	III	III	III	II-III	II-III

Note: Fuzzy comprehensive assessment method-FCA; Projection pursuit model-PPM; Set-Pair Analysis method-SPA; Artificial Neural Network- ANN; Variable Fuzzy Theory-VFT.

6 CONCLUSION

In the present study, variable fuzzy theory is developed to establish a new evaluation method for river ecological vulnerability. The method showed its great advantages in the case study of the Minjiang River in Yangtze River Basin.

The model proposed in this study determines relative membership degrees and functions between the simple index and the standard interval of each level by varying the parameters of the mode. Comparison with the traditional methods shows that the proposed model improves the reliability of the assessment process and effectively diagnoses river ecological vulnerability status.

ACKNOWLEDGMENTS

This study was supported by the National Natural Science Fund of China (No. 51309131, 51679118, 41571017), The National Key Research and Development Program of China (No. 2016YFC0401501-03), Hunan Province Water Conservancy Science and Technology Major Project - River evolution trend analysis of Chenglingji out flow reaches in Dongting Lake (Xiang Water Science and Technology 2016).

REFERENCES

Chen, S.Y., Guo, Y. 2006. Variable fuzzy sets and its application in comprehensive risk evaluation for flood-control engineering system. Fuzzy. Optim. Decis. Ma., 5, 153–162.

Donevska, K.R., Gorsevski, P.V., Jovanovski, M., Pesˇevski I. 2012. Regional non-hazardous landfill site selection by integrating fuzzy logic, AHP and geographic information systems. Environ Earth Sci 67: 121–131.

Eakin, H., Luers, A.L. 2006. Assessing the vulnerability of social-environmental systems. Annu. Rev. Environ. Resour. 31, 365–394.

Ford, J.D., Smit, B., Wandel, J. 2006. Vulnerability to climate change in the arctic: a case study from arctic bay canada. Glob. Environ. Change 16 (2), 145–160.

Kia, M.B., Pirasteh, S., Pradhan, B., Mahmud, A.R., Sulaiman, W.A., Moradi, A. 2012. An artificial neural network model for flood simulation using GIS: Johor River Basin, Malaysia. Environ Earth Sci 67: 251–264.

Li, A.N., Wang, A.S., Liang, S.L., Zhou, W.C., 2006. Eco-environmental vulnerability evaluation in mountainous region using remote sensing and GIS/A case study in the upper reaches of Minjiang River, China. Ecol. Model 192, 175–187.

Li, Q., Zhou, J., Liu, D., Jiang, X. 2012. Research on flood risk analysis and evaluation method based on variable fuzzy sets and information diffusion. Safety Science, 50(5), 1275–1283.

Mao, X.Y., Meng, J.J., Xiang, Y.Y. 2013. Cellular automata-based model for developing land use ecological security patterns in semi-arid areas: a case study of Ordos, Inner Mongolia, China. Environ Earth Sci 70: 269–279.

Mortberg, U.M., Balfors, B., Knol, W.C. 2007. Landscape ecological assessment: a tool for integrating biodiversity issues in strategic environmental assessment and planning. J Environ Manag 82: 457–470.

Navas, J.M., Telfer, T.C., Ross, L.G. 2012. Separability indexes and accuracy of neuro-fuzzy classification in geographic information systems for assessment of coastal environmental vulnerability. Ecol Inform 12: 43–49.

Wang, X.D., Zhong, X.H., Gao, P. 2010. A GIS-based decision support system for regional eco-security assessment and its application on the Tibetan Plateau. J Environ Manag 91: 1981–1990.

Wang, Y.K., Wang, D., Wu, J.C. 2011. A Variable Fuzzy Set Assessment Model for Water Shortage Risk: Two Case Studies from China. Hum Ecol Risk Assess 17: 631–645.

Wang, Y.K., Wang, D.W., Ji, C. 2013. A fuzzy assessment model for river ecological vulnerability: a case study in Yangtze river basin. In *Proceedings of the 4th International Conference on Risk Analysis and Crisis Response*, 187–192.

Yao, J., Ding, J., Ai, N. 2004. assessment of ecological vulnerability in upper reaches of mingjiang river. Resources and Environment in the Yangtze Basin, 13(4), 380–383.

Zadeh, L.A. 1965. Fuzzy sets. Inf Control 8(3): 338–53.

Risk Analysis and Management – Trends, Challenges and Emerging Issues – Bernatik, Huang & Salvi (Eds)
© 2017 Taylor & Francis Group, London, ISBN 978-1-138-03359-7

Using game theory for scheduling patrols in an urban environment

Cheng-Kuang Wu
Department of Computer Science and Engineering, Xiamen Institute Technology, Xiamen City, Fujian Province, P.R. China

ABSTRACT: This study proposes a framework for police patrol services that incorporates two game theory models for the allocation of police officers when the threat level is raised. First, the interactions between the criminal activity and the response agent dispatched by the police department are modeled as a non-cooperative game, after which a mixed strategy Nash equilibrium is used to derive the Threat-Vulnerability-Consequence (TVC) value for each district patrolled. The TVC value represents the threat, vulnerability, and consequence of crimes for each district. All TVC values for all districts are collected and each district's Shapley value computed for three shifts. Then, a simulation is carried out with a minimum of three sets of police patrol shifts per day for twelve districts. The experimental results show that proposal model is feasible as a method for deciding on the deployment polices for patrolling the districts in a city.

1 INTRODUCTION

For a Police Department (PD) to provide full security coverage at all times is a challenging task anywhere in the world, particularly in an urban environment when the available police resources are limited. The increasing randomization of criminal activity makes the scheduling of police patrols or rostering ever more difficult and the construction of optimized work timetables for personnel that will deter potential aggressors and prevent the occurrence of crimes is almost impossible (Zhang et al., 2015). In addition, the more security resources the police have to deploy, and the more districts that need to be patrolled, the greater the growth in the costs of policing. Mass deployment tends to waste resources, while the deployment of too few security resources makes it easier for vulnerabilities to be exploited, causing the occurrence of more crimes. Thus, the robustness of a complete police patrol schedule depends upon a balance between the policing requirements and the risk of criminal activity. This creates for the PD the dilemma between the reduction of the crime rate and ever expanding policing costs (Livio 2014). Specific measures for rational decision-making are lacking, and mathematical models have not been applied to capture the interactions between police resources and law breakers in a protected district. The response agent of the PD should have at hand a tool to measure the strength of the criminal activity (i.e., crime rate) and the resistance capability of the police forces (i.e., arrest rate). The utilities of the moves available to the police and the criminals need to be considered. Randomized patrol strategies are needed to avoid the possibility that criminals will observe and be able to predict vulnerabilities in patrol scheduling (Pita et al. 2009).

A model is proposed and applied for the deployment of police personnel, specifically for the development of PD patrol schedules. A simplified workflow chart describing the principles of optimal police deployment is shown in Figure 1. Two game-theoretic models are constructed, representing the two stages needed for economical police officer deployment. In the first step, the interactive movements between the criminal and the PD response agent (or police officers) are modeled and analyzed as a non-cooperative and zero-sum game, after which the TVC value for each patrol district is derived from a mixed strategy Nash

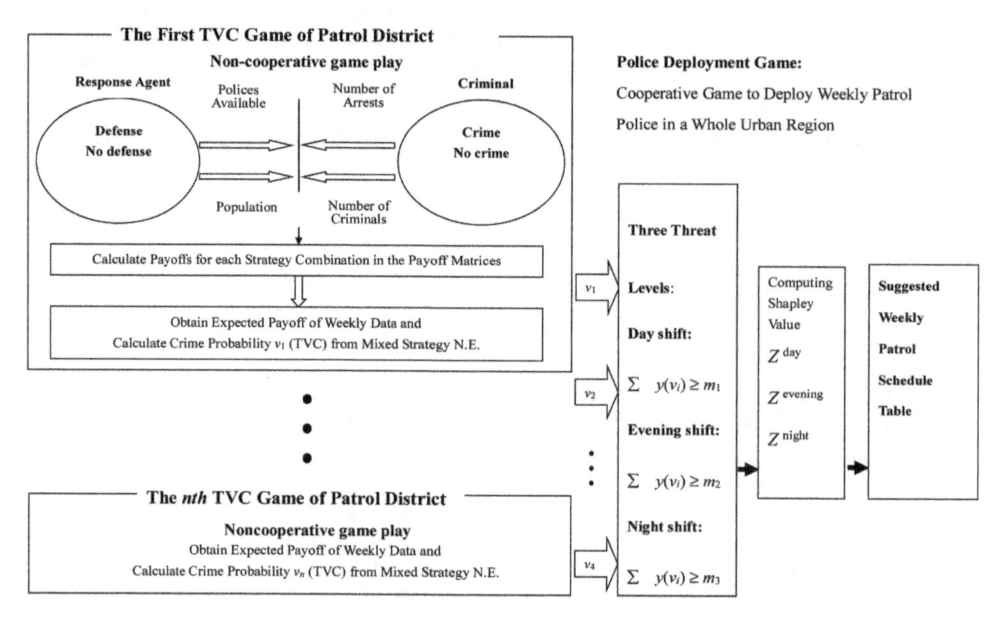

Figure 1. Workflow of an optimal deployment model for an urban environment.

equilibrium. This value quantifies the Threat, Vulnerability, and Consequence (TVC) of criminal activities within each patrol district. In the second step, the interactions of the twelve protective districts within the whole urban area are likened to the playing of a cooperative game. In the case study, twelve TVC values are utilized to compute the Shapley value for each patrol district for three different times of day. The Shapley value assists in setting priorities for the planning of police patrol schedules for three shifts (e.g., day, night, and evening) so as to maximize the effectiveness of patrolling duties within the whole region.

2 THE PROPOSED MODEL

Certain assumptions must be made when treating criminal activities and police patrol scenarios in the first game. The assumptions are detailed below.

a. With multiple criminal events in different districts, patrol actions will overlap in time, and the PD manager needs to arrange schedules for each patrol district simultaneously. In the TVC game, the two game players simultaneously use a single static step. This is played out in each patrol district as a noncooperative zero-sum game. Player 1 is the criminal or perpetrator who could carry out any type of crime, such as robbery, sexual assault, vehicle theft, identity fraud, weapons violations, terrorist activity, and so on. Player 2 is the response agent who is responsible for enforcing the law, responding to complaints, and patrolling the district.
b. The first model is a normal form game for modeling the interactions between two players. There are a specific number of criminals living or active within each district. And each district has a different number of police officers per one thousand of the population. The ratio between police officers and criminals C_i is the proportion of the number of criminals w_i divided by the available police officers o_i. In this game, there is complete information regarding the ratio of the number of criminals and police officers, and the population of the patrol districts (as shown in Table 1).
c. The information concerning the number of crimes, number of arrests, arrest rate and crime success rate per week in the twelve patrol districts is given in Table 2.

Table 1. The number of criminals, the number of police officers per one thousand population, and the police arrest and crime rate in the twelve patrol districts (from A to L).

District	Population $\delta_i \times 1000$	Number of criminals w_i	Number of police officers o_i	Ratio between criminals and police officers C_i
A	236	20	225	0.089
B	235	30	226	0.133
C	234	40	227	0.176
D	233	50	228	0.219
E	232	60	229	0.262
F	231	70	230	0.304
G	230	80	231	0.346
H	229	90	232	0.388
I	228	100	233	0.429
J	227	110	234	0.470
K	226	120	235	0.511
L	225	130	236	0.551

Table 2. The number of crimes, number of arrests, arrest rate and crime success rate in the twelve patrol districts.

District	Number of crimes	Number of arrests	Arrest rate α_i	Crime success rate $1-\alpha_i$
A	137	30	0.22	0.78
B	136	31	0.23	0.77
C	135	32	0.24	0.76
D	134	33	0.25	0.75
E	133	34	0.26	0.74
F	132	35	0.27	0.73
G	131	36	0.27	0.73
H	130	37	0.28	0.72
I	129	38	0.29	0.71
J	128	39	0.30	0.70
K	127	40	0.31	0.69
L	126	41	0.33	0.67

2.1 *The TVC game in the patrol district*

The game in the first step of the proposed model is designed to obtain the TVC value of the criminal risk in each district. This study assumes that the behavior of the criminal (Player 1) and response agent (Player 2) can be captured in a two-person game theory model. The parameters for determining the threat, vulnerabilities, and consequence measures are defined in the following paragraphs.

Threat: The criminal (Player 1) has two choices that could happen in the patrol district: there could be a criminal event or no criminal event. S_1 denotes the set of player 1's strategies: $S_1 = \{u_1, u_2\}$ = {criminal offence, no criminal offence}. The greater the number of criminals, the more police officers are needed. W denotes the set of police required for resisting criminals for the whole administration region. $W = \{w_1, w_2,...,w_n\}$. The variable w_i denotes the number of criminals in the ith district. The district response agent is denoted as player 2. In a TVC game, each patrol district has a response agent, comprised of police officers with which to prevent multiple criminal offences. In this game, the response agent dispatches police offers on a patrol or in response to the occurrence of criminal offences. He/she may choose a defense or a no defense strategy. S_2 denotes the sets of player 2's strategies:

$S_2 = \{d_1, d_2\}$ = {defense, no defense}. O denotes the set of police available to response agents in the whole region. $O = \{o_1, o_2, \ldots, o_n\}$. o_i denotes the number of polices available in the ith district. It is assumed in this study that the number of police officers is greater than the number of criminals in each district.

Vulnerability: Levitt (1998) indicated that when the police report an increase in the arrest rate for a specific crime, the overall crime rate will be reduced. Thus it is assumed in this study that when the arrest rate increases in one district, then the crime rate will decrease and the security situation will improve. The arrest rate is calculated by

$$\alpha_i = \frac{\text{The number of arrests}}{\text{The number of crimes}} \tag{1}$$

Consequence: A criminal event in a densely populated district increases the number of crimes and thus gains greater benefit for the criminal (player 1), while the response agent (player 2) must pay a higher police service cost. Criminal offenses lead to an increase in human casualties, monetary loss and economic damage, which necessitates improvement of the police services. Here, δ_i denotes the population of the ith district in the whole region.

The two players make strategic decisions simultaneously in this game. A 2×2 payoff matrix is created based on the two players' strategies and interactions; see Table 3. The payoff to player 1 for choosing a particular strategy when player 2 makes their selection can be represented either as a gain by player 1 or a loss for player 2. In this model, it is depicted as a summation of the gains of player 1, while player 2 tries to minimize this gain. Thus, this is a zero-sum game. The payoff for each outcome is given as a 2×2 payoff matrix, as shown in Table 3.

The proposed matrix has no pure strategy Nash equilibrium (N.E.). A mixed Nash equilibrium pair (p, q) exists in the normal form game if the game has no pure strategy N.E., which is an optimal strategy (Lemke and Howson, 1964). Player 2's expected payoff is computed when player 1 and player 2 play mixed strategies p and q, respectively. The mixed N.E. for the probability vector is $p^* = \{p^*(u_1), p^*(u_2)\}$, with actions $\{u_1, u_2\}$ for the criminal and the vector $q^* = \{q^*(d_1), q^*(d_2)\}$ with actions $\{d_1, d_2\}$ for the response agent. The optimal strategies are always mixed, and player 2 makes a defense with the probability:

$$q^*(d_1) = \frac{C_i}{1 + \alpha_i C_i} \tag{2}$$

while player 1 makes a criminal offence with the probability

$$p^*(u_1) = \frac{1}{1 + \alpha_i C_i} \tag{3}$$

This study defines the v_i as the TVC value of ith district, given by:

$$p^*(u_1) = v_i \tag{4}$$

Where vi represents the probability of player 1 committing a criminal offence in a TVC game, which represents the TVC value in the ith district. The next model applies the TVC values of all districts for computation of the Shapley value of each district within the cooperative game.

Table 3. Payoff matrix for the TVC game in the ith district.

	Response agent (Police officer)	
Criminal	defense q	No defense $1-q$
Crime offence p	$(1-\alpha_i)\ \delta_1 C_i$	δ_1
No Crime offence $1-p$	$\delta_1 C_i$	0

2.2 *Police deployment game*

The interactions of all response agents in an urban region are likened unto the playing of a cooperative game. An efficient method is needed to decide on the number and priority of the deployment of police officers for the three different shifts in one day (e.g., day shift, evening shift, and night shift) where police officers patrol multiple district simultaneously. The Shapley value is a power index for cost allocation. The cooperative game provides a suitable model for the design and analysis of resource allocation. It has been shown that the well-known Shapley value rule satisfies many of the properties of fairness, thus is applied in the proposed model to create an optimal cost allocation for the deployment of PD police patrol schedules.

In this study y: $V \rightarrow R^{+}$ is defined as a one-to-one function where a positive real number is assigned to each element of v and $y(0) = 0$, $V = \{v_1, v_2, v_3\}$. There are twelve response agent districts for deployment of police officers based on the threat level h. The three threat levels in one day (e.g., day shift, evening shift, and night shift) are $H = \{h_1, h_2, h_3\}$, where $0 < m_1 < m_2 < m_3$ represent the corresponding threshold values. Given the output vector of twelve TVCs, the threat level L of the response region is equal to h_f if the sum of the TVCs of the districts is greater than or equal to m_f:

$$L = \begin{cases} h_1 \text{ if} \sum_{i=1}^{N} y(v_i) \geq m_1 \\ h_2 \text{ if} \sum_{i=1}^{N} y(v_i) \geq m_2 \\ h_3 \text{ if} \sum_{i=1}^{N} y(v_i) \geq m_3 \end{cases} \tag{5}$$

Where $m_1 = v_{Mini} + m_1$, $m_2 = m_1 + m_{in}$, $m_3 = m_2 + m_{in,}$ $m_{in} = (v_{Max} - v_{Mini})/4$.

The twelve TVCs for all the districts can be grouped into three threat levels according to the average value of the interval m_{in} for the threshold. This is divided by three threat levels from the maximum TVC v_{Max} to the minimum v_{Min}.

The TVCs for the twelve districts can be modeled as a 12-person game where $X = \{1, 2, 3, \ldots, 12\}$, which includes the set of players and each subset $V \subset N$, where $v_i \neq 0 \; \forall_i \in V$ is called a coalition. The coalition of X groups in the mth threshold for a specific threat level and each subset of the X (coalition) represents the observed threat pattern for different levels of H. The aggregate value of the coalition is defined as the sum of all the TVCs for the entire district, $y(C) = \Sigma_{i \in C}\, y(v_i)$ and is called a coalition function. Each TVC coincides with one or another given m_f threshold of the threat level. The priorities for deployment in the twelve districts are derived from the three thresholds (e.g., m_1, m_2, m_3). The Shapley value, which is derived based on the emergency threat level for each district with respect to the others, and the effect of the threshold values given the three threat levels, is used to represent the relative importance of each district. Now let $y(C) = \Sigma_{i \in C}\, y(v_i)$, $v_i \in V$, $C \subset X$ be the value of coalition C with cardinality c. The Shapley value of the ith district of the emergency vector is defined by:

$$Sh(i) = \sum_{\substack{C \subset X \\ i \in C}} \frac{(c-1)!(n-c)!}{n!} [y(C) - y(C - \{i\})] \Rightarrow Sh(i) = \sum_{\substack{C' \subset X \\ i \in C'}} \frac{(c-1)!(n-c)!}{n!} \tag{6}$$

Equation (6) can be simplified, because the term $y(C)$-$y(C$-$\{i\})$ will always have a value of 0 or 1, taking the value 1 whenever C' is a winning coalition. If C' is not a winning coalition, the terms C-$\{i\}$ and $y(C)$ are 0 (Owen, 2001). Hence, the Shapley value is $Sh(i)$, where C' denotes the winning coalitions with $\Sigma v_i \geq m_f$, $i \in C'$. The Shapley value of the ith response agent output indicates the relative TVC value for the thresholds m_f (i.e., threat levels). Therefore, a Shapley value represents the strength of the criminal offense which the PD should consider when assigning police patrols and security service requirements for one day. The Shapley values for the ith district are applied to compute the number of officers patrolling for the h_f threat level. Given m_f thresholds of the threat level, the number of police officers allocated to patrol the ith district is defined by

$$e(i) = Sh(i) \times O_{all} \tag{7}$$

The number of police officers patrolling the ith district $e(i)$ is derived from the Shapley value of the ith district $Sh(i)$ multiplied by the total number of police officers available in the whole region O_{all}. Finally, the administrator of the PD can generate weekly police patrol schedules that will suitably allocate police officers so as to improve efficiency and mitigate criminal offenses.

2.3 *Simulation experiments*

This study hypothesizes that the PD administration possesses different numbers of police officers available for twelve districts. Their goal is to effectively manage the number of police patrol shifts each day given the three threat levels (e.g., day shift, evening shift, and night shift). The information employed in the simulation regarding the ratio of number of criminals and police officers, arrest rate per week, and population of the twelve patrol districts is summarized in Tables 1–2.

The experiment is first modeled as a noncooperative game which generates simulated sets of TVC measures for the twelve districts for a specified period of time (e.g., a week). The payoff matrix of the zero-sum TVC game for each patrol district is modeled according to Equation (1) appears in Table 3. From the numerical examples, a Nash equilibrium is created by enumerating the mixed strategies in the strategy game. Then, the TVCs are calculated for the twelve districts according to Equations (3–4); see Figure 2.

Second, the twelve TVCs for the different districts are utilized to compute a Shapley value for each district (see Figure 3). The three threat levels in the police roster correlated to the night shift, evening shift, and day shift. The Matlab tool is adopted to compute the Shapley value of each district for the three shifts. In the simulation, three minimum sets of police

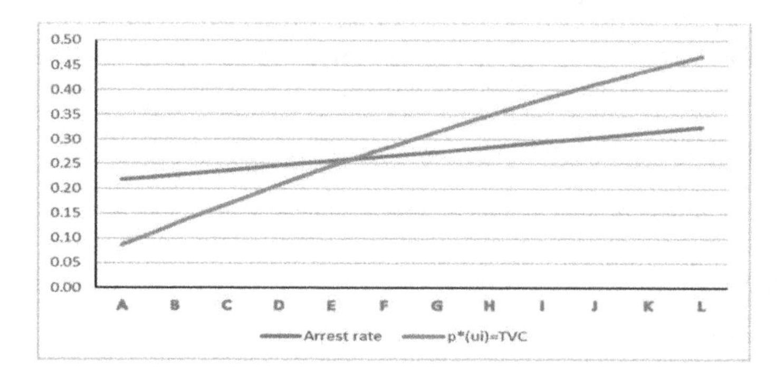

Figure 2. TVCs and arrest rates in the twelve patrol districts.

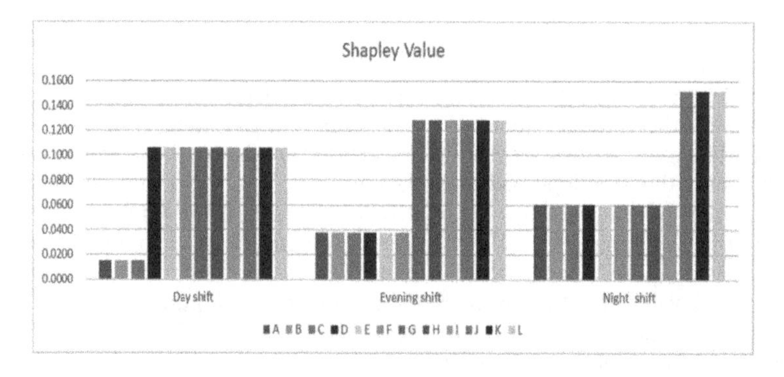

Figure 3. Shapley values for the twelve districts for the three shifts.

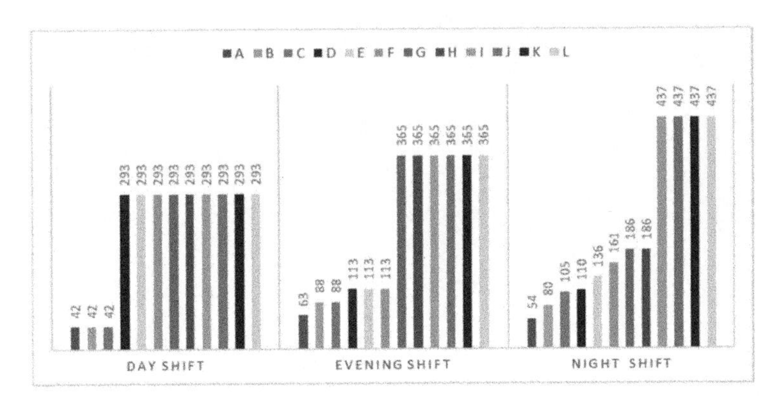

Figure 4. Police patrol schedules for the three shifts.

officer patrol shifts per day for the twelve districts are computed using Equations (6) - (7); see Figure 4.

3 CONCLUSIONS

In the proposed framework, two game theoretic models are applied to develop rational police patrol schedules for three threat levels. In the first step, a game strategy is created which uses the mixed strategy Nash equilibrium to find the TVC value for each district for each response agent. In the second step the TVC values of all the districts are used to compute the Shapley value for each response agent given the three threat levels. The Shapley values are then used for deployment of police officers to patrol the entire protected region during all three shifts. The simulation results demonstrate that the proposed model is able to improve patrol efforts to counteract multiple criminal offenses.

REFERENCES

Lemke, C.E., Howson, J.T. 1964. Equilibrium Points of Bimatrix Games, SIAM J. Applied Math., 12: 413–423.

Levitt, S.D. (1998). Why do increased arrest rates appear to reduce crime: deterrence, incapacitation, or measurement error?. *Economic inquiry, 36*(3), 353–372.

Livio, Di Matteo (2014). Police and Crime Rates in Canada: A Comparison of Resources and Outcomes. Fraser Institute. <http://www.fraserinstitute.org>.

Owen, G. (2001). *Game Theory*, 3rd Ed. New York, NY: Academic Press.

Pita, J., et al. 2009. "Using game theory for Los Angeles airport security." AI Magazine 30.1 (2009): 43.

Zhang, C., Sinha, A., Tambe, M. (2015). Keeping pace with criminals: Designing patrol allocation against adaptive opportunistic criminals. In *Proceedings of the 2015 international conference on Autonomous agents and multiagent systems* (pp. 1351–1359). International Foundation for Autonomous Agents and Multiagent Systems.

Risk Analysis and Management – Trends, Challenges and Emerging Issues – Bernatik, Huang & Salvi (Eds)
© 2017 Taylor & Francis Group, London, ISBN 978-1-138-03359-7

A new approach to the streamflow gauge network design

Pengcheng Xu & Dong Wang
Key Laboratory of Surficial Geochemistry, Ministry of Education, Department of Hydrosciences, School of Earth Sciences and Engineering, State Key Laboratory of Pollution, Control and Resources Reuse, Nanjing University, Nanjing, China

ABSTRACT: Due to an increasing risk of climate change, the uncertainties of hydrologic events (e.g., precipitation and streamflow) have a great impact on the optimization design of the hydrological network. The copula entropy is recommended as a substitute for mutual information estimation. The approach employs a measure of the information flow between gauging stations in the network, which is referred to as the Intensity of Directional Information Transfer (IDIT). The Maximum Pseudo-Likelihood estimator (MPL) is used to approximate the multivariate probability density functions required in the copula entropy calculation. The potential application of the approach is illustrated by using monthly streamflow data from a collection of gauging stations located in the Yiluo river basin of China.

1 INTRODUCTION

The collected hydrometric data are indispensable for the effective planning, decision-making and operation of exactly all water resources systems, such as the optimal design of water reservoirs, flood forecasting, risk assessment of regional freshwater resources, and establishing of water distribution systems, to name but a few. It recommends a variety of users of hydrometric data ranging from hydrologists, water resources managers, environmentalists, hydrogeologists, and researchers of numerous studying fields. An optimally designed network should provide sufficient information for various users and industries. In addition, a lot of reasons including the increasing and conflicting water demands, climate change, long-term benefits, and influence of network density on the precision of hydrological models have promoted the necessities of the network design (Mishra and Coulibaly, 2009). It is thus urgent for a comprehensive approach to be utilized in hydrometric network designing.

The concept of entropy has been very popular in the scientific literature over the last several decades and has several applications in hydrology and water resources engineering (Singh, 1997). The fundamental basis in designing networks based on the entropy concept is that the stations should have as little transinformation as possible, which implies that the stations should be independent of each other. Transinformation is defined as mutual information in the entropy theory. Husain (1987) derived bivariate and multivariate continuous distributions to derive entropy terms. Information transmission between station pairs was calculated for different cases of probability distributions and on the basis of information maximization principles, and the optimum locations of the stations to be retained were identified. Yang and Burn (1994) proposed a measure of the information flow between gauging stations in the network, which is referred to as the Directional Information Transfer index (DIT) based on entropy. The entropy of a model output can be computed from historical data and thus characterizes the variability inherent in the process. Al-Zahrani and Husain (1998) used Shannon's information measure to study the existing hydrological network in hydrological area III, which is located in the southwestern region of the Kingdom of Saudi Arabia, to examine their suitability toward providing maximum hydrological information. In recent years, hybrid methods are often adopted. Markus et al. (2003) created a hybrid combination of GLS and entropy by including a function of the negative net information as a penalty function in the GLS. The weights of the combined model were determined to maximize the average correlation with the results of GLS and entropy. Chen et al. (2008) presented a method by combining kriging with entropy that can determine the optimum number and spatial distribution of rain gauge stations in catchments. Mahmoudi-Meimand et al. (2016) established an optimization model based on entropy and kriging by using GIS for determining the number and location of the rain gauges. The candidate stations are those with maximum variance of Kriging error and minimum information entropy.

Although the establishment of networks based on information theory has progressed considerably, the calculated mutual information is decided by the multivariate distribution models. It should be noted, however, that the distributions of hydrological processes in the real world may not obey certain probability distributions. Multivariate distribution models cannot avoid estimating the joint distribution function. Due to the difficulty in estimating the joint Probability Density Function (PDF) of an unknown distributed sample. Copula, which is a natural substitute for digging the correlation of two variables with the product of the associated copula density function, which offers an extraordinary perspective for estimating mutual information (Zeng and Durrani, 2011). The main advantage of copula functions over classical multivariate hydrologic modeling is that the marginal distributions and multivariate dependence modeling can be determined in two separate processes, thereby giving additional flexibility to the practitioner in choosing different marginal and joint probability functions (Zhang and Singh 2006; Genest and Favre 2007; Song and Singh 2010; Sraj et al. 2015). Previous studies using copula usually focus on multivariate variables of single hydrological stations, and therefore studies using copula rarely analyze the arbitrary station pair.

The objectives of this paper are designed as follows: (1) understand the correlation of different station pairs in certain catchments by establishing respective copula functions; (2) calculate the copula entropy instead of mutual information; (3) evaluate the optimal network design based on the modified IDIT (the Intensity of Direction Information Transfer) index considering the distance between station pairs.

2 METHODS

The methods used in this paper including the entropy calculation of a single site, estimation method for the copula parameter, goodness-of-fit test for copulas, and the calculation of copula entropy. The details of these methods can be found in the literature (Press et al., 1995; Genest and Favre, 2007; Nelsen, 2006; L. Zhang et al., 2007).

Inspired by the idea of directional information transfer (Yang and Burn,1994), the Intensity of Direction Information Transfer (IDIT) is combined with the distance between each station pair:

$$IDIT(X,Y) = T(X,Y)/[H(X)^{*}d(X,Y)] \tag{1}$$

where, $T(X, Y)$ is calculated by the copula entropy, $d(X, Y)$ is the distance between station X and Y. It is noticed that $IDIT$ is not symmetrical. Therefore, $IDIT_{yx} = T / [H(Y)*d]$ while $IDIT_{xy} = T / [H(X)*d]$.

When all remaining stations in the group have strong mutual connections with each other (i.e. both $IDIT_{yx}$, and $IDIT_{yx}$ are high), these can be further selected according to a criterion of $S(i)$, $R(i)$, and $N(i)$, which is defined as follows:

$$\mathrm{S}(i) = \sum\nolimits_{j=1,j\neq 1}^{m} IDIT_{ij} \tag{2}$$

$$\mathrm{R}(i) = \sum\nolimits_{j=1,\,j\neq 1}^{m} IDIT_{ji} \tag{3}$$

$$N(i) = (S - IDIT_{i}) - (R - IDIT) \tag{4}$$

3 CASE STUDY

The related monthly streamflows of 2001–2013 from the Yiluo catchment of the Yellow River of China are collected. Thirteen potential gauge stations are analyzed, and 8 station pairs are chosen to prove the feasibility of MPL methods and copula entropy, thereby estimating the mutual information.

3.1 *Test of the parameter-estimated methods*

It may be found that the copula entropy method that is used to estimate the mutual information only relies on the copula density function, which is determined by the parameter. Therefore, it is indispensable to test the parameter-estimated method of MPL and IKT.

By using the three types of copula to establish joint distributions between pairs of stations and compare the results of Goodness of Fit (GOF), a choice of θ and copula, which brings the best GOF, is made. In Table 1, the results are presented. The highlighted minimum AIC represents the best fitting situation. Thus, it can be deduced that each station pair of the selected 8 pairs can belong to a certain copula function, which has been marked in bold typeface. In Table 1, the method of Inversion of Kendall's Tau (IKT) has been restricted because of overvaluing the boundary. Therefore, in most circumstances, the parameter estimated by MPL performs better than that through IKT.

3.2 *Classification of the network*

Based on Table 2 of the IDIT value, the typical application of the IDIT values in the hydrological network design process is in classification or grouping of potential stations. If both

Table 1. The GOF results of different copulas.

Station pair		MPL	IKT	RMSE
6–7	Clayton	−132.3022	−125.5473	0.01255
	Frank	−132.8218	−132.6469	0.01157
	G-H	−116.5486	−103.1404	0.01457
1–6	Clayton	−43.0414	−27.9522	0.02154
	Frank	−41.4736	−41.4732	0.02587
	G-H	−51.47845	−51.3583	0.01597
3–7	*Clayton*	−92.9563	−92.1834	0.00934
	Frank	−57.7171	−57.7102	0.02547
	G-H	−85.5352	−84.1190	0.01194
2–5	Clayton	−2.8074		0.01835
	Frank	−2.9796	−2.8758	0.01578
	G-H	−2.7607		0.02015
1–4	*Clayton*	−79.3201	9.6340	0.01235
	Frank	−26.0210	−26.0056	0.01578
	G-H	−61.2533	−57.6474	0.01958
5–12	*Clayton*	−10.3150	−10.2706	0.01089
	Frank	−6.2846	−6.2791	0.01618
	G-H	−9.2023	−9.2016	0.01238
11–12	Clayton	−18.0697	−8.6298	0.01792
	Frank	−24.4088	−24.3808	0.01137
	G-H	−21.4711	−24.2425	0.01689
4–9	*Clayton*	−28.0254		0.01235
	Frank	−6.2563	−6.2451	0.01578
	G-H	−25.1238	−22.6405	0.02015

Table 2. The Intensity of Direction Information Transfer (IDIT) matrix.

	1	2	3	4	5	6	7	8	9	10	11	12	13
1	1	0.15	0.29	0.29	0.21	0.24	0.29	0.25	0.26	0.22	0.2	0.15	0.14
2	0.05	1	0.07	0.02	0.05	0.06	0.05	0.05	0.04	0.04	0.02	0.04	0.03
3	0.3	0.24	1	0.17	0.34	0.19	0.32	0.31	0.26	0.27	0.18	0.17	0.21
4	0.27	0.05	0.15	1	0.11	0.13	0.24	0.15	0.15	0.12	0.1	0.17	0.08
5	0.16	0.12	0.24	0.09	1	0.04	0.15	0.12	0.08	0.1	0.04	0.06	0.09
6	0.22	0.15	0.16	0.12	0.04	1	0.27	0.2	0.21	0.22	0.14	0.13	0.1
7	0.26	0.14	0.27	0.23	0.18	0.27	1	0.3	0.3	0.3	0.18	0.24	0.22
8	0.18	0.11	0.22	0.12	0.12	0.17	0.24	1	0.3	0.26	0.15	0.13	0.18
9	0.23	0.1	0.22	0.14	0.09	0.21	0.3	0.37	1	0.33	0.18	0.16	0.22
10	0.19	0.11	0.22	0.11	0.11	0.21	0.28	0.3	0.32	1	0.19	0.18	0.27
11	0.07	0.02	0.06	0.04	0.02	0.05	0.07	0.07	0.07	0.08	1	0.04	0.06
12	0.05	0.04	0.05	0.06	0.03	0.05	0.09	0.06	0.06	0.07	0.04	1	0.04
13	0.11	0.09	0.17	0.07	0.1	0.1	0.2	0.21	0.21	0.26	0.15	0.1	1

Table 3. The criterion of $S(i)$, $R(i)$, and $N(i)$ for group 1.

$N(i)$	1	3	4	6	7	8	9	10	13	$S(i)$
1	0.22	0.29	0.29	0.24	0.29	0.25	0.26	0.22	0.14	1.98
3	0.3	0.33	0.17	0.19	0.32	0.31	0.26	0.27	0.21	2.03
4	0.27	0.15	0.04	0.13	0.24	0.15	0.15	0.12	0.08	1.29
6	0.22	0.16	0.12	–0.02	0.27	0.2	0.21	0.22	0.1	1.50
7	0.26	0.27	0.23	0.27	0.01	0.3	0.3	0.3	0.22	2.15
8	0.18	0.22	0.12	0.17	0.24	–0.42	0.3	0.26	0.18	1.67
9	0.23	0.22	0.14	0.21	0.3	0.37	0.01	0.33	0.22	2.02
10	0.19	0.22	0.11	0.21	0.28	0.3	0.32	–0.08	0.27	1.90
13	0.11	0.17	0.07	0.1	0.2	0.21	0.21	0.26	–0.09	1.33
$R(i)$	1.76	1.70	1.25	1.52	2.14	2.09	2.01	1.98	1.42	

IDIT values are higher than a given threshold value, this station pair would be classified as one group. Others are not shown in this figure because the information transfer intensity is too weak. Therefore, in Figure 9, a remarkable information relation is marked by using an arrow representing the information transmission. If the threshold value for IDIT is set as 0.25, then 13 stations can be classified as four groups according to the previous grouping principles. Stations 2, 11, and 12 are divided into three groups respectively, because IDIT of the station with others is smaller than the threshold value. That is to say that, these should be separated from the others. Similarly, stations 1, 3, 4, 6, 7, 8, 9, 10, and 13 should be classified into the same group. Station 5 can be predicted by station 3 ($IDIT_{35} = 0.34$, $IDIT_{53} = 0.24$), and so it is deleted from the network.

3.3 *Station selection for the network*

Since all of the stations in each group are mutually strongly related (both IDIT values of every pair are above the threshold value), further selection may be accomplished by the criterion of $S(i)$, $R(i)$, and $N(i)$. Consider, as an example, the group 1 that is composed of stations 1, 3, 4, 6, 7, 8, 9, 10, and 13.

The results of the three criteria and their rankings are displayed in Tables 3 and 4 and Figure 1. The results recommend that the information $S(i)$ sent from station i to station j correlates positively with the information $R(i)$ received from station j to station i. The received information indicates a small negative correlation with $N(i)$. Higher ranks ($r = 9$ or $r = 8$)

Table 4. Ranking based on the criterion of $S(i)$, $R(i)$, and $N(i)$ for group 1.

	Information			Rank(r)		
Station no.	$S(i)$	$R(i)$	$N(i)$	$S(i)$	$R(i)$	$N(i)$
1	1.98	1.76	0.22	6	5	8
3	2.03	1.70	0.33	8	4	9
4	1.29	1.25	0.04	1	1	6
6	1.50	1.52	–0.02	3	3	3
7	2.15	2.14	0.01	9	9	5
8	1.67	2.09	–0.42	4	8	1
9	2.02	2.01	0.01	7	7	4
10	1.90	1.98	0.08	5	6	7
13	1.33	1.42	–0.09	2	2	2

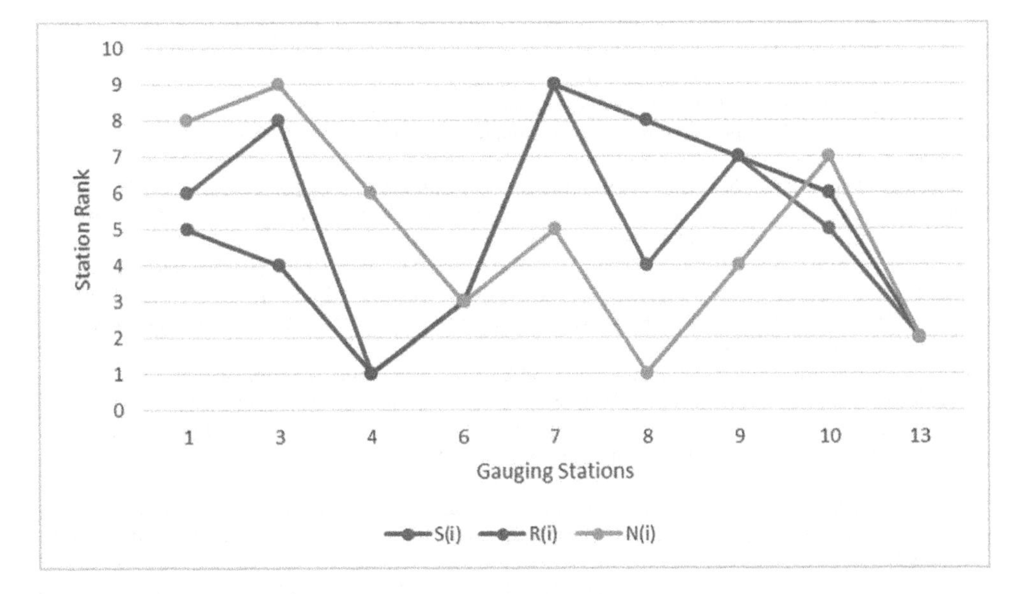

Figure 1. Station rankings based on $S(i)$, $R(i)$, and $N(i)$. The gauging numbers are given in Table 4.

present the station with higher information value that should be remained in the network. On the contrary, lower ranks recommend that gauges with little regional value will be discontinued if the number of stations in the group needs to be cut down. For example, the station 4 should be the first to be removed from the network on the basis of $S(i)$ and $R(i)$, while station 8 would be the first to be discontinued according the ranking of $N(i)$. And station 3, 7, and 10 should be retained with the highest priority.

4 CONCLUSIONS

In this paper, a transinformation analysis has been conducted between different station pairs based on the copula entropy method through the criterion of IDIT; the results recommended that the choice of the MPL method as a semi-parametric estimation for selecting the best copula was better than the parametric estimation. And IDIT (Intensity of Direction Information Transfer) index has two applications: it helps to confirm the certain station to be removed or retained in the gauge station network design and is a good tool for classifying the network into certain groups. Furthermore, the mutual information estimated by using the copula entropy method proved to be rational, because the method of copula entropy avoids the assumption of the density function of a single station.

ACKNOWLEDGMENT

This study was supported by the National Natural Science Foundation of China (Grant no. 41571017).

REFERENCES

Al-Zahrani, M., Husain, T. 1998. An algorithm for designing a precipitation network in the south-western region of Saudi Arabia, *Journal of Hydrology*, 205: 205–216.

Chen, Y.C., Wei, C., Yeh, H.C. (2008), Rainfall network design using kriging and entropy, *Hydrological process*, 22(3): 340–346.

Genest, C., Favre, A.C. 2007. Everything you always wanted to know about copula modeling but were afraid to ask. *Journal of Hydrological Engineering*, 12(4): 347–368.

Husain, T. 1987. Hydrologic network design formulation, *Canadian Water Resources Journal*, 12(1): 44–59.

Mahmoudi-Meimand, H., Nazif, S., Abbaspour, R.A., Sabokbar, H.F. 2016. An algorithm for optimization of a rain gauge network based on geostatistics and entropy concepts using GIS, *Journal of Spatial Science*, 61(1): 233–252.

Markus, M., Knapp, H.V., Tasker, G.D. 2003. Entropy and generalized least square methods in assessment of the regional value of streamgages, *Journal of Hydrology*, 283: 107–121.

Mishra, A.K. & Coulibaly P. 2009. Developments in hydrometric network design: A review. *Reviews of Geophysics*, 47(2): 2415–2440.

Nelsen, R.B. An Introduction to Copulas [M]. 2nd Edition. New York: Springer, 2006.

Press, W.H., Teukolsky S.A., Vetterling, W.T., Flannery, B.P. 1995. Numerical recipes in Fortran 77, the art of scientific computing. *Statistical Description of Data, Cambridge University Press*, 14: 626–630.

Singh, V.P. 1997. The use of entropy in hydrology and water resources, *Hydrological Processes*, 11, 587–626.

Song, S.B., Singh, V.P. 2010. Frequency analysis of droughts using the Plackett copula and parameter estimation by genetic algorithm. *Stochastic Environmental Research and Risk Assessment, 24*(5): 783–805.

Sraj, M., Bezak, N., Brillly, M. 2015. Bivariate flood frequency analysis using the copula function: a case study of the Litija station on the Sava River. *Hydrological Processes*, 29(2): 225–238.

Yang, Y., Burn, D.H. 1994. An entropy approach to data collection network design, *Journal of Hydrology*, 157: 307–324.

Zeng, X., Durrani, T.S. 2011. Estimation of mutual information using copula density function. *Electronics Letters*, 47(8): 493–494.

Zhang, L., Singh, V.P. 2006. Bivariate flood frequency analysis using the copula method, *Journal of Hydrologic Engineering*, 11(2): 150–164.

Zhang, L., Singh, V.P. 2007. Bivariate rainfall frequency distributions using Archimedean copulas. *Journal of Hydrology*. 332: 93–109.

Risk Analysis and Management – Trends, Challenges and Emerging Issues – Bernatik, Huang & Salvi (Eds)
 ISBN 978-1-138-03359-7

A study on the evaluation of tourism environmental quality in Huangshan city

Lizhong Yao
School of Tourism, Huangshan University, Huangshan, China

ABSTRACT: With the development of the society and economy, more and more people pay attention to traveling. Traveling has become a kind of popular lifestyle. The rapid development of outbound tourism brings about not only the expansion of various factors in tourism, but also many problems. In order to protect the tourism resources, maintain the healthy, orderly and sustainable development, and improve the economic, social and environmental benefits, in this paper, on the basis of relevant experts' opinions, Huangshan is set as an example, related studies are combined, and a city tourism environmental quality evaluation index system is constructed including Tourist Destination Sightseeing environment, tourism environment, tourism service life social environment, and tourism ecological environment. It evaluates quantitatively on the Huangshan tourism environment by using the pattern of tourists' evaluation, which is based on the needs of tourists' aesthetic psychology.

Keywords: tourism environment; environmental quality; evaluation system; Huangshan city

1 THE CONCEPT AND CHARACTERISTICS OF AN URBAN TOURISM ENVIRONMENT

1.1 *The concept of urban tourism environment*

There is no unified definition of tourism environment in the academia as different cities, scenic spots, and administrative geographical districts have different versions of it. For example, Zhou Yanting holds that tourism environment refers to the physical environment formed by the natural and human factors in all areas with tourism value. Others such as Chang Fengchi believe that this concept refers to a kind of environment including the natural environment, social environment, economic environment, and political environment, where people can enjoy tourism activities with aesthetic feelings as well as the pleasure of obtaining certain knowledge (Wan Xu-cai et al. 2003). By making a comparison between the two definitions mentioned above, we eventually define the tourism environment as follows: it is an objective existence which has a direct effect on the quality of tourists' journey, thereby containing four aspects such as the general environment of scenic spots, the environment of service system, the social environment, and natural ecological environment.

1.2 *The characteristics of urban tourism environment*

1.2.1 *The complexity of influencing factors*

According to the definition of tourism environment, we can find that it is a system with a complex structure and concerning a large scale of aspects. Therefore, the factors which can influence the urban tourism environment turn to be quite varied and complicated, especially for Huangshan, which is a young tourism city with tourism as its pillar industry and which has a close connection with other industries. Therefore, the influencing factors are just more complex.

1.2.2 *The difficulty of assessment*

Besides some mandatory quota that can be easily measured, these toughly measurable flexible targets account for a lion's share in the various factors affecting the tourism environment, such as service level, hygiene level, ornamental value, cultural atmosphere etc. (Wang Qun et al. 2006). To a large extent, the assessment of these flexible targets depends on tourists' feeling and the satisfaction of the aesthetic psychology, and therefore it has difficulty of assessment.

1.2.3 *The dynamics of evaluation*

Based on the difficulty of assessment, we can figure out that the tourists' aesthetic values and aesthetic feeling can largely determine the value of the tourism environment's quality. For instance, both people's material and spiritual requirements will be greater and greater with an improvement in people's livelihood. As a result, the evaluation of tourists will not be constant, but dynamic. Meanwhile, an urban environment involves all aspects of social life, any of which can influence tourists' psychological feeling. As a result, the assessment of the tourism environment's quality will be more and more dynamic (Qian Yi-chun 2007).

2 THE BASIC RESEARCH METHODS OF THE URBAN TOURISM ENVIRONMENTAL QUALITY

On the basis of the value of environmental quality and the previously reported work, we adopted the way of questionnaire and consulting relevant reference books to select the evaluation index through the method of tourists' evaluation (Jiang Zong-hao 1996). We adopted the opinions of some experts, selected 22 indexes and established a set of systems that was suitable for Huangshan's tourism environmental quality. We have also adopted a 10-point system and set a rule that the score above 8 points including 8 is considered as good, 6–8 points including 6 is considered as medium, and the rest as poor. All the numbers are accurate to two decimal places to evaluate every index of Huangshan city.

2.1 *The technical route of evaluation*

Construction of an index system; the weight of determining the evaluation index factor; obtaining factor scores; and performing a comprehensive evaluation (the scores of the calculated comprehensive evaluation) are the four steps of the technical route of evaluation. According to the principle of evaluation, the first three steps mainly involve the tourist evaluation method (Figure 1).

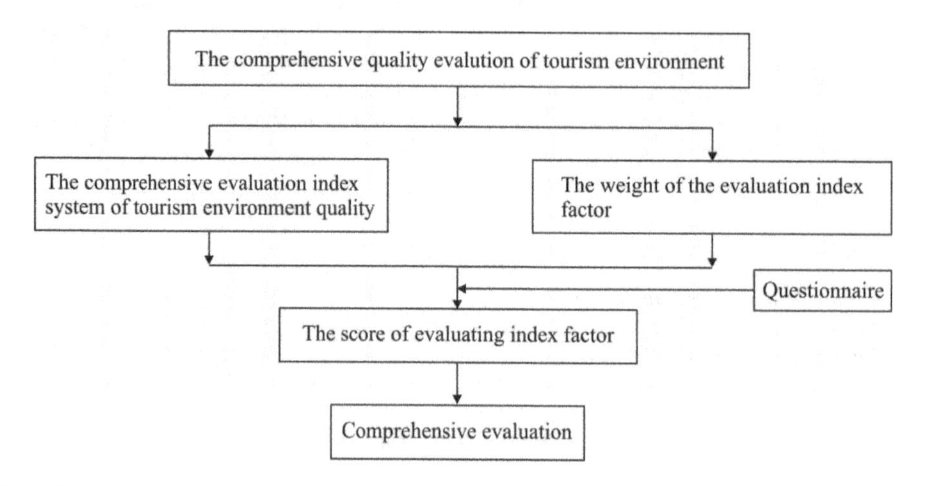

Figure 1. The technical route of evaluation.

2.2 *To build the index system*

Starting from the aesthetic psychological needs of the tourists, on the principles of tourists-oriented assessment, guidance, timeliness and conciseness, and on the basis of reference of many evaluation systems and opinions of relevant experts, we selected the evaluation index by making an assessment of the tourists and then we built the index system framework.

2.2.1 *Questionnaire to tourists*

According to the characteristics of the urban tourism environmental quality of Huangshan and the factors influencing tourist satisfaction, we have chosen 22 indexes as the main researching areas on the basis of the reference of the related data and opinions of some experts during the summer vacation in 2011 and searched for dozens of visitors from different cities and different ages to fill in our questionnaire.

2.2.2 *To select indexes*

According to the different characteristics of tourists' indexes on the degree of tourism environmental perception, in this paper, on the basis of predecessors' work, assesses the importance of the evaluation indexes, and we also invite some experts and one who knows how to compare and analyze the importance among the indexes, and then calculate each index. Finally, we obtain the importance percentage of the evaluation system to tourism environmental quality.

2.2.3 *To build an index system framework*

The system, including tourism environment, life service environment, social environment, and nature ecological environment, can be divided into goal layer, eight big indexes, and 22 indexes by using the top–down method.

2.3 *To determine the weighted calculation of the evaluation index factor*

It is very important to determine the weighted calculation of evaluation indicators in the evaluation of urban tourism environmental quality, because the scientificity and rationality of the evaluation results will be directly affected based on whether the weighted calculation is reasonable or not (Cui Feng-jun 1998).

There are many methods to achieve determination of index weighted calculation, such as analytic hierarchy process, fuzzy evaluation method, Delphi method, and expert scoring method.

In this paper, on the basis of previous work, the importance of each index respectively according to the different characteristics of tourists' perceived degree of the tourism environment is judged.

In this paper, the tourist evaluation method and AHP method are adopted, and some experts and insiders are invited to compare and analyze once again, and then recalculate the weighted calculation of each index. Finally, we obtain the weighted calculation system (for specific numbers, please see Table 1).

2.4 *To obtain the evaluation index factor score values*

Tourism environment evaluation should meet the physiological needs of tourists firstly, but above all, it should emphasize the satisfaction degree of tourists' aesthetic psychology (Liu Xiao-bing et al. 1996).

In order to ensure that the evaluation result deeply reflects the personal experience of tourists, we add the tourists' evaluation points and take the average one as each index score by means of a questionnaire on the principles of tourists-oriented assessment, guidance, timeliness, and conciseness (for specific numbers, please see Table 1).

2.5 *Comprehensive evaluation*

In order to reflect the quality of tourism environment accurately and objectively, we determined to draw a comprehensive score through a single index evaluation system by using a mathematical model.

Table 1. The weighted calculation and value index of tourism environment in Huangshan city.

Target layer	Weight	General index layer	Weight	Elaborate index layer	Weight	Score
The tour environment	0.38	Scenery	0.31	Diversity	0.11	8.7
				Characteristics	0.14	8.5
				Ornamental value	0.06	7.8
		Equipment	0.07	Safety facilities	0.05	8
				Guiding marker	0.02	6.4
Living service environment	0.38	Traffic	0.11	Convenience	0.08	6.9
				The traffic order	0.03	5.6
		Accommodation catering and entertainment	0.15	Characteristics	0.04	8.3
				Service level	0.06	6.6
				Reasonable price	0.03	7.1
				Material supply	0.02	7.1
		Public health	0.11	Public facilities	0.03	6.7
				Cleanliness and tidiness	0.08	7.3
Social environment	0.12	Tourism policies and regulations	0.06	Security	0.04	8.6
				Service attitude	0.01	7.1
				Service efficiency	0.01	7.1
		Humanistic atmosphere	0.06	Culture atmosphere	0.02	8.9
				Warm and friendly people	0.02	7.6
				The quality of residents	0.02	6.5
The natural ecological environment	0.12	The ecological environment	0.12	Make green	0.04	9.1
				Air quality	0.04	8.6
				Water quality	0.04	8.2

It can help us comprehensively know the overall level of tourism environment quality in Huangshan city by making a necessary evaluation. In order to carry out the comprehensive evaluation simply, conveniently, and accurately, in this work, we use the weighted calculation comprehensive index method to establish the mathematical model:

$$Y = \sum\left(\sum_{i=1}^{m}\left(\sum_{j=1}^{n} I_j * R_j\right) * W_i\right) X_i$$

In the above-mentioned formula, Y stands for the comprehensive score of the tourism environmental quality value in Huangshan city, I_i stands for the single index evaluation score, R_j stands for the weighted calculation of single index, W_i stands for the weighted calculation of eight indicators, and X_i stands for the weighted calculation of four aspects.

3 THE SURVEY OF TOURISM ENVIRONMENTAL QUALITY IN HUANGSHAN CITY

3.1 *The survey of tourism environmental quality in Huangshan city*

3.1.1 *Tourism environment*

Huangshan, which is a city in the south of Anhui Province, is located between East Longitude 117.0" to 118.55" and North Latitude 29.24" to 30.24". As the world's cultural and natural heritage site and a national key scenic spot, Huangshan city is an important scenic area in Anhui Province and one of the major tourism cities of China. It is a famous hilly city with a total area of 9807 square kilometers and a population of 1.5 million. The well-known Huangshan scenic area is located in Huangshan District. The "Five Wonders", grotesquely shaped pines, queer rocks, sea of clouds, hot springs and winter snow, make Mt Huangshan a world-renowned location. Traditional crafts with a history of hundreds of years, such as Hui Ink-stick, inkstone, Hui Sculpture, Hui Bonsai, Qimen Porcelain, Wan'an Compass and Hui

Lacquered Engraving, are now bursting with youthful energy. Graceful natural landscape and Huizhou culture create favorable conditions for the local tourism industry. Besides Huangshan scenic areas, there are many other scattered scenic spots, just like all the stars twinkling around the bright moon, such as the ancient villages Hongcun and Xidi, the world cultural heritage site and national 5A-class scenic spot; Huanshan Mysterious Grottoes, which is regarded as the ninth wonder of the world; the Jade Valley, where the film "Crouching Tiger & Hidden Dragon" was shot; the mysterious "Eight Diagrams Village" Chenkan, which is the miniature of the west lake Tangmo; Mount Qiyun, which is one of the four famous Taoism mountains; Taiping Lake, to which the title "Huangshan love" is given etc. Meanwhile, Hui studies, Tibetology, and Dunhuangology are considered as three region studies. Nowadays, Hui trip has sparked an excitement that can be felt here and there.

3.1.2 *Life service environment*

At present, a complete system of the travel service is basically formed in Huangshan city, encompassing food service, lodging, transportation, sightseeing, shopping, and entertainment. Besides, Huangshan city is trying to improve its scenic spots, enrich its cultural connotation, and perfect its supporting facilities by continually transforming and completing sightseeing resources and developed spots inside it. Moreover, it continually improves the professional quality and service skills of staff in the tourism industry and offers training courses to tourism employees working at different star-level hotels, travel agencies, and tourist attractions, thereby enhancing their service skills on the purpose of improving the overall management level and service quality.

3.1.3 *Social environment*

Huangshan city designates the tourism industry as its new economic growth point and pillar industry, thereby paying special attention to tourism administration and environment optimization. Responding to the call "to make full use of Mt. Huangshan and Hui culture", the local government is actively exploiting, managing, and conducting publicity measures; meanwhile, it is carrying out the construction of spiritual civilization for scenic regions and tourism enterprises, administering the industry in accordance with the law system, and strengthening functions of the tourism quality supervision organization. The government is dedicated to creating a friendly and civilized social environment for tourists all over the world.

3.1.4 *Natural eco-environment*

Huangshan city, which belongs to the humid sub-tropical monsoon climate, has always enjoyed the reputation of being a "natural park" for its ample sunshine, covered trees, and rich negative oxygen ions. The great people's educator, Tao Xingzhi once admired the city as the Land of Utopia. Huangshan experiences four district seasons and a mild and humid climate; it is an ideal paradise for people.

3.2 *The assessment results and brief analysis*

3.2.1 *The assessment results*

Huangshan is a typical tourism city; in order to obtain the tourists' satisfaction with its tourism environmental quality, we adopted the method of conducting a questionnaire survey and organized managers or employees related to evaluate all aspects of the tourism environment, and then we arrived at a conclusion about its scores in each index of tourism environmental quality evaluation.

3.2.2 *The brief analysis*

It can be seen from Table 1 that the overall standard of Huangshan city tourism environmental quality is middle-level. Among the 22 detailed travels, nine of them are ranked excellent, including greening, cultural atmosphere, landscape diversity, public security, air quality, landscape characteristics, features of food, accommodation and entertainment, water quality, and safety facilities, and the other 11 aspects, such as values of landscape viewing, residents cordiality and friendliness, tidiness, materials supply, service efficiency, attitude towards tourism policies and regulations, degree of transportation convenience, public facilities, reasonable price, residents quality, and guiding marks, are rated middle-level; while the traffic order and service level are considered to be poor. In addition, the assessment results correspond to the actual situation of tourism environmental quality in Huangshan city. Therefore, this, to some extent, illustrates the rationality of the evaluation system.

4 MEASURES OF OPTIMIZING HUANGSHAN TOURISM ENVIRONMENT

4.1 *Strengthen the management of tourism infrastructure and enhance accessibility*

According to the actual situation of Huangshan, we should strengthen the construction of the following facilities, such as the internal and external traffic facilities of the town, the post and telecommunications, entertainment, shopping, public environmental sanitation, and so on. The degree of accessibility should also be increased greatly, and this can be illustrated by the opening of the traffic of Huihang Expressway, the construction of Hefei–Tongling–Huangshan Expressway, the effort to the construction of a highway project from Mt Huangshan to Wuyuan and the merger of Mt Huangshan and Kaihua, some substantial progress on the early construction of the expressway from Tunxi to Jingdezhen, and the endeavor to make the expressway from Huangshan to Xuancheng and Qiandao Lake become a provincial plan. Besides, actively implement the reconstruction project of the national and provincial road network and county roads. What is more, we should make an effort to strengthen the supervision and management of emissions of tourist vehicles and control waste pollution along the way and tourism destination to make the tourism environment and tourism facilities more complete (Zhao Hong 2000).

4.2 *Improve the quality of working staff and service level*

With respect to the service level of Huangshan's service industry, we suggest taking active measures to cultivate and introduce some corresponding professionals, expanding the scope of training, strengthening the tourism professionals and related industry working practitioners' training and qualification training, and continuously improving the level of the service industry, especially the tour guide and those who have direct communication with tourists on the basis of dealing well with the qualification training of the tour guide, travel agency, hotel, and other professional tourism management measures. At the same time, we should deepen the importance of the protection of the tourism environment so as to improve the quality of the employees and the awareness of environmental protection.

4.3 *Strengthen the publicity of the tourism environment*

We should forcefully strengthen the publicity of Huangshan's tourism by adopting a variety of ways and channels and traditional media such as newspapers, television, broadcast, and the Internet, which can always introduce the tourism environment. Some tourism festival activities can be organized, such as the festival organized by International Tourism Festival of Mt Huangshan and we can widely participate in various kinds of tourism trades and expositions, and at the same time, we can invite the business representatives and the media representatives to pay a visit to Mt Huangshan to investigate the tourism environment, to strengthen the publicity and promotion, to make more foreign visitors understand the cultural and geographical significance of Mt Huangshan, and to ensure that the reputation of Mt Huangshan spreads to all parts of the world.

4.4 *Establish and perfect the legal system*

A perfect legal system is the effective guarantee measure to protect the tourism environment. The government should establish a perfect legal system and ensure its powerful implementation (Wan Xu-cai et al. 2002). Besides the strict legal system, we should also have the strict law enforcement personnel. The government should strictly control the emissions of factory sewage waste and supervise the measures of the discharge and treatment of the sewage waste. The government should also strictly control the flow of people to various scenic spots and assess the carrying capacity of scenic spots to ensure sustainable development of the tourism environment.

4.5 *Speed up the adjustment of the industrial structure of important tourist attractions*

Low pollution and high efficiency of the third industry is the inevitable outcome and developing trend of social development (Wang Xiang 2001). The government should move the highly polluted and highly discharged industries and factories from the scenic spots and the surrounding areas to the industrial zone. At the same time, the government should strengthen the supervision of the hotel around the scenic areas and industries inside the scenic areas and strictly examine the reconstruction of buildings to reduce the destruction of some harmful behavior to the tourism environment to further improve the city planning and industrial layout (Qian Wei 1998).

5 CONCLUSION

We evaluate the tourism environmental quality of Huangshan by evaluation of tourists and consultancy to related experts and also the on-the-spot questionnaire investigation, which show that the tourism environmental quality of Huangshan and tourism carrying rationality are quite good, but the traffic order should be improved. The tourism environment can appeal to tourists and meet the needs of market and tourists and the purpose of tourism environmental protection and administration is the motivation of tourism. In this paper, we make important evaluations on each index based on the previous work and the different characteristics of the degree of tourists' perception of the tourism environmental index. We obtain this coefficient of weight of the evaluation system at all levels by using tourists' evaluation and AHP methods and some experts and related professionals' analysis and comparison of the importance between each index and the calculation of each index. In this paper, the tourists' aesthetic demands of all times are demonstrated and, at the same time, the subjectivity of evaluation is avoided. Thus, in this paper, the tourists' demand of the tourism environmental quality is reflected more accurately and truly and also the demand of the market. We should pay attention to the tourism environmental quality and develop the tourism environment reasonably, in order to promote the sustainable development of tourism in the near future.

ACKNOWLEDGMENTS

This work is supported by the Anhui Social Sciences Key Projects (SK2015A532; SK2015A168) and Huangshan Social Science Project (A2015009).

REFERENCES

Cui, Feng-jun. Issues of Tourism Environment Research [J]. Tourism Tribune, 1998, (5): 35–39.

Jiang, Zong-hao. Study on Tourism Carring Capacity and Environmental Qualities of Mount Huangshan. [J]. Rural Eco-Environment, 1996, (2): 22–25.

Liu, Xiao-bing, Bao Ji-gang. Development of the Researches on the Environmental Impacts on Tourist Development. [J]. Geographical Research, 1996, 15 (04): 92–99.

Qian, Wei. Bring advantages into play and grasp the opportunity to provide new products and services for Japan and Singapore Tourist for "twenty-first Century international tourism and tourism development seminar in Hangzhou" [J] *Journal Beijing Second Foreign Language Institute,* 1998, (03): 48–51.

Qian, Yi-chun. Quality Evaluation of Tourism Enviroment of hangjiajie National Forest Park [J]. *Journal of Northeast Forestry University,* 2007, (1).

Wan, Xu-cai, Bao, Hao-sheng. Study on the Integrated Evaluation for Tourism Environmental Quality in the Tourist Destination of Mountain A Case Study of Huang Mountain and Tianzhu Mountain in Anhui Province. [J] *Journal of Nanjing Agricultural University*, 2002, (01): 48–52.

Wan, Xu-cai, Zhang, An, Li Gang, Xu Fei-fei. Study on Synthetical Evaluation for Tourism Environmental Quality of City in View of Tourists a Case Study of Nanjing and Suzhou [J]. Economic Geography, 2003, (1): 133–138.

Wang, Qun, Ding, Zu-rong, Zhang Jin-he, Yang Xing-zhu. Study on the Model of Tourist Satisfaction Index about Tourism Environment: a Case Study of Huangshan Mountain [J]. Geographical Research, 2006, (1): 171–175.

Wang, Xiang. On the Assess of the Quality of Tourism Environments. [J] *Journal of Beijing Union University,* 2001, (02): 35–37.

Zhao, Hong. Ecological and environmental problems and Countermeasures for the sustainable development of tourism industry in China [J]. Shandong Environment, 2000, (06): 4–5.

Risk Analysis and Management – Trends, Challenges and Emerging Issues – Bernatik, Huang & Salvi (Eds)
 ISBN 978-1-138-03359-7

An evaluation method for regional disaster loss based on the internet of intelligence technology

Fanlei Zeng
China Meteorological Administration Training Center, Beijing, China

Shujun Guo
Handan Meteorological Bureau, Handan, China

Chongfu Huang
Ministry of Civil Affairs and Ministry of Education, Academy of Disaster Reduction and Emergency Management, Beijing, China

ABSTRACT: It is very important to evaluate the loss of a disaster rapidly, which is the basis for disaster relief and recovery. The traditional evaluation method based on expert's statistical research personally is relatively low in timeliness. This paper presents a method of disaster loss assessment using the Internet of intelligence technology. First, we divide the disaster area into a number of subunits according to geographical distribution; second, we select a number of staff in each unit as agents who will report the unit's disaster information with a network interactive way to a service platform; finally, with the information processing model and the loss evaluation model of the platform, we make the information fusion and the comprehensive loss evaluation process of the disaster region. The proposed method demonstrates a case of flood disaster.

Keywords: Internet of intelligence; regional disaster; loss evaluation; information fusion process

1 INTRODUCTION

The occurrence of natural disasters is usually regional, sudden, and serious (Pan Xiaohong et al. 2009), and evaluating the loss of a disaster is an important basis for disaster release, rehabilitation, and reconstruction. In general, emergency work, such as disaster relief or rehabilitation, requires much more in timeliness and accuracy of the disaster loss assessment. However, for the limitation of space, the traditional disaster assessment method is time-consuming and labor-intensive, and it is difficult to guarantee the validity of the evaluation results. The emergence of the Internet technology allows people to get rid of the constraints of time and space and to transmit a variety of information in real time, which facilitated the improvement of the efficiency of the disaster loss assessment. At the same time, it allows choosing the stakeholders of disaster loss to provide the basic data of a disaster so that the basic data can be collected and transmitted quickly and efficiently. On the basis of this, in this paper, we proposed a disaster loss assessment method based on the Internet of Intelligence (IOI) technology, introduced the method's basic principle, and demonstrated its feasibility by taking an area's flood disaster loss evaluation as example.

Disaster loss usually refers to economic losses, casualties, social effects, and ecological damage, among which economic losses are generally divided into direct economic losses and indirect economic losses (Zheng Dawei 2015). As with social impact, ecological damage is difficult to quantify, and researchers involved in the study of disaster losses mainly focus on the direct economic losses of disasters, casualties, and so on. The purpose of disaster loss assessment is to calculate the actual loss or loss of risk. There are several studies on the assessment of disaster losses conducted by different experts, scholars, and organizations. According to the difference of research ideas or disaster types, the research methods have different divisions. For example, according to the assessment time, which is quicker or later than the occurrence of the disaster, it can be divided into pre-disaster risk assessment, disaster assessment, and post-disaster assessment (Xu Feiqiong 1998; Liu Gaofeng, Li Na 2008). According to the disaster's type difference, it can be divided into floods, earthquakes, typhoons, drought, storm surge, and others. According to the differences in research ideas, it can be divided into mathematical statistics method, index system method, and physical mechanism model method, and so on (Pan Xiaohong et al. 2009). With the continuous development of computer science, satellite remote sensing, Internet technology, and their application in disaster research, the level of disaster loss assessment technology has been improved greatly, and related software packages for disaster loss assessment have been developed. For example, HAZUS-MH is a multihazard assessment model developed by the National Emergency Management Agency and the National Academy of Architectural Sciences in the 1990s to assess the loss of three disasters in the United States, including earthquakes, floods, and hurricanes. SLOSH is a tropical storm (tide) model developed by the National Oceanic Meteorological Agency. RADIUS is a tool specifically designed for urban earthquake disaster risk assessment. In order to solve the problem of earthquake disaster loss assessment, Chen Hongfu developed HAZ-China software preliminary (Chen Hongfu 2012), Wang and Yanyan developed the loss model of flood disaster in Shanghai on the basis of GIS technology (Wang Yanyan et al., 2001), and Zhang Fei studied the method of agricultural flood disaster loss assessment on the basis of dynamic input–output loss model (Zhang Fei 2013). With the maturity of mobile Internet technology and the popularity of mobile communication terminals, people can easily receive and send a variety of information, and with the idea of group participation and the tool of Internet, it can help people solve many practical problems. On the basis of the above ideas, some scholars put forward the concept of Internet of intelligence.

IOI is a technology of collecting problem-related information from a number of intelligent agents on the basis of the Internet technique and processing the collected information with an effective mathematic model to form a problem-related solution. From the IOI conception was first proposed in 2011, until now, the IOI research has experienced a process from theoretical research to practical application. From theoretical viewpoint, Chongfu Huang first proposed the concept of IOI in 2011 and explained its feasibility on online risk analysis (Chongfu Huang 2011). In the follow-up study, Fanlei Zeng further explored IOI technology's workflow, proposed two kinds of screening strategies to avoid malicious information, and put forward an incentive method, which aimed to encourage agents' participation (Zeng F.L. et al. 2013). Fuli Ai proposed a compound participation model for dealing with unstructured flexible information (Huang C.F. 2013; Ai Fuli 2013). From the viewpoint of application, several experimental IOI service platforms have been explored, including the "IOI platform of risk assessment of college candidate voluntary", which aimed to solve the problem of risk in college candidate voluntary filling; "Zhejiang Wenzhou aquaculture typhoon disaster insurance IOI services platform", which aimed to solve the problem of an aquaculture typhoon insurance's feasibility; "Seismic group test group anti-IOI service platform", which aimed to study the mechanism of the reporting of seismic anomaly information; and the "Community risk radar IOI service platform", which aimed to study the community security problem. In this paper, we applied the IOI technology to the regional disaster loss evaluation problem, with the participation of corresponding person within the disaster-affected area to provide basic data of disaster loss, combined with specific loss evaluation model, and to achieve a rapid loss evaluation for the disaster-affected area.

In the second part of this paper, we first analyzed the contents of IOI technology and disaster loss assessment problem and then discussed a way of realizing the disaster loss assessment by using the IOI technology. In the third part, we demonstrated the implementation of the proposed method with a flood disaster case.

2 DISASTER LOSS ASSESSMENT METHOD BASED ON THE IOI TECHNOLOGY

2.1 *Definition of internet of intelligence*

The concept of Internet of intelligence has been briefly described in the previous section. Here, we define the Internet of intelligence and explain some basic concepts, including Intelligent Agents, Internet Servers, and Information Processing Models to better define IOI:

Intelligent Agents refers to a group of certain individuals who have the ability of observation, interpretation, and logical reasoning and have the functions of monitoring and feedback.

Internet Server refers to a platform that can support the Internet service for the IOI's running.

Information Processing Model refers to some mathematical models that can process basic data into certain useful information.

The strict definition of the Internet of intelligence can be described as:

Definition 1: Let A be a set of *intelligent* agents, N is an Internet server platform used by A, and M is the information processing model; ternary object $<A, N, M>$ was named an Internet of intelligence system, which can be marked as Φ.

And function (1) is the formula for an IOI system:

$$\Phi = <A,N,M> \tag{1}$$

Figure 1 is the topology diagram of a simple IOI system (Chongfu Huang 2015).

2.2 *Modules and processing steps for a disaster loss evaluation IOI system*

Disaster loss assessment refers to an evaluation of the damage of a disaster using some methods, which served the purposes of disaster relief, recovery, reconstruction, and so on. This paper applied the IOI technology into the issue of disaster evaluation. The methods can be described simply as below.

First, to choose the stakeholders within the disaster-affected area to be intelligent agents, who should provide basic data with some certain standard into the IOI system; then, to process the provided information with certain mathematic model in the IOI system, and finally, to evaluate the disaster loss, as shown in Figure 2.

There are four main modules in the system, and the whole process can be divided into four steps. The following subsections will introduce the proposed method from the viewpoints of the modules and the process.

2.2.1 *Main modules*

There are four modules in an IOI-based disaster loss evaluation platform. They are (1) Intelligent agent module, who is the basic information provider. It is joined in the system through a user registration action; (2) Data-inputting module, which provides

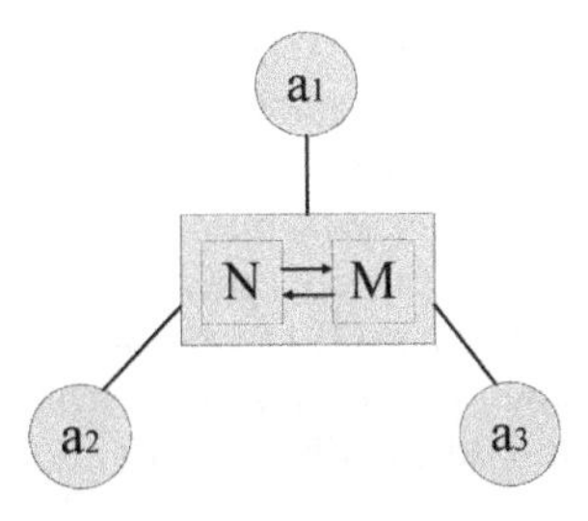

Figure 1. Topology diagram of a simple IOI system. N refers to Internet server, M refers to information processing model, and a_1, a_2, and a_3 are basic intelligent agents.

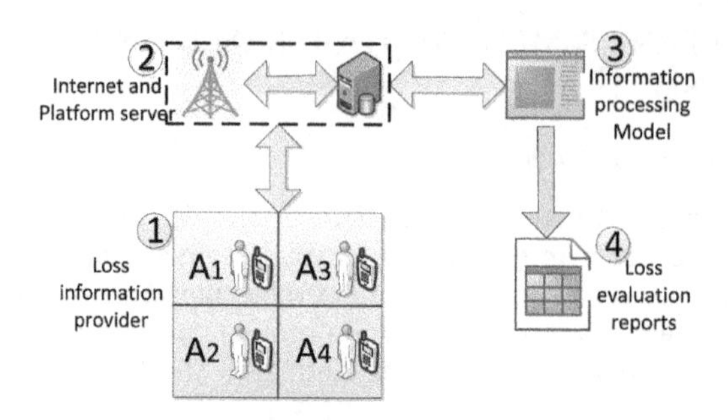

Figure 2. A simple schematic diagram for the loss evaluation process of a regional disaster using IOI technology.

a visual interface for the user to input basic loss information with certain data format. It is the data entrance of the system that is operated by intelligent agents; (3) Information processing module: as the basic data is provided by agent, the data format may not be standardized, and the data contents may be inconsistent, there should be some model to process such issues; (4) Disaster loss computing module, which will compute the final disaster losses with certain mathematic statistical formula or other mechanism formula.

2.2.2 *Evaluation process*

The whole process of the IOI-based disaster losses evaluation system involves the following four steps:

1. Choose the intelligent agents. To divide several key sections according to the situation of the disaster-affected area, we selected a number of stakeholders in different key sections as the information provider. In order to verify the provided information, we can employ the staff of the grassroots administrative institutions in the affected area as information providers. For example, China can be divided into five administrative units: province, city, county, township, and village. Thus, staff at different administrative levels can be selected as an agent to provide information.
2. Collecting disaster information. After the disaster, the selected agents within the affected areas should provide specific disaster information, which generally includes casualties, housing collapse, agriculture and animal husbandry, and infrastructure.
3. Processing basic data. To process the basic data with the corresponding model according to the data format.
4. Computing damage loss. To input the processed data into certain loss-computing model to calculate the final disaster loss.

3 AN APPLICATION EXAMPLE

Since July 18, 2016, several areas of China, including North China, Huanghuai, Jianghan, and other places, have experienced a large range of heavy rainfall, with rainstorm areas receiving more than 50 mm per 0.635 million km^2. Precipitation occurred in some local areas of Hebei and Henan provinces after nearly a century. And Handan City suffered severe storm floods, and its western mountainous areas, such as Wu'an and other areas, had casualties, housing damage, crops flooding, and other serious consequences. In order to assess the disaster loss in Handan as soon as possible, using the IOI technology, relying on Handan City's multiple level of meteorological departments, we organized the grassroots staff of the affected areas to carry out disaster investigation and rapidly collect and obtain the basic loss information of a number of disaster areas.

The specific assessment process is as follows.

3.1 *Choose agent*

Several counties including Wu'an, Shexian, Jize, and Yongnian were seriously affected by this disaster. When collecting the disaster information, we first divide each affected county into several key sections according its township-level administrative department and specify an agent in each administrative unit to be responsible for disaster verification in the area. Figure 3 shows China's five-level administrative division, and Figure 4 shows the administrative divisions in Handan City and the distribution of IOI agents chosen in Wu'an county.

3.2 *Collect and process data*

For the consequences of floods, the index system of disaster loss evaluation is summarized into four first-level indicators, including "affected population", "crop disaster situation", "housing collapse situation", and "direct economic loss", each of which is divided into several second-level indicators, as shown in Table 1. To describe the similarity in the process of agent's information feedback, in this paper, we only take Wu'an county as an example, to verify the information collection process and the loss evaluation process.

There are 22 townships and 502 administrative villages under the administration of Wu'an county. Considering the complexity of the operation and the normativity of the management, we set the township as the data feedback unit, and all results should be fed back to the IOI system. Tables 2 and 3 show the basic disaster data reported by villages under the Bei An

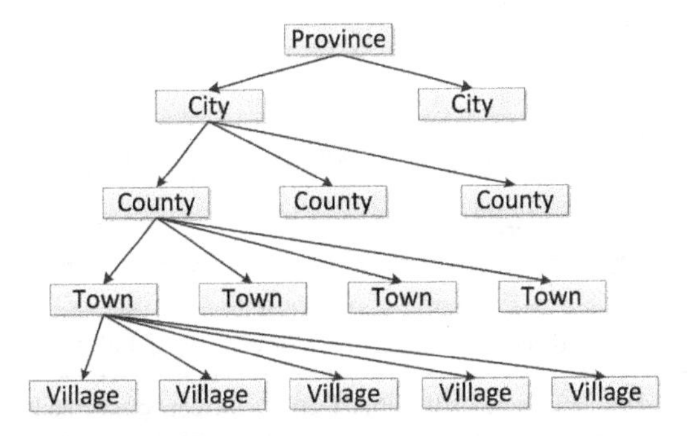

Figure 3. Graph of China's five-level administrative division.

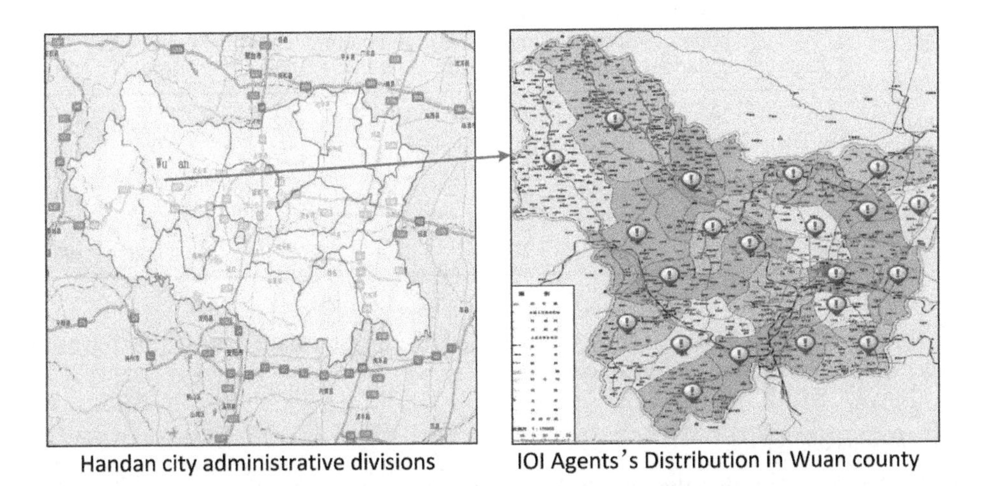

Figure 4. Handan City's administrative divisions and the distribution of IOI agents in Wu'an county.

Table 1. An indicator system for flood disaster loss evaluation.

First level	Second level
Population-affected situation (person)	Affected population Decentralized population Death population
Crop-affected situation (acres)	Affected area Disaster area Death area
House-affected situation (family)	Houses collapsed number Serious damaged house number General damaged house number
Direct economic loss (million)	Agriculture loss Industrial and mining enterprises loss Family property loss

Table 2. Damage situation of population and agriculture in villages of Bei An Le Township.

		Population-affected situation (person)		Crop-affected situation (acres)		
Bei An Le Township's damage situation	Village	Affected population	Decentralized population	Affected area	Disaster area	Death area
	Nan Tian	4026	189	491.6	16.15	304.78
	Bei Tian	1450	10	41.35	18.29	23.06
	Nan An Le	1900	23	97.69	42.67	52.88
	Kang Su	5600	187	155.85	40.2	90.12
	…	…	…	…	…	…
	Total	31666	1132	2030.8	683.16	840.5

Table 3. Housing damage and direct economic losses in villages of Bei An Le Township.

		House-affected situation (family)			Direct economic loss (million yuan)			
Bei An Le Township's damage situation	Village	No. of houses collapsed	No. of houses seriously damaged	No. of houses moderately damaged	Agriculture loss	Industrial and mining enterprises loss	Family property loss	Total
	Nan Tian	14	9	2	11	15	13	42.4
	Bei Tian	0	0	1	1	0	0.09	1.29
	Nan An Le	0	0	1	2	0.3	0.5	2.805
	Kang Su	2	5	0	1.07	30.17	1.6	32.95
	…	…	…	…	…	…	…	…
	Total	44	48	21	25.4	110.55	23.56	196.705

Le Township administration. Tables 4 and 5 show the disaster data reported by townships under the administration of Wu'an county.

3.3 *Results analysis*

Tables 2 and 3 show that the overall disaster situation of Bei An Le Township is realized through the disaster situation statistic of villages in Bei An Le Township. The specific damage situation is as follows: the total affected population is 31666, total decentralized population

Table 4. Damage situation of population and agriculture in the townships of Wu'an county.

		Population-affected situation (person)		Crop-affected situation (acres)		
	Township	Affected population	Decentralized population	Affected area	Disaster area	Death area
Wu An county's damage situation	Bei An Le	31666	1132	2030.77	683.16	840.53
	Bo Yan	21367	160	562.68	370.64	313.36
	Bei An Zhuang	7230	12	849.09	640.86	334.43
	Ci Shan	28567	2186	1750.68	578.25	548.6
	...	...	...	...	...	...
	Total	411468	19343	55365.14	29571.25	16254.04

Table 5. Housing damage and direct economic losses in townships of Wu'an county.

		House-affected situation (family)			Direct economic loss (million yuan)			
	Township	No. of houses collapsed	No. of houses seriously damaged	No. of houses moderately damaged	Agriculture loss	Industrial and mining enterprises loss	Family property loss	Total
Wu An county's damage situation	Bei An Le	44	48	21	25.4	110.55	23.56	196.705
	Bo Yan	12	14	0	8.5055	0	22.7207	74.9857
	Bei An Zhuang	0	11	4	3.974	0	1.312	12.0562
	Ci Shan	95	125	430	8.5646	0	110.5435	131.1081
	...	...	...	...	...	...	...	...
	Total	3693	1893	4806	378.6457	251.5975	629.7777	1644.7

is 1132, crop-affected area is 2030.8 acres, disaster area is 683.16 acres, crop death area is 840.5 acres, the number of collapsed houses is 44, the number of seriously damaged houses is 48, the number of moderately damaged houses is 21, direct economic losses is 196.705 million yuan, agricultural loss is 25.4 million yuan, industrial and mining enterprises loss is 110.55 million yuan, and family property loss is 23.56 million yuan.

Tables 4 and 5 show that the overall disaster situation of Wu'an county is realized through the disaster situation statistic of townships in the county. The specific damage situation is as follows: the total affected population is 411468, total decentralized population is 19343, crop-affected area is 55365.14 acres, disaster area is 29571.25 acres, crop death area is 16254.04 acres, the number of collapsed houses is 3693, the number of seriously damaged houses is 1893, the number of moderately damaged houses is 4806, direct economic loss is 1644.7 million yuan, agricultural loss is 378.6457 million yuan, industrial and mining enterprises loss is 251.5975 million yuan, and family property loss is 629.7777 million yuan.

4 CONCLUSION

In this paper, we proposed a disaster loss evaluation method on the basis of the Internet of Intelligence technology. There are four modules in the method's principle, including intelligent agent module, data inputting module, information processing module, and disaster loss comprehensive computing module, and there are four steps to implement the method, including choosing the information providing agent, collecting disaster information, processing basic data, and computing final loss results. Then, we demonstrated the feasibility of method taking the flood disaster evaluation in Wu'an county, Handan, China, as an example. Results show that the proposed method can take the advantage of the Internet and have a high precision in timeliness and efficiency.

The disadvantage of the method is that the accuracy and normativity of the data are difficult to guarantee, as the basic data are provided by the agents. Therefore, when conditions permit, it should be as far as possible to use higher credible staff to be the agent to provide basic information.

REFERENCES

Ai, F.L. 2012. A service platform for risk assessment of college candidate voluntary supported by Internet of intelligences. In Huang C.F. and Zhai G. F (eds), *Innovative Theories and Methods for Risk Analysis and Crisis Response:* 44–49. Paris: Atlantis Press. 2012.

Ai, Fuli 2013. A Research on Building and Applying the Risk Analysis of Natural Disaster of Internet of Intelligences Service Platform [D], Beijing Normal University, 2013.

Chen, Hongfu 2012. General Development and Design of HAz-China Earthquake Disaster Loss Estimation System [D], instinute of engineering mechanics, China Earthquake Administration Harbin. China, 2012.

Chongfu, Huang 2011. Internet of Intelligences in Risk Analysis for Online Services [J]. *Journal of Risk Analysis and Crisis Response*, 2011, 1(2): 110–117.

Chongfu, Huang 2015. Internet of Intelligences Can Be a Platform for Risk Analysis and Management [J]. Human and Ecological Risk Assessment, 2015, 21(5): 1395–1409.

Huang, C.F. 2013. The measurement of effective knowledge in multiple Internet of intelligences and application in risk assessment, Intelligent Systems and Decision Making for Risk Analysis and Crisis Response — *Proceedings of the 4th International Conference on Risk Analysis and Crisis Response* (Istanbul, Turkey, August 27–29, 2013), Eds. Chongfu Huang, Cengiz Kahraman, Boca Raton, USA: CRC Press, 1–8.

Liu, Gaofeng, Li, Na 2008. Construction of Loss Evaluation Index System for Urban Flood Disaster [J]. Modern Agricultural Science and Technology, 2008, No.492(22): 268–270.

Pan, Xiaohong, Jia, Tiefei, Wen, Jiahong ect. 2009. The Application and Commentary of Mulit-hazard Loss Estimation Model [J]. J. of Institute of Disaster-Prevention Science and Technology, 2009, 11(2): 77–82.

Wang, Yanyan, Lu, Jikang, Zheng, Xiaoyang, etc. 2001. Development of flood damage assessment system of Shanghai [J]. *Journal of catastrophology*, 2001, 16(2): 8–14.

Weidan, Wang, Chongfu, Huang, Fuli, Ai 2012. A Tentative Exploration of Earthquake Prediction Strategy Based on Internet of Intelligences [C] In Huang C.F. and Zhai G. F (eds), *Innovative Theories and Methods for Risk Analysis and Crisis Response*: 44–49. Paris: Atlantis Press. 2012.

Wu, T., Huang, C.F, Ai, F.L, Zeng, F.L, Guo, J. 2013. A discussion on using internet of intelligences to improve risk radar, Intelligent Systems and Decision Making for Risk Analysis and Crisis Response — *Proceedings of the 4th International Conference on Risk Analysis and Crisis Response* (Istanbul, Turkey, August 27–29, 2013), Eds. Chongfu Huang, Cengiz Kahraman, Boca Raton, USA: CRC Press, 831–837.

Xu, Feiqiong 1998. Disaster loss evaluation and structure of the system [J]. *Journal of catastrophology*, 1998, 13(3): 80–83.

Zeng, F.L., Huang, C.F., Ai, F.L. 2013. Operation mechanism and interfered information's screening strategy of internet of intelligence, Intelligent Systems and Decision Making for Risk Analysis and Crisis Response *Proceedings of the 4th International Conference on Risk Analysis and Crisis Response* (Istanbul, Turkey, August 27–29, 2013), Eds. Chongfu Huang, Cengiz Kahraman, Boca Raton, USA: CRC Press, 313–318.

Zhang, Fei 2013. An Economic Loss Estimation on Agricultural Natural Disaster Based on Input-Output Model [D]. Central South University of Forestry and Technology, 2013.

Zheng, Dawei 2015. Basic Catastrophology [M]. Beijing, Peking University Press, 2015.

Risk Analysis and Management – Trends, Challenges and Emerging Issues – Bernatik, Huang & Salvi (Eds)

ISBN 978-1-138-03359-7

Parameter sensitivity analysis of a seawater intrusion model

Xiankui Zeng, Dong Wang, Jichun Wu & Xiaobin Zhu

Key Laboratory of Surficial Geochemistry, Ministry of Education, Department of Hydrosciences, School of Earth Sciences and Engineering, State Key Laboratory of Pollution Control and Resource Reuse, Nanjing University, Nanjing, China

ABSTRACT: Seawater intrusion is a natural phenomenon caused by the difference of densities between seawater and freshwater, which is influenced by many factors from ground surface to underground and from groundwater to seawater, for example, precipitation, evapotranspiration, and pumping. In general, for the numerical simulation of seawater intrusion, some model parameters are always specified by personal experiences of model users or reference value in the literature, such as the ratio of hydraulic conductivity along columns to rows and the ratio of horizontal to vertical hydraulic conductivity. In addition, the parameters representing the groundwater stage and seawater level are critical for the process of seawater intrusion. In this paper, on the basis of a seawater intrusion program, SEAWAT4, the sensitivities of model parameters of an artificial seawater intrusion model are analyzed by a stepwise regression analysis. The most sensitive variables of the seawater intrusion model are seawater level and the boundary head of the groundwater model. In addition, aquifer's hydraulic conductivity and the ratio of horizontal to vertical hydraulic conductivity also have an important impact on seawater intrusion.

Keywords: Seawater intrusion, SEAWAT, Model, Sensitivity analysis

1 INTRODUCTION

Seawater intrusion is a natural phenomenon caused by the difference of densities between seawater and freshwater, which is influenced by many factors, such as regional hydrogeological conditions, precipitation, evapotranspiration, runoff conditions, storm surge, and climate change (Shi and Jiao, 2014). Studies of seawater intrusion in coastal areas are always focused on determining saltwater distribution, transport of dissolved salts, movement of fluid density front, groundwater–seawater interaction, climate change effects, and numerical modeling and prediction (Tam et al., 2014).

The seawater intrusion process is complex, and is influenced by many factors from ground surface to underground and from groundwater to seawater that can interact with each other (Zeng et al., 2014). Sensitivity analysis has been frequently used to identify the complicated input–output relationship in groundwater models. The factors affecting output variable can be identified by sensitivity analysis. Recognizing the characteristics of model parameters will help improve the model structure and provide feedback for data-collecting activities relating to model prediction and uncertainty analysis (Wang et al., 2009; Zeng et al., 2012).

In general, the sensitivity analysis methods can be divided into two broad categories: local sensitivity analysis and global sensitivity analysis. The important uncertainty variables that affect the output of the seawater intrusion model can be identified by the global sensitivity analysis. The most important parameters affecting seawater intrusion modeling or groundwater solute transport modeling are hydraulic conductivity, longitudinal dispersivity, and some boundary conditions (e.g., boundary head and recharge rate) (Nassar and Ginn, 2014). Considering the influences of human activities and seawater level rising (climate change), sensitivities of the groundwater stage and the seawater level are analyzed in this study.

The rest of this paper is organized as follows. The methods used for sensitivity analysis will be described in the second section. Next, an artificial groundwater model is described briefly. The important sensitivity factors of seawater intrusion model are presented in the Results and Discussion section. Finally, the main conclusions drawn from the analysis are presented in last section.

2 METHODS

Stepwise regression analysis is a common approach for global sensitivity analysis. The basic idea for regression analysis is to fit the input and output variables with a linear regression model (Mishra et al., 2009; Pappenberger et al., 2008). The model generated at every step is tested to ensure that all the regression variables are important to the model. The *t*-test is applied to test the importance of a variable. If some variables are found to be insignificant, then the most insignificant variable is removed from the model. The stepwise regression process will continue until each variable in the regression model becomes significant and the variables outside of the model are insignificant (Mishra et al., 2009). After that, the uncertainty importance of the input variable can be defined as Standardized Regression Coefficient (*SRC*):

$$SRC = b_j \sigma(x_j) / \sigma(y) \tag{1}$$

where $\sigma(x_j)$ is the standard deviation of x_j, $\sigma(y)$ is the standard deviation of y, and b_j is the coefficient of x_j in the regression model.

3 MODEL DESCRIPTION

3.1 *Artificial seawater intrusion model*

For the purpose of sensitivity analysis, we constructed an artificial three-dimensional steady-state groundwater flow model (Figure 1). The model domain is 5000 m in the x direction, 3000 m in the y direction, and 53 m in the z direction (thickness).

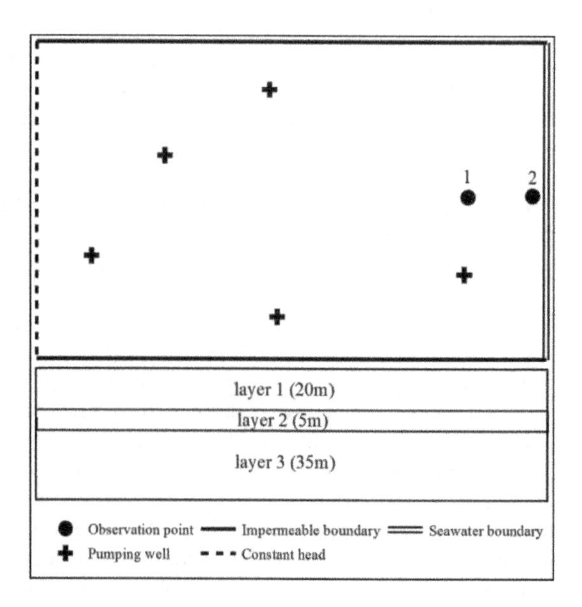

Figure 1. Schematic diagram of the seawater intrusion model.

The model parameters used for sensitivity analysis include the hydraulic conductivity of layer 1 (K1), longitudinal dispersivity (D), the ratio of hydraulic conductivity along columns to hydraulic conductivity along rows (R1), the ratio of horizontal to vertical hydraulic conductivity (R2), the ratios of horizontal transverse dispersivity or vertical transverse dispersivity to the longitudinal dispersivity (R3, R4), the head of west boundary (H), and the seawater level of east boundary (L), which are numbered from 1 to 8, respectively.

3.2 *Implementation of methods*

The numerical model of seawater intrusion was built using SEAWAT4 (Langevin et al., 2007), involving the following procedures:

1. Constructing the model.
2. Setting the boundary conditions, including precipitation rate and pumping rate.
3. Setting the model parameters, including hydraulic conductivity, dispersivity, groundwater head of west boundary, and seawater level at east boundary, whose values are set by sampling a value uniformly from the corresponding range.
4. Running the established model and collecting model outputs.
5. Repeating steps 3 and 4 for 500 times.
6. Conducting sensitivity analysis for model output and model parameters by a stepwise regression analysis and a mutual entropy analysis.

4 RESULTS AND DISCUSSION

Figure 2 shows the results of stepwise regression analysis. The output variable is the groundwater concentration at observation 1, and the input variables are eight model parameters. It is obvious to find that variable L (No. 8) obtains the largest coefficient, and variables K1, R1, and R3 (No. 1, 3, and 5, respectively) are excluded from stepwise regression. Moreover, the largest regression coefficient (L) is 0.0015, followed by R4 and H (0.001 and 0.0008, respectively). According to the basic consensus from groundwater modeling studies, the parameters representing the groundwater head and seawater level are the most important factors affecting groundwater flow and solute transport, and the hydraulic conductance and dispersivity also have important influence. However, parameter K1 could not be identified by the stepwise regression model.

Figure 3 plots the results of stepwise regression analysis, and the output variable is the groundwater concentration at observation 2. It is easy to find that parameters R2, R3, and

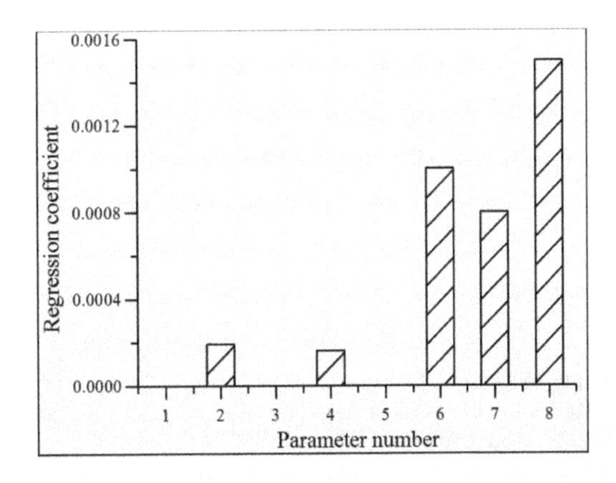

Figure 2. Regression coefficients of model parameters and the output variable is the groundwater concentration at observation 1.

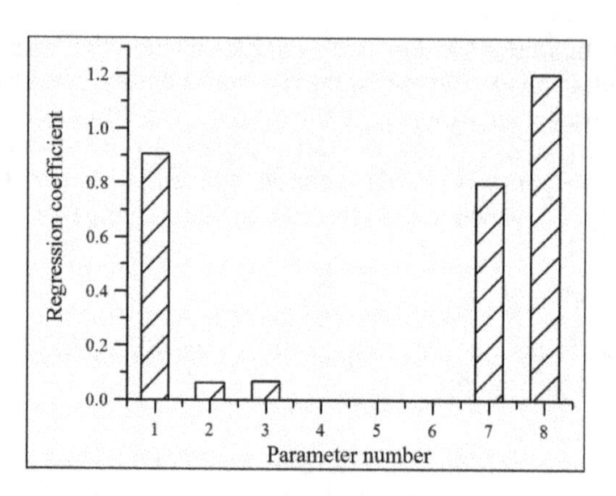

Figure 3. Regression coefficients of model parameters, and the output variable is the groundwater concentration at observation 2.

R4 (No. 4, 5, and 6, respectively) are excluded from the stepwise regression model, and parameter L (No. 8) has the largest regression coefficient. In addition, the coefficients of L, K1, H, D, and R1 are 1.2, 0.9015, 0.8, 0.06179, and 0.06818, respectively.

On the basis of the results above, the parameters representing the groundwater boundary head and the seawater level have the largest sensitivities to groundwater concentration. Thus, the seawater intrusion process is mainly influenced by the groundwater head and seawater head. This result means that the solute transport from seawater to groundwater is controlled by the difference between the heads of groundwater and seawater.

5 CONCLUSION

In this paper, using SEAWAT4, the sensitivities of an artificial seawater intrusion model are analyzed by stepwise regression analysis. We can find that the most sensitive variables of the seawater intrusion model are the seawater level of east boundary (L) and the groundwater head of west boundary (H). Furthermore, the hydraulic conductivity of the aquifer (K1) and the ratio of horizontal to vertical hydraulic conductivity (R2) also have significant sensitivities.

ACKNOWLEDGMENTS

This study was supported by the National Natural Science Fund of China (Nos. 41302181, 41172207, 41030746, 51190091, and 41071018) and the Program for New Century Excellent Talents in University (NCET-12-0262).

REFERENCES

Langevin, C.D., Thorne, D.T., Jr., D., et al., 2007. SEAWAT Version 4: A Computer Program for Simulation of Multi-Species Solute and Heat Transport: U.S. Geological Survey Techniques and Methods Book 6, Chapter A22.

Mishra, S., Deeds, N., Ruskauff, G., 2009. Global Sensitivity Analysis Techniques for Probabilistic Ground Water Modeling. Ground Water, 47(5): 730–747.

Nassar, M.K., Ginn, T.R., 2014. Impact of numerical artifact of the forward model in the inverse solution of density-dependent flow problem. Water Resour Res, 50(8): 6322–6338.

Pappenberger, F., Beven, K.J., Ratto, M., et al., 2008. Multi-method global sensitivity analysis of flood inundation models. Adv Water Resour, 31(1): 1–14.
Shi, L., Jiao, J.J., 2014. Seawater intrusion and coastal aquifer management in China: a review. Environ Earth Sci, 72(8): 2811–2819.
Tam, V.T., Batelaan, O., Le, T.T., et al., 2014. Three-dimensional hydrostratigraphical modelling to support evaluation of recharge and saltwater intrusion in a coastal groundwater system in Vietnam. Hydrogeology Journal, 22(8): 1749–1762.
Wang, D., Singh, V.P., Zhu, Y.S., et al., 2009. Stochastic observation error and uncertainty in water quality evaluation. Adv Water Resour, 32(10): 1526–1534.
Zeng, X.K., Wang, D., Wu, J.C., 2012. Sensitivity analysis of the probability distribution of groundwater level series based on information entropy. Stoch Env Res Risk A, 26(3): 345–356.
Zeng, X.K., Wu, J.C., Wang, D., Zhu, X.B., 2014. The characteristics of probability distribution of groundwater model output based on sensitivity analysis. [J] Hydroinform, 16(1): 130–143.

Risk Analysis and Management – Trends, Challenges and Emerging Issues – Bernatik, Huang & Salvi (Eds)
 ISBN 978-1-138-03359-7

Heavy rain hazard risk assessment and analysis in Mount Huangshan scenic area, China

Junxiang Zhang, Shanfeng Hu & Hongbing Zhu
School of Tourism, Huangshan University, Huangshan, China

Qinghua Gong
Guangzhou Institute of Geography, Guangzhou, China
Guangdong Open Laboratory of Geo-spatial Information Technology and Application, Guangzhou, China

ABSTRACT: On the basis of the data of heavy rains in Mount Huangshan scenic area during 1961–2010, this paper aims to quantify the temporal characteristics and evaluate the hazard risk of heavy rains by using the fuzzy information diffusion technique in Huangshan scenic area. The analysis results are depicted by exceeding probability curves. This study can be applied to the heavy rain disaster prevention planning of Mount Huangshan scenic area or provide follow-up research and disaster insurance and other related planning in the future.

Keywords: Heavy rain, risk assessment, information diffusion technique, exceeding probability, Huangshan scenic area

1 INTRODUCTION

At the dawn of the 21st century, the tourism industry has been rocked by a series of events that forced it to completely change its perspective on crises and disasters (Alexandros Paraskevas). The 9/11 incident; SARS; tourism-targeted terrorist attacks in Bali, Jakarta, Amman, and Sharm-el-Sheik; the Southeast Asia tsunami; and the hurricanes Katrina, Rita, and Wilma in the United States are only some of the recent events, which apart from their horrific death and destruction toll, have had a dramatic impact on international travel and tourism. Mountains are among the most popular areas for outdoor recreation and tourism in China. However, there is growing scientific evidence that many mountain tourism destinations have become increasingly disaster-prone in recent decades because of increased vulnerability and climatic extremes. However, there is usually no consideration of disaster characteristics of mountainous scenic areas by the existing national and local disaster prevention and mitigation and emergency mechanism. Therefore, many mountainous scenic areas are almost undefended in the face of natural disasters. At the same time, studies conducted on the natural disasters in scenic areas worldwide are often focused on the disaster responses and recovery (Richie, B. 2008). There are few established policies and plans for disaster risk reduction in mountainous scenic areas.

Mount Huangshan in Anhui Province, East China, is one of the country's 10 best-known scenic spots. It is characterized by the four wonders: odd-shaped pines, grotesque rock formation, seas of clouds, and crystal-clear hot springs. The Huangshan scenic area is frequently hit by heavy rains, which have tangible and intangible impacts on the tourism industry. Because of the "risk" characteristics of natural disasters, such as uncertainty and loss, the concept of risk is mostly used worldwide to analyze the occurrence of natural disasters and achieve a comprehensive disaster prevention. The concept of risk was applied to the study of heavy rain hazard in this paper, and the fuzzy information diffusion technique (Chongfu H., Yong S. 2002) was used to evaluate the hazard risk of heavy rains in the Huangshan scenic area.

2 STUDY AREA

Huangshan is a mountain range in southern Anhui Province in East China (see Figure 1). It is a UNESCO World Heritage Site and is classified as an AAAAA scenic area by the China National Tourism Administration. The Huangshan mountain range has many peaks, some more than 1,000 m (3,250 ft) high. The World Heritage Site covers a core area of 154 km² and a buffer zone of 142 km². Having at least 140 sections open to visitors, Huangshan is a major tourist destination in China. In 1985, Huangshan was selected as one of the top 10 natural and cultural attractions of China. The mountain is increasingly targeted by tourists, and foreigners are being encouraged to visit it. There were only 460,000 visitors in 1985, which rose between 1995 and 2015 from 831,000 to 3.182 million. Tourism has been one of the sources of income of the local people. However, rise in a variety of natural disasters, such as heavy rain, torrential flood, and slope hazards, with a warming climate put the future of the tourism in Huangshan scenic area at some risk, which is the subject of this study.

3 ANALYSIS OF THE CHARACTERISTICS OF HISTORICAL HEAVY RAINS DISASTER IN HUANGSHAN SCENIC AREA

The meteorological data used in this paper are daily precipitation from 1961 to 2010 in Mount Huangshan meteorological station. The number of heavy rain days between 1961 and 2010

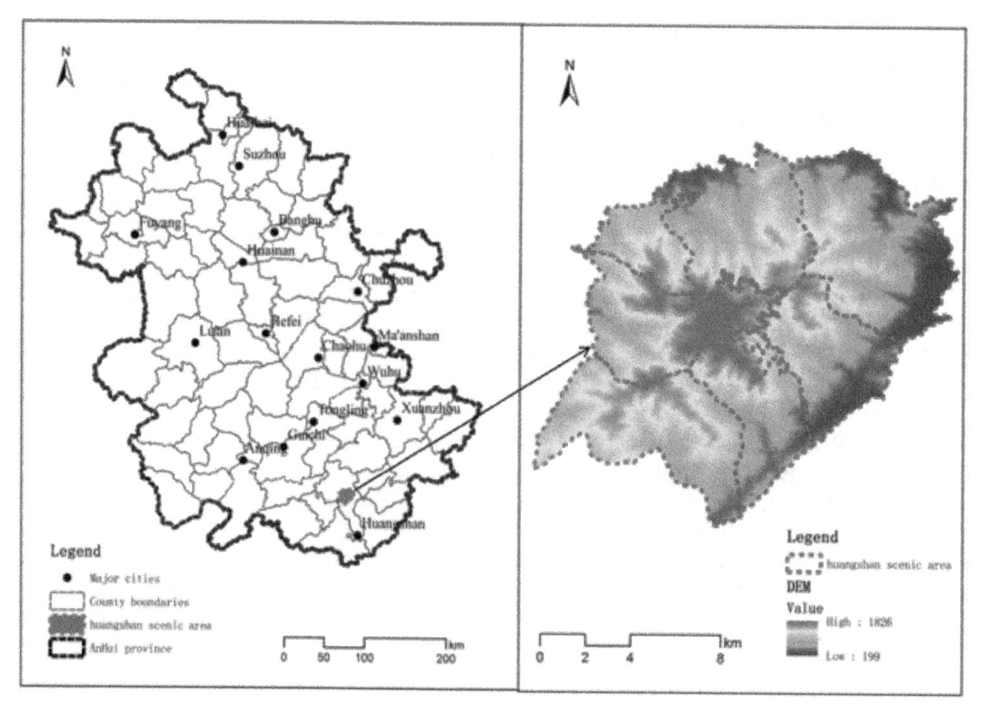

Figure 1. Study area site.

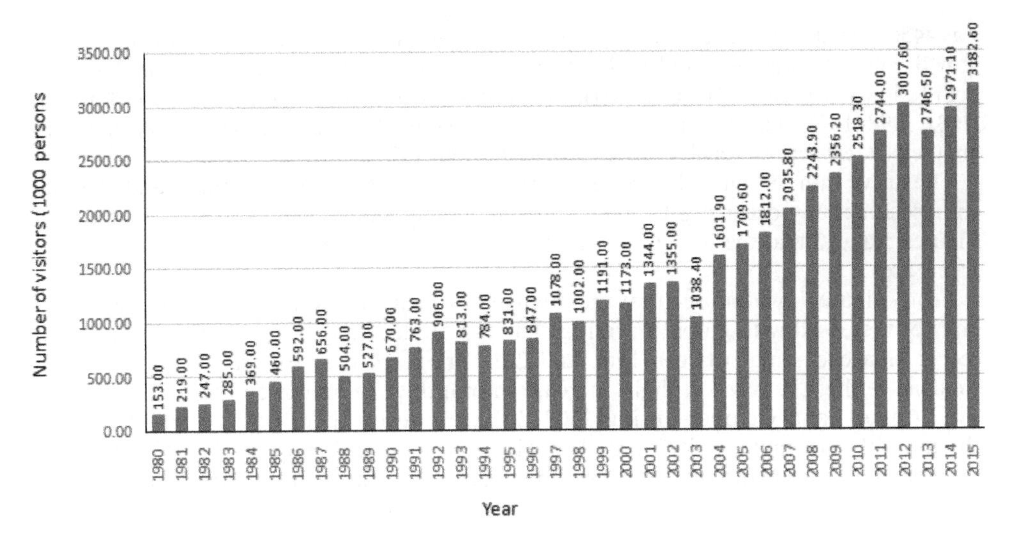

Figure 2. Numbers of visitors to Huangshan scenic area during 2003–2015.

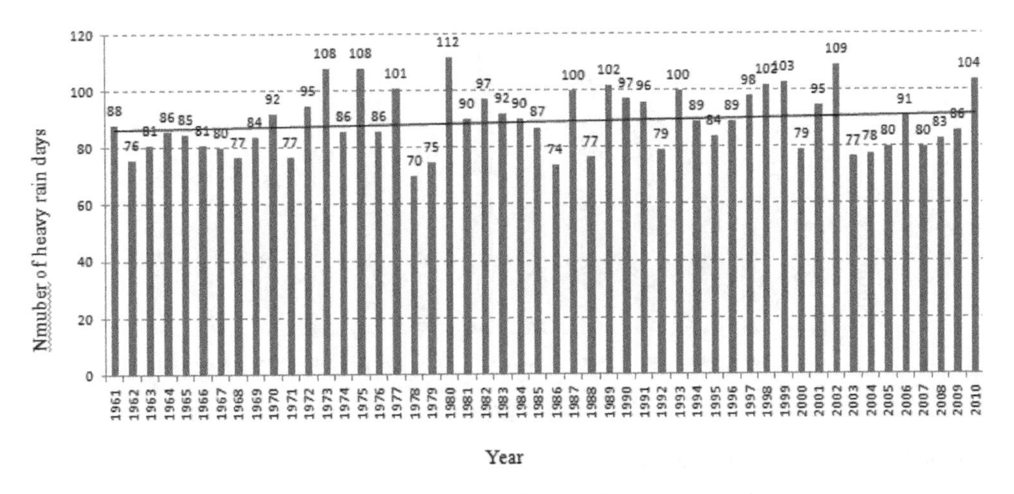

Figure 3. Number of heavy rain days during 1961–2010 in Huangshan scenic area.

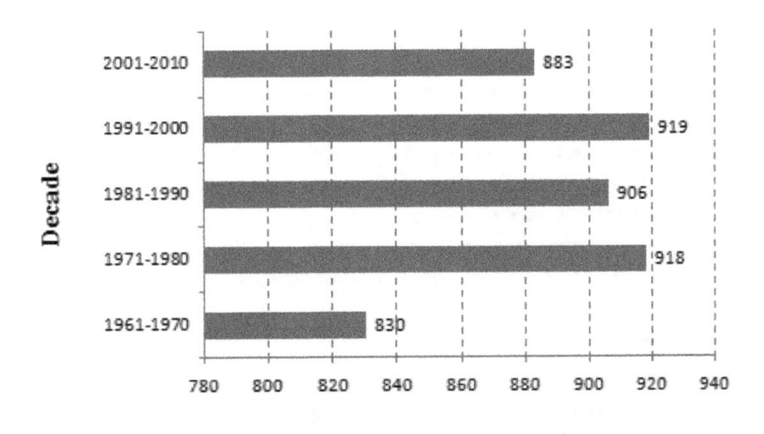

Figure 4. Number of heavy rain days in each decade during 1961–2010.

was 4456, giving an average of around 89 per year. The number of heavy rain days has a weak growth trend (see Figures 3 and 4). The top 10 years with most frequent heavy rain days from high to low are 1980 (112), 2002 (109), 1973 (108), 1975 (108), 2010 (104), 1999 (103), 1989 (103), 1998 (102), and 1977 (101). It is interesting to note that most frequent heavy rain years are the last few years of each decade except 1973 and 1975.

The heavy rain season in Huangshan Scenic area spans between February and October, with peak activity occurring between April and August (see Figure 5). We can see from Figures 5 and 6 that the heavy rain season is also the tourism-peak season in Huangshan scenic area.

From Figure 7, we can see that the intensity of heavy rains in Huangshan scenic area has not increased with global warming. The top 10 deadliest heavy rains in Huangshan scenic area occurring between 1961 and 2010 are marked in Figure 7. By comparing, we can see that there was a slight increase in monthly maximum daily rainfall in April, July, August, and October during 2003–2010. However, there was a slight decline in the monthly maximum daily rainfall in March and June during 2003–2010 and a significant decrease in the monthly maximum daily rainfall in May and September (see Figures 8–15).

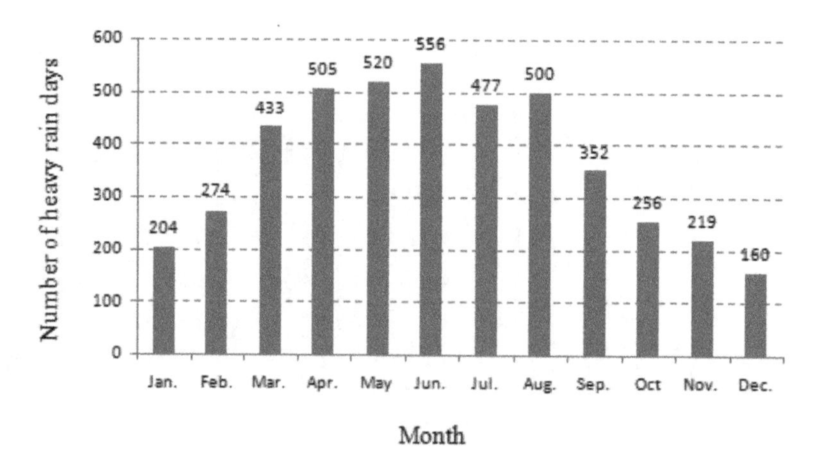

Figure 5. Number of heavy rain days in each month during 1961–2010 in Huangshan scenic area.

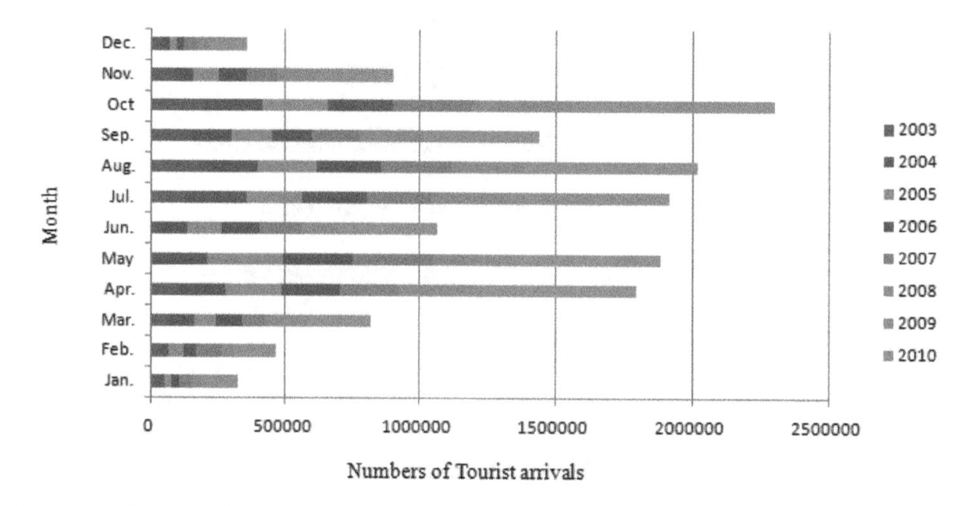

Figure 6. Numbers of tourist arrivals in Huangshan scenic area during 2003–2010.

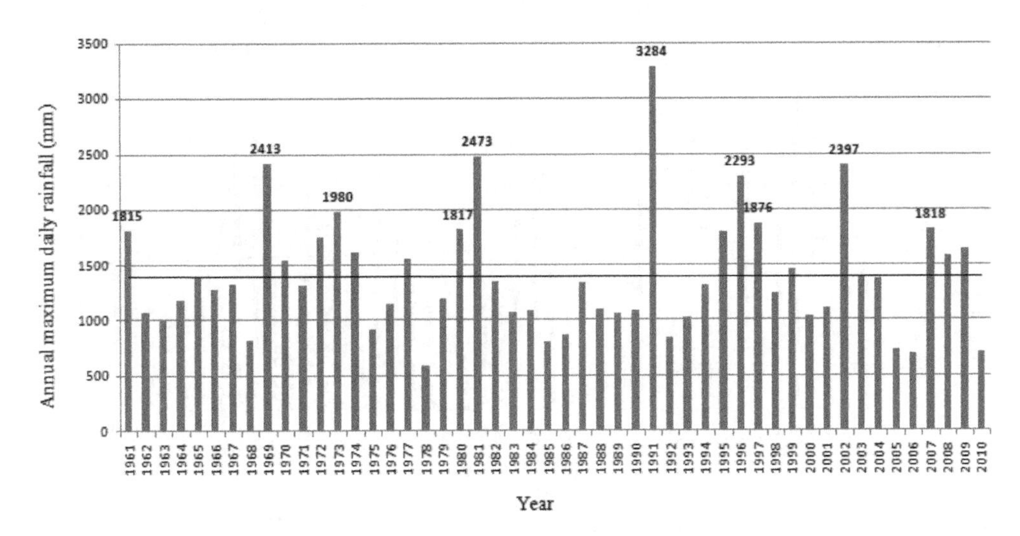

Figure 7. Trend of annual maximum daily rainfall in Huangshan scenic area during 2003–2010.

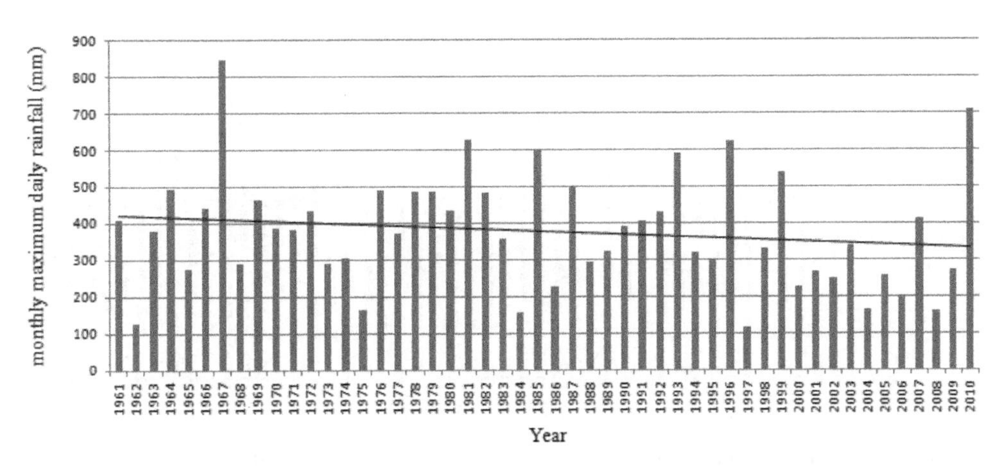

Figure 8. Trend of monthly maximum daily rainfall in March during 2003–2010.

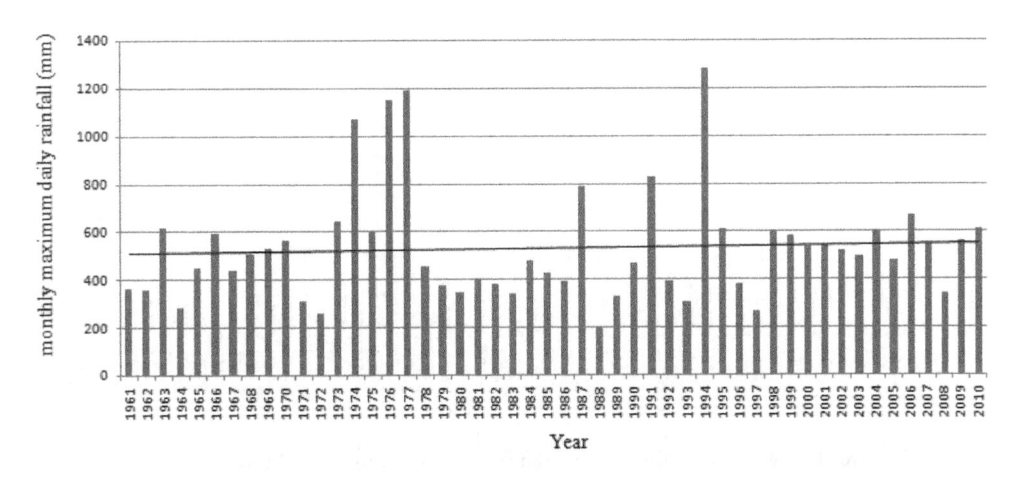

Figure 9. Trend of monthly maximum daily rainfall in April during 2003–2010.

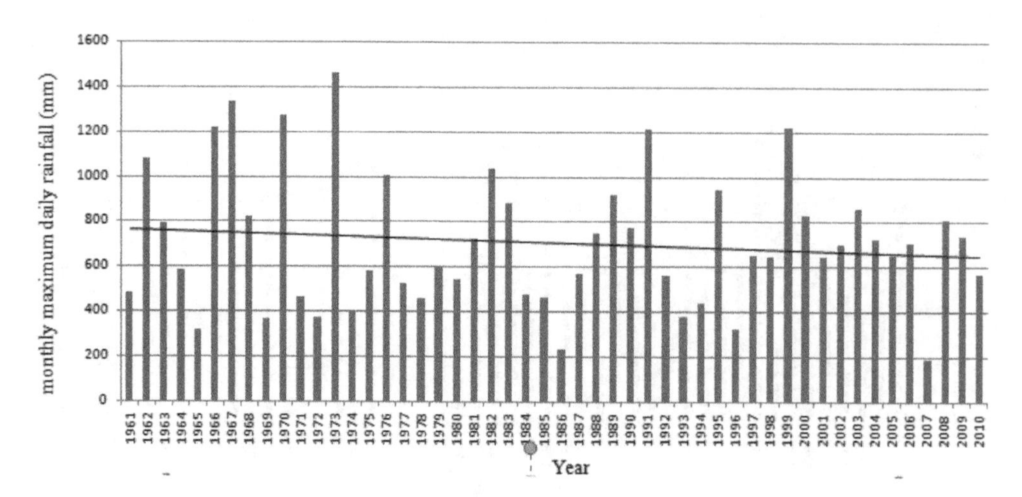

Figure 10. Trend of monthly maximum daily rainfall in May during 2003–2010.

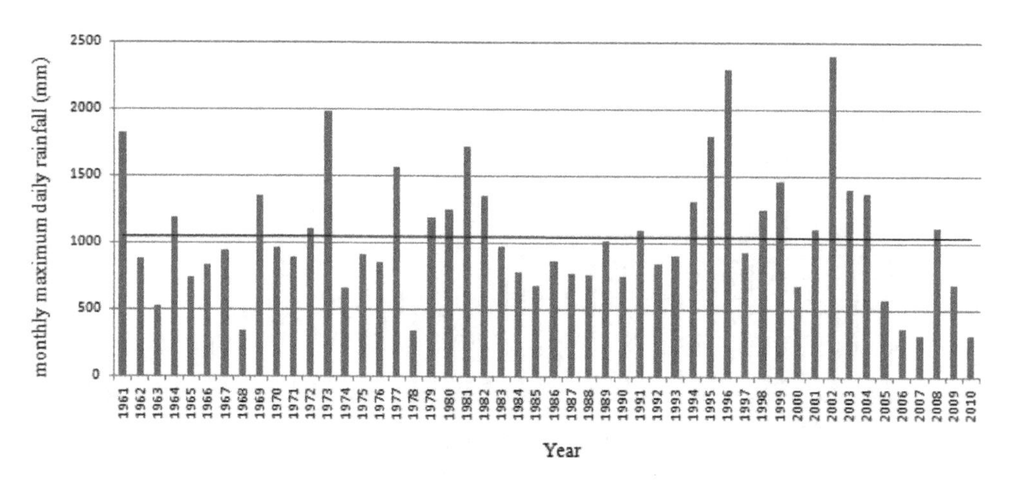

Figure 11. Trend of monthly maximum daily rainfall in June during 2003–2010.

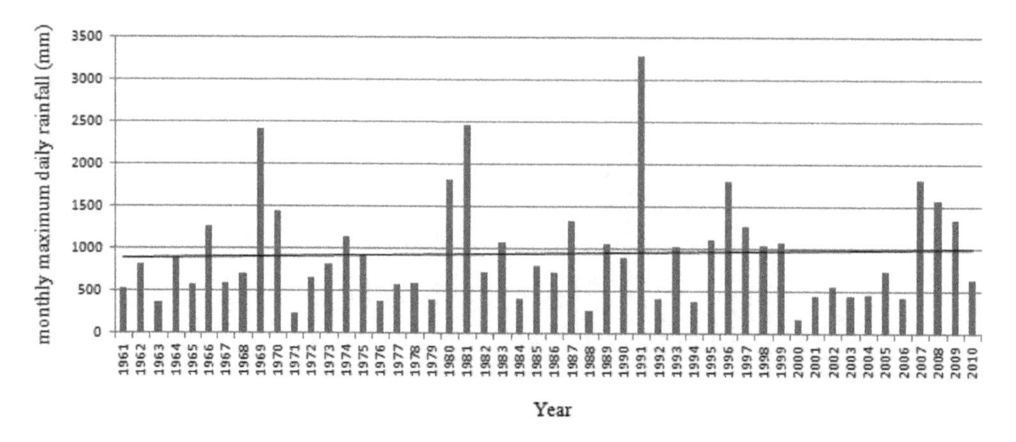

Figure 12. Trend of monthly maximum daily rainfall in July during 2003–2010.

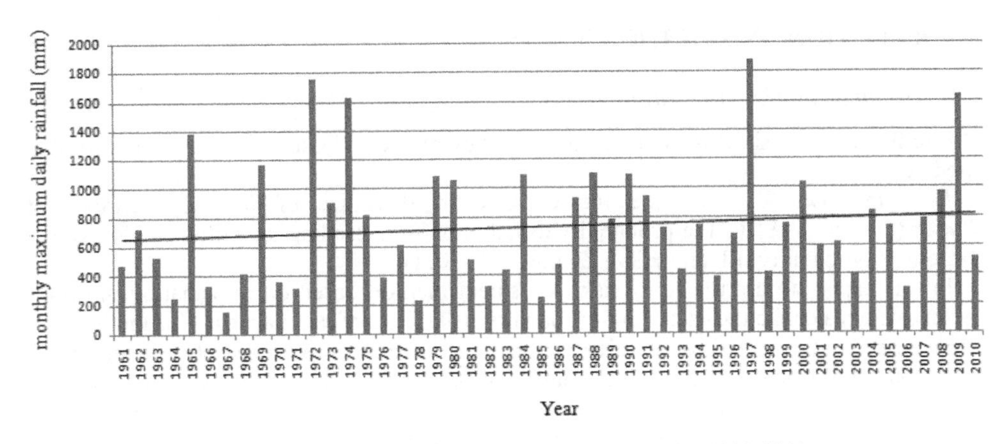

Figure 13. Trend of monthly maximum daily rainfall in August during 2003–2010.

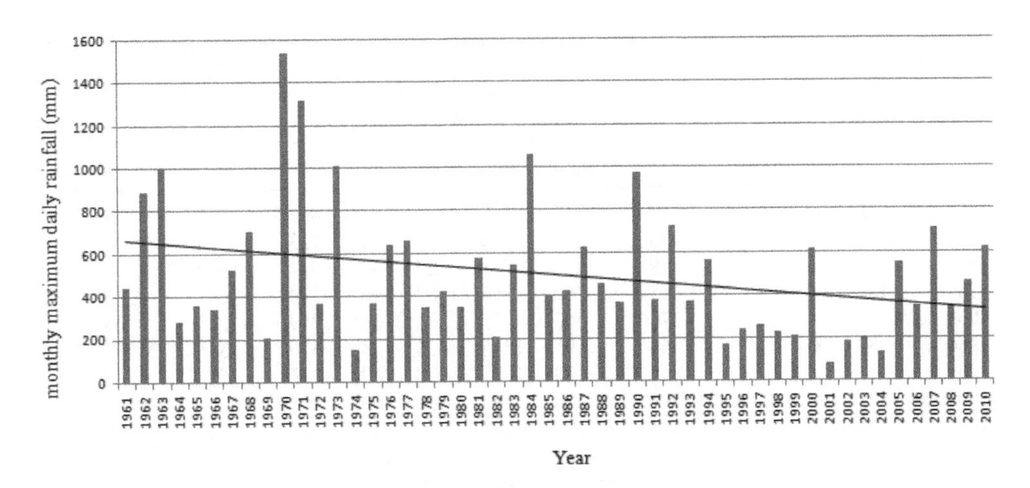

Figure 14. Trend of monthly maximum daily rainfall in September during 2003–2010.

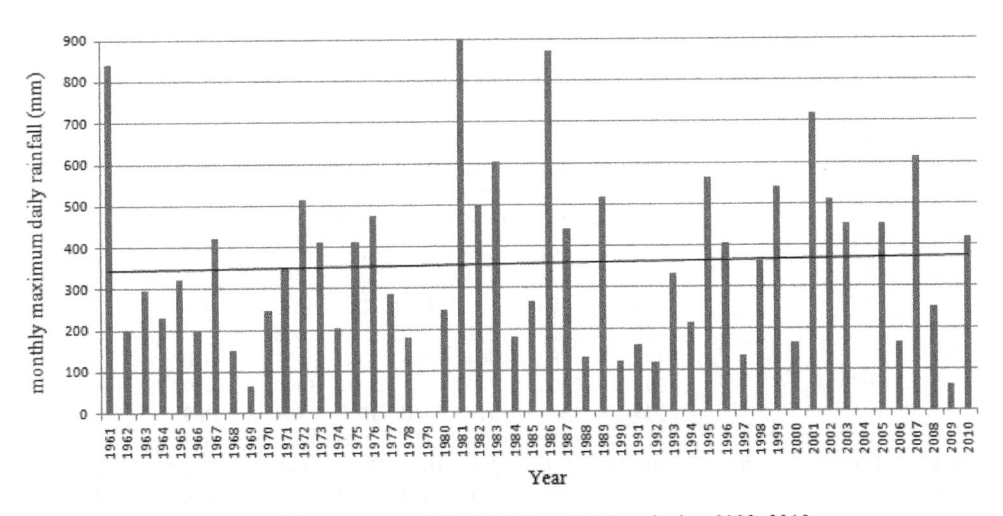

Figure 15. Trend of monthly maximum daily rainfall in October during 2003–2010.

4 DEFINITION AND METHODOLOGY

4.1 *Definition of heavy rain hazard risk*

Before further analysis, it is essential to define how the term "risk" of heavy rain hazard is used here.

Definition. Let $X = \{x\}$ be the universe of discourse of natural hazards and $P = \{p\ (\xi \geq x) | x \in X\}$ be the probability distribution of exceeding magnitude x. P is called exceeding-probability risk (Junxiang Zhang et al., 2011).

In this paper, we use the fuzzy risk assessment method based on information diffusion technique to calculate the exceeding-probability risk of heavy rain hazard. This method has been applied to the analysis of a variety of hazard risks, such as earthquake (Junxiang Zhang et al. 2009) and tropical cyclone (Junxiang Zhang et al. 2007). The main feature of information diffusion technology is to change a traditional sample point into a fuzzy set to partially fill the gaps caused by incompleteness of data (Huang 2002). For a detailed description

Table 1. Statistical of annual maximum daily rainfall and monthly maximum daily rainfall during 2003–2010 in Huangshan scenic area.

Year	Annual maximum daily rainfall (mm)	Monthly maximum daily rainfall (mm)							
		March	April	May	June	July	August	September	October
1961	1815	410	366	481	1815	529	473	443	841
1962	1078	126	360	1078	879	821	723	890	201
1963	1009	379	615	789	523	379	524	1009	297
1964	1187	495	287	586	1187	906	244	287	231
1965	1383	276	453	315	738	585	1383	363	322
1966	1282	444	595	1215	833	1282	330	345	201
1967	1331	848	438	1331	937	600	146	527	421
1968	822	290	508	822	339	707	414	708	151
1969	2413	464	533	367	1350	2413	1163	207	66
1970	1544	386	565	1270	962	1444	360	1544	247
1971	1316	382	316	465	886	241	307	1316	348
1972	1754	434	262	372	1107	669	1754	370	515
1973	1980	290	648	1458	1980	830	904	1010	413
1974	1620	304	1071	396	655	1142	1620	151	205
1975	917	162	598	577	911	917	814	366	411
1976	1151	492	1151	1003	851	382	385	643	475
1977	1560	373	1187	522	1560	586	606	661	285
1978	592	486	456	459	337	592	229	351	180
1979	1189	488	375	596	1189	398	1077	427	-
1980	1817	436	349	540	1244	1817	1051	347	247
1981	2473	630	405	724	1719	2473	504	579	910
1982	1350	483	383	1035	1350	732	316	211	498
1983	1073	358	341	880	972	1073	432	550	605
1984	1085	156	478	473	778	419	1085	1063	180
1985	811	599	426	461	672	811	248	400	266
1986	872	226	395	231	857	718	468	426	872
1987	1337	499	787	564	773	1337	927	630	442
1988	1097	292	206	749	761	274	1097	457	131
1989	1067	325	334	919	1010	1067	783	369	519
1990	1087	391	471	772	750	900	1087	977	122
1991	3284	405	830	1209	1090	3284	940	382	161
1992	839	433	396	561	839	424	725	726	117
1993	1030	590	311	376	902	1030	435	372	334

(*Continued*)

Table 1. (*Continued*)

Year	Annual maximum daily rainfall (mm)	Monthly maximum daily rainfall (mm)							
		March	April	May	June	July	August	September	October
1995	1794	303	613	940	1794	1117	386	171	563
1996	2293	623	385	321	2293	1812	680	241	405
1997	1876	116	269	650	929	1273	1876	260	133
1998	1247	332	607	647	1247	1043	415	230	361
1999	1459	539	582	1216	1459	1087	753	207	542
2000	1032	228	538	828	678	173	1032	617	165
2001	1107	266	544	642	1107	449	599	79	720
2002	2397	249	522	699	2397	567	616	182	511
2003	1398	340	497	855	1398	448	403	204	450
2004	1371	165	604	726	1371	464	834	136	-
2005	734	257	479	653	570	734	731	555	453
2006	703	200	668	703	362	440	297	350	163
2007	1818	412	551	189	311	1818	784	715	614
2008	1582	159	342	809	1110	1582	971	341	249
2009	1638	271	559	737	690	1357	1638	466	63
2010	711	711	613	565	311	641	514	625	420

"-" indicates no heavy rain records.

of the method, the readers are referred to the book, Towards Efficient Fuzzy Information Processing-Using the Principle of Information Diffusion (Buckle 2000).

4.2 *Hazard risk assessment of heavy rain in Huangshan scenic area*

Let the annual maximum daily rainfall be the intensity index of heavy rain. We choose the records of maximum daily rainfall ≥ 50 mm to constitute a sample of heavy rainfall. Let $x_i = (i = 1,2,\ldots, 50)$ be the annual maximum daily rainfall of heavy rains over the years. The data used are listed in the first column of Table 1.

According to the variation range of the annual maximum daily rainfall, let the set [450, 3300] in the space of one dimension be the domain of x_i. Then, the continuous domain [450, 3300] was transformed into the discrete domain by the equal distance.

Taking into account the accuracy of the calculation, 20 control points ($n = 20$) were constructed to form the discrete domain as follows.

$$U = \{u_1, u_2, \ldots, u_{20}\} = \{450, 600, 750, 900, 1050, 1200, 1350, 1500, 1650, 1800, 1950, 2100, 2250, 2400, 2550, 2700, 2850, 3000, 3150, 3300\}$$

Here, the sample number (m) is 50, the maximum value of this sample (b) is 3284, and the minimum value of this sample (a) is 592. We get the diffusion coefficient ($h = 78.057012$) by using formula (6). Then, the estimated risk value of heavy rain in Huangshan scenic area can be drawn by using formulas (9)–(15), as shown in Table 2.

We can draw out the exceeding-probability distribution curve of the annual maximum daily rainfall in Huangshan scenic area by using the data in Table 2, as shown in Figure 16.

Figure 16 shows that the exceeding-probability is 0.1 if the annual maximum daily rainfall is greater than or equal to 2250. In other words, heavy rain with the annual maximum daily rainfall is greater than or equal to 2250 for a 10-year return period. Similarly, heavy rain with the annual maximum daily rainfall is greater than or equal to 2700 for a 50-year return period. According to this curve and the elevation and terrain characteristics of the scenic area, the high-risk area of heavy rain disaster can be determined.

The timescale used in the above calculation is annual. For the natural hazard management of scenic areas, the smaller the timescales of the analysis, the more meaningful the results of the analysis. Next, we will use the same method to calculate the heavy rain hazard risk based on monthly data. Because the tourism-peak season in Huangshan scenic area is from March to October of each year, this paper analyzes the heavy rain hazard risk in these months.

According to the variation range of the monthly maximum daily rainfall shown in Table 1 and the accuracy of calculation, the discrete domains of x_i (i = March, April, May, June, July, August, September, October) are as follows:

$U_{Mar.}$ = {100,150,200,250,300,350,400,450,500,550,600,650,700,750,800,850,900},
$U_{Apr.}$ = {200,250,300,350,400,450,500,550,600,650,700,750,800,850,900,950, 1000,1050,1100,1150,1200,1250,1300},
U_{May} = {150,250,350,450,550,650,750,850,950,1050,1150,1250,1350,1450,1550},
$U_{Jun.}$ = {250,400,550,700,850,1000,1150,1300,1450,1600,1750,1900,2050,2200,2350,2500},
$U_{Jul.}$ = {150,300,450,600,750,900,1050,1200,1350,1500,1650,1800,1950,2100,2250,2400,2550, 2700,2850,3000,3150,3300},

Table 2. Estimated risk values of annual maximum daily rainfall.

Risk level	Exceeding-probability	Risk level	Exceeding-probability
450	1	1950	0.143
600	0.997	2100	0.105
750	0.965	2250	0.099
900	0887	2400	0.082
1050	0.805	2550	0.035
1200	0.647	2700	0.02
1350	0.519	2850	0.02
1500	0.376	3000	0.02
1650	0.298	3150	0.02
1800	0.229	3300	0.016

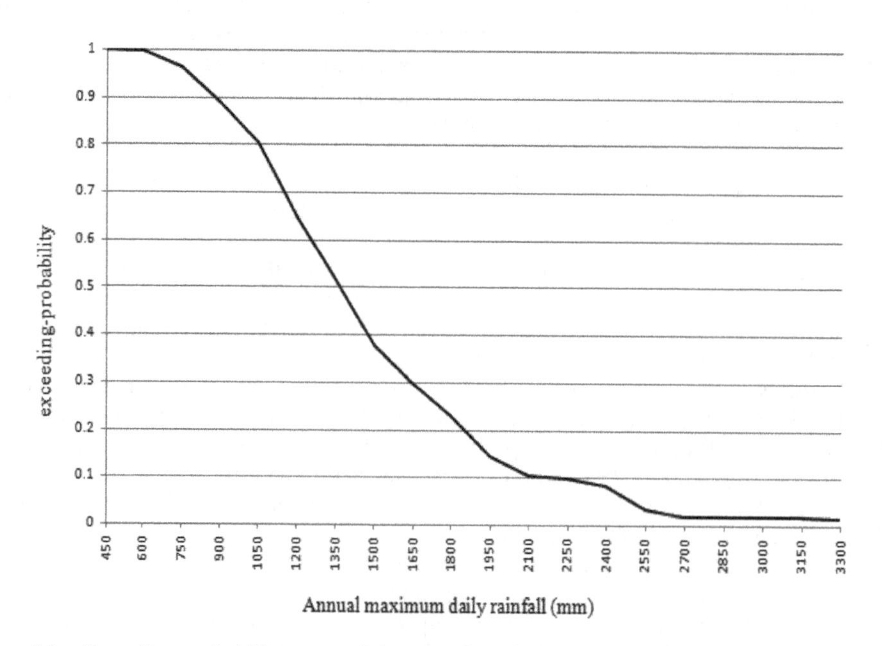

Figure 16. Exceeding-probability curve of the annual maximum daily rainfall.

Table 3. Concerned parameters of monthly hazard risk assessment of heavy rain.

Month	Sample X_{ij}	No. of Sample m	Maximum b	Minimum a	Diffusion coefficient h
Mar.	$X_{Mar.j}$	50	848	116	21.225012
Apr.	$X_{Apr.j}$	50	1281	206	31.170612
May	$X_{May.j}$	50	1458	189	36.795820
Jun.	$X_{Jun.j}$	50	2397	311	78.520947
Jul.	$X_{Jul.j}$	50	3284	173	100.238898
Aug.	$X_{Aug.j}$	50	1876	146	50.162939
Sep.	$X_{Sep.j}$	50	1544	79	42.479020
Oct.	$X_{Oct.j}$	48	910	63	25.604630

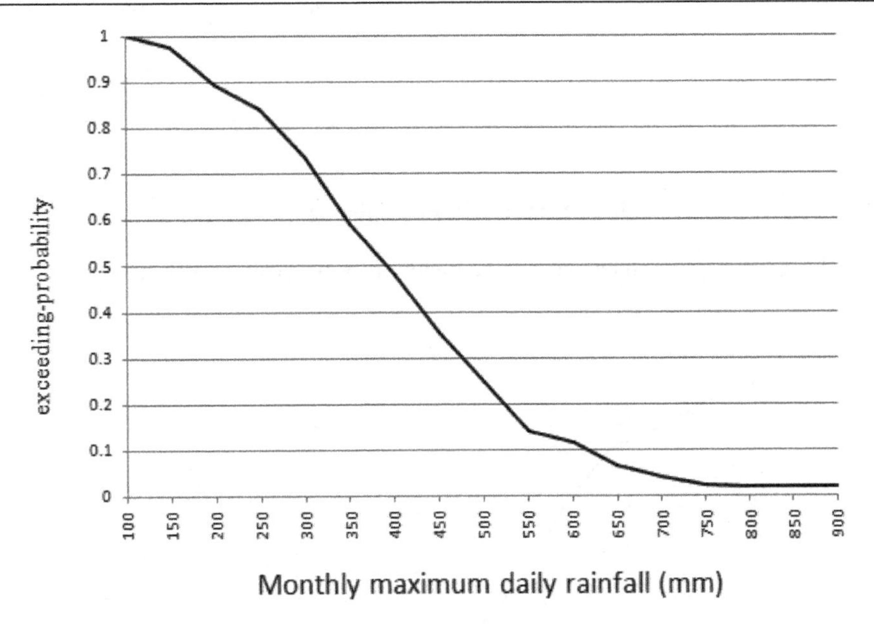

Figure 17. Exceeding-probability curve of the monthly maximum daily rainfall in March.

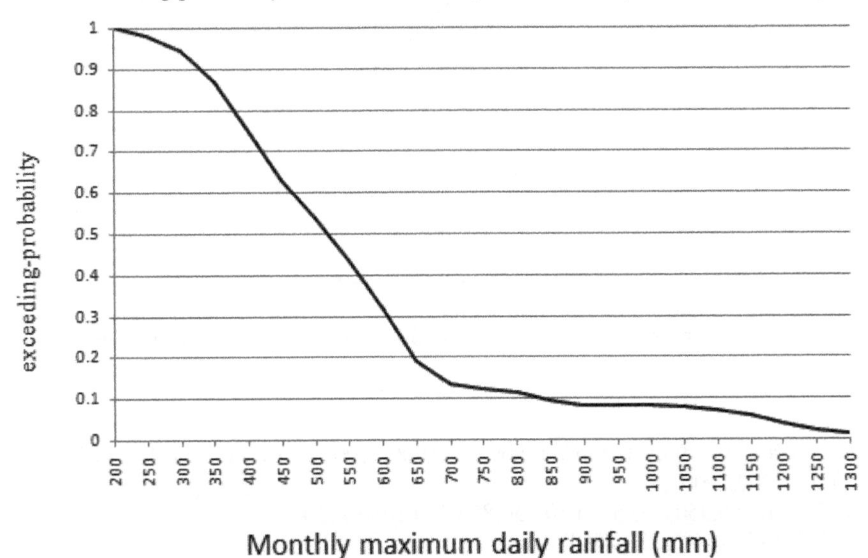

Figure 18. Exceeding-probability curve of the monthly maximum daily rainfall in April.

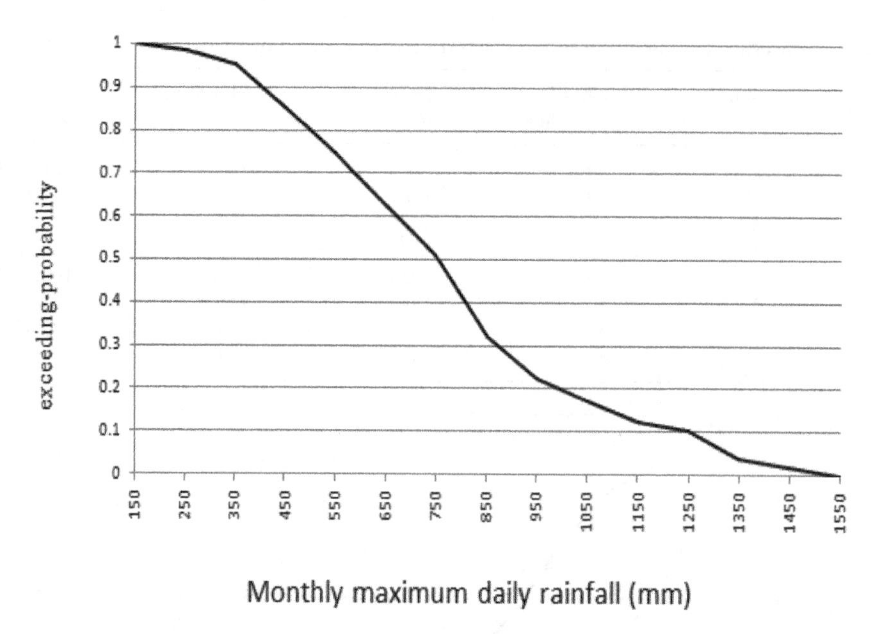

Figure 19. Exceeding-probability curve of the monthly maximum daily rainfall in May.

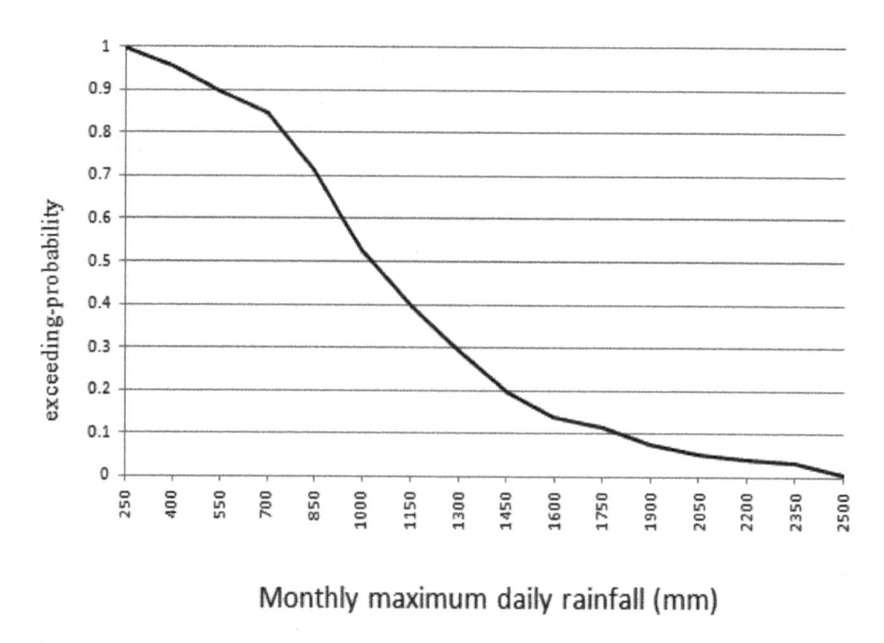

Figure 20. Exceeding-probability curve of the monthly maximum daily rainfall in June.

$$U_{Aug.} = \{100,200,300,400,500,600,700,800,900,1000,1100,1200,1300,1400,1500,1600,1700,1800,1900\},$$
$$U_{Sep.} = \{50,150,250,350,450,550,650,750,850,950,1050,1150,1250,1350,1450,1550,1650\},$$
$$U_{Oct.} = \{50,100,150,200,250,300,350,400,450,500,550,600,650,700,750,800,850,900,950,1000\}.$$

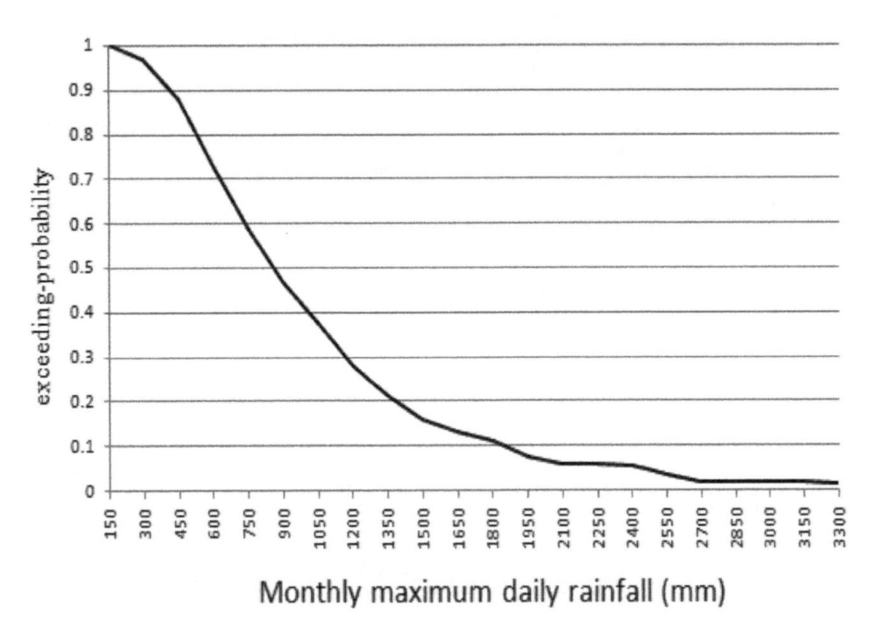

Figure 21. Exceeding-probability curve of the monthly maximum daily rainfall in July.

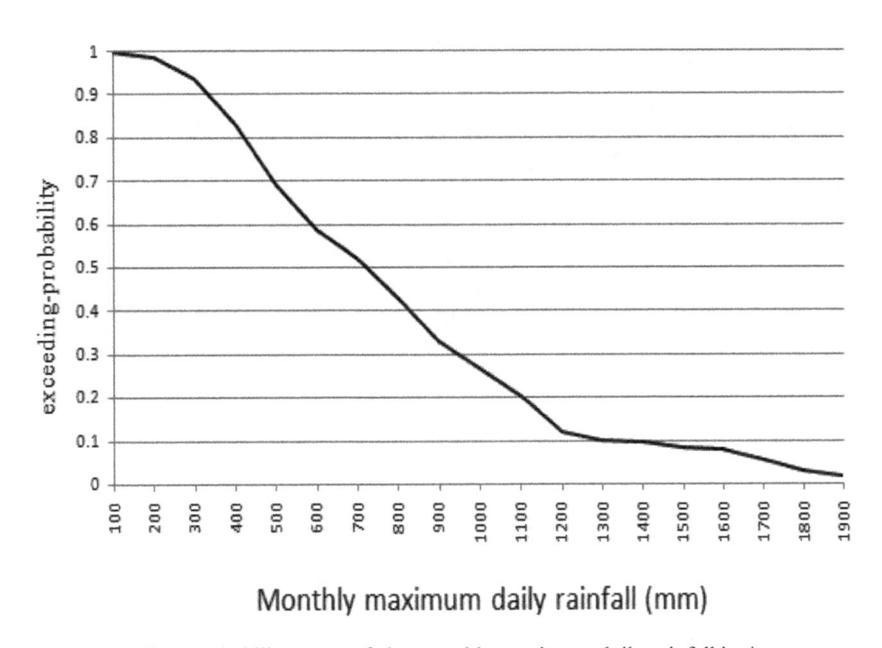

Figure 22. Exceeding-probability curve of the monthly maximum daily rainfall in August.

Table 3 shows the concerned parameters of monthly hazard risk assessment of heavy rain, including the number of samples m, the maximum b, the minimum a, and the diffusion coefficient h.

Then, the estimated risk value of heavy rain from March to October in Huangshan scenic area can be calculated by using formula (9)–(15), respectively. Finally, we can draw out the exceeding-probability distribution curve of the monthly maximum daily rainfall by using the result as shown from Figures 17 to 24. These exceeding probability curves show different shapes, which indicate the difference of heavy rain hazard risks in each month. For example,

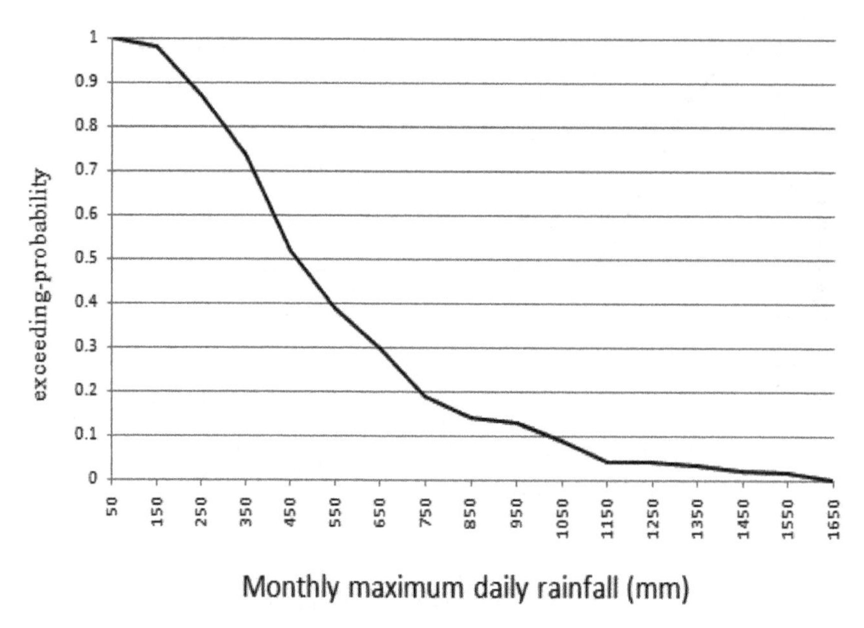

Figure 23. Exceeding-probability curve of the monthly maximum daily rainfall in September.

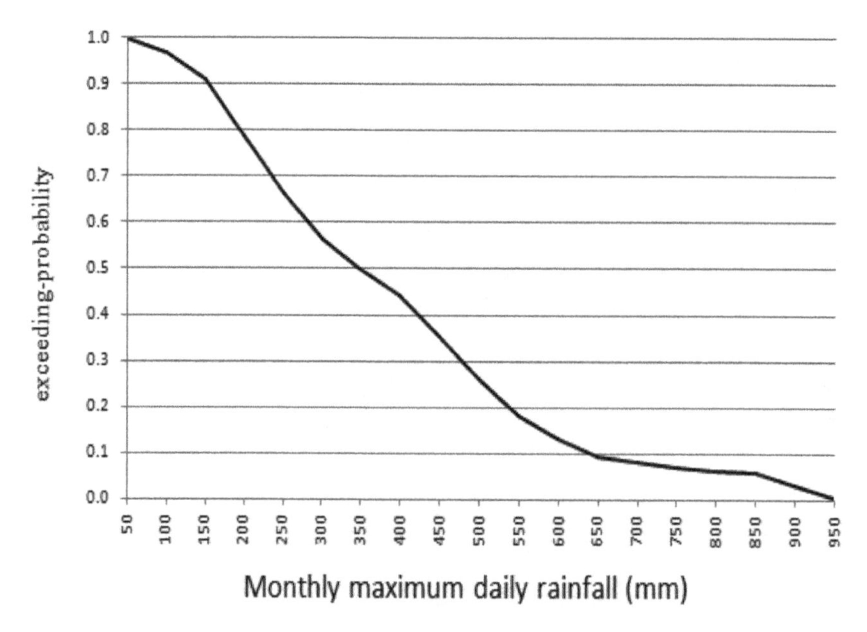

Figure 24. Exceeding-probability curve of the monthly maximum daily rainfall in October.

when the exceeding probability was 0.5, the risk levels of heavy rain from March to October were [350,400], [500,550], [700,750], [1000,1050], [800,850], [650,700], [400,500], [300,350], and [325,375], respectively.

5 CONCLUSIONS AND DISCUSSIONS

On the basis of the annual data of heavy rains in Mount Huangshan scenic area during 1961–2010, this paper analyzes the characteristics of historical heavy rain. The results show that the number of heavy rain days has a weak growth trend, and the intensity of heavy rains in Huangshan scenic area has not increased with global warming. Analysis based on monthly data shows that there are obvious differences in the development trend of heavy rain in different months from March to October. The heavy rain season in Huangshan scenic area runs between February and October, with peak activity occurring between April and August. However, this period is the tourism-peak season in Huangshan scenic area.

In this paper, we evaluate the hazard risks of heavy rains by using the fuzzy information diffusion technique based on the annual data and monthly data of heavy rains in Huangshan scenic area. The analysis results are depicted by exceeding-probability curves. This study can be applied to the heavy rain disaster prevention planning of Huangshan scenic area or provide follow-up research and disaster insurance and other related planning in the future.

The timescale of this study is annual and monthly. For the natural hazard management of scenic areas, the smaller the timescales of the analysis, the more meaningful the results of the analysis. In future works, we will use the daily data to quantitatively analyze the relationship between heavy rain hazard and tourists amount with the aim to provide reference for heavy rain hazard risk management in Huangshan scenic area.

ACKNOWLEDGMENT

The authors acknowledge the financial support by the National Natural Science Foundation of China (No. 41671506) and the 2012 annual Huangshan University scientific research project (No. 2012xkjq005).

REFERENCES

Alexandros, Paraskevas. Introduction of Tourism and Disaster Research Group. http://www.atlas-euro.org/sig_disaster.aspx.

Chongfu, H., Yong, S. 2002. Towards Efficient Fuzzy Information Processing-Using the Principle of Information Diffusion. Physica-Verlag (Springer), Heidelberg, Germany.

Junxiang, Zhang, Chongfu, Huang, Xulong, Liu, Qinghua, Gong 2011. Risk assessment of land-falling trapical cyclones in Guangdong Province, China. Human and Ecological Risk Assessment, 2011, 17(03), pp, 732–744.

Junxiang, Zhang, Chongfu, Huang 2009. Risk Analysis of Earthquake in Sichuan Province. Tropical Geography, 2009, 29(3): 280–284.

Junxiang, Zhang, Pingri, Li, Guangqing, Huang, et al. 2007. Risk Assessment of Storm Surge Disaster in Coastal Area of China Induced by Typhoon Based on Information Diddusion Method. Tropical Geography, 2007, 27(1): 11–14.

Richie, B. 2008. Tourism Disaster Planning and Management: From Response and Recovery to Reduction and Readiness. Current Issues in Tourism, 2008, 11(5): 315–48.

Risk Analysis and Management – Trends, Challenges and Emerging Issues – Bernatik, Huang & Salvi (Eds)
© 2017 Taylor & Francis Group, London, ISBN 978-1-138-03359-7

Path optimization in typhoon scenarios based on an improved ripple-spreading algorithm

Ming-Kong Zhang, Xiao-Bing Hu & Hang Li
Academy of Disaster Reduction and Emergency Management, Beijing Normal University, Beijing, China

Jain-Qin Liao
MiidShare Technology Ltd., Chengdu, China

ABSTRACT: We all know that adverse weather conditions often have a serious impact on our travelling behaviors. In this paper, how to plan the travelling path under typhoon events, in order to make our travelling trajectory such that the typhoon-covered area is avoided in the most cost-efficient way, and therefore to reduce the risk and to improve the safety during the travelling period are the subjects of focus. This problem is actually a Dynamical Path Optimization (DPO) problem, as the typhoon covered area usually keeps changing during the travelling period. Traditional DPO methods need to conduct real-time online path optimization based on the current typhoon-covered area, but they can hardly achieve the optimal actual travelling trajectory under a given typhoon scenario. To perform path optimization in a dynamical routing environment, we have recently proposed the concept of Co-Evolutionary Path Optimization (CEPO) and then reported a Ripple Spreading Algorithm (RSA) to resolve the CEPO. In this study, we aim to apply CEPO and RSA to the path optimization in typhoon scenarios by considering some realistic factors, i.e., there are some different categories of routes and different wind strengths in typhoon-covered areas, and different route categories have different speed limits. In different wind strength areas, the travelling speed is slower than that under normal weather conditions. Therefore, the RSA needs to be modified to allow different ripple spreading speeds in different areas and different routes, in order to find the optimal path. The modified RSA can make the result of path optimization more reliable in typhoon scenarios. This is demonstrated by a case study on two historical typhoon events in the Hainan Province of China.

1 INTRODUCTION

By analyzing the global Emergency Events Database (EM-DTA), which is obtained from the center for research on the epidemiology of disasters of Leuven University, an increasing trend in typhoon frequency can be observed in recent years (Wu, et al, 2014). A typhoon event usually has a severe impact on our daily travelling life, because it can not only significantly reduce the comfortableness of travelling (Weisser, et al, 1997; Lam, et al, 2008), but also seriously jeopardize the safety of travelling (Keay, et al, 2008; Keay, et al, 2006; Maarten and Aalst, 2005). Therefore, it will be very useful to take advantage of the information and technologies of typhoon forecast, in order to optimize the travelling trajectory in typhoon situations. By optimizing the travelling trajectory, we can not only avoid the area affected by the typhoon, but also reduce the exposure time under disaster. In the traditional path optimization, travelling cost may be measured in terms of path length, but under typhoon scenarios, travelling time usually has the highest priority, as travelling time is related to exposure time, and reducing exposure time means lowering risk, which leads to a safer journey. If the travelling plan cannot be cancelled under typhoon scenarios, we need to complete the journey as soon as possible; and so, it is important to optimize the travelling trajectory and reduce the exposure time. Actually, in the field of disaster risk science, minimizing travelling time is usually very important for evacuation, rescue, and emergency management. Therefore, path optimization in adverse weathers has long been studied in the field of disaster reduction and risk management (Krozel, et al, 1999).

Apparently, path planning under typhoon scenarios is called as Dynamic Path Optimization (DPO), because the typhoon keeps moving during the travelling period. When a typhoon comes, it will bring strong winds and heavy rains, which can reduce the safety of travelling. For example, a strong wind will cause windborne debris to damage vehicles and injure travelers, and heavy rain can reduce visibility and worsen road conditions. Now, there are many methods to resolve DPO (Ramalingam and Reps, 1996). When traditional DPO methods are used to plan the path under typhoon scenarios, these mainly conduct on-line optimization according to the current typhoon condition. When the typhoon moves and the route network changes, traditional DPO methods have to conduct on-line recalculation for new optimal paths. In a single on-line run of optimization, traditional DPO methods treat the road network as static. Traditional DPO methods pay close attention to the searching efficiency of on-line recalculation. Rather than starting from scratch in each on-line recalculation, traditional DPO methods focus on modifying an existing plan by exploring a set of promising nodes and links as small as possible. Even so, traditional DPO methods still cannot guarantee to achieve the actual optimal travelling trajectory under a given typhoon scenario. The main reason is because the given typhoon dynamics is ignored in the on-line recalculation of DPO.

In contrast, we have recently proposed a new path optimization method in dynamical routing environments, the so-called Co-Evolutionary Path Optimization (CEPO) in Hu, et al, 2016a. A fundamental difference between the CEPO and traditional DPO is that, in a single run of optimization, the DPO treats weather conditions as static, while the CEPO allows weather conditions to co-evolve with the searching process. To achieve CEPO, a Ripple-Spreading Algorithm (RSA) was developed. However, the CEPO and RSA reported in Hu, et al, 2016a and Zhang, et al, 2016 are highly simplified. In this paper, we will improve the CEPO and RSA by taking into account some realistic factors in typhoon scenarios, i.e., there are different route categories with different speed limits even under normal weather conditions; such speed limits will reduce in typhoon scenarios, and the reduction on the speed limit largely depends on how far away the route is from the typhoon center. With these realistic factors under consideration, we need to modify the CEPO and RSA accordingly, and as a result of new modifications, we present a more practical CEPO and RSA to resolve the path optimization problem under typhoon scenarios in this paper.

2 IMPROVING CEPO AND RSA

In this paper, our optimization objective is to minimize travelling time. To achieve the actual optimal travelling trajectory under a given typhoon dynamic, we follow the methodology of Co-Evolutionary Path Optimization (CEPO) proposed in the literature (Hu, et al, 2016a and Hu, et al, 2016b). In CEPO, the routing environment keeps changing (e.g., due to typhoon dynamics) during a single run of optimization. Traditional DPO methods can hardly solve the CEPO problem. Therefore, Hu, et al, (2016a) developed a new Ripple-Spreading Algorithm (RSA), which can provide a solution to CEPO by using just a single offline optimization.

Actually, Ripple-Spreading Algorithm (RSA) is a newly reported nature-inspired method that mimics the natural ripple-spreading phenomenon occurring on a water surface (Hu, et al, 2013 and Hu, et al, 2016b). The natural ripple-spreading phenomenon reflects a simple optimization principle, which is, on a water surface, a ripple spreads out at the same speed in all directions, and therefore it reaches the closest spatial point firstly; based on this simple optimization principle, the RSA was developed as a simulated ripple relay race on a route network. Basically, in the RSA, an initial ripple is started from the source node. When the ripple reaches an unvisited node, the node will be activated to generate its own ripple. When all of the nodes that are directly connected to the epicenter node of a ripple have generated their own ripples, then the ripple will stop and be eliminated. When the destination node is visited for the first time, the first shortest path is found and the ripple relay race will be terminated. During the ripple relay race, all ripples always travel at the same preset constant speed.

The change in typhoon conditions is time-unit-oriented. The searching procedure of traditional DPO methods is link-oriented; and so, it is very difficult, if not impossible, to integrate the change in typhoon conditions into the searching procedure. In contrast, the behaviors of ripples and nodes in the RSA are simulated-time-unit-oriented operations; and so, it is easy to integrate the change in typhoon conditions into the ripple relay race of RSA. Therefore, RSA makes it possible to address CEPO by using a single run of optimization (Hu, et al, 2016a).

However, the RSA cannot be directly used to solve the CEPO problem; therefore, some necessary modifications were introduced to the RSA by Hu, et al, (2016a). First, the change in the route network was integrated into the ripple relay race of the RSA. Second, a waiting behavior was introduced to ripples, because when a node/link is closed, waiting before the closed node/link might save more travelling time than going around the closed node/link. With these two major modifications, the RSA can be extended to resolve CEPO. Fig. 1 gives an illustration about how the modified RSA can achieve the actual optimal traveling trajectory in a dynamic routing environment.

In this study, we want to apply the CEPO and RSA in Hu, et al's (2016a) work to find the actual optimal traveling trajectory under a typhoon scenario. The study of Hu, et al, (2016a) assumed that all routes are of the same category, and the impact of disruptive factors on routes within an obstacle area is the same. In this study of typhoon scenarios, we need to take into account different route categories and different impacts within an obstacle area. Simply speaking, different route categories have different speed limits, and the impact of the typhoon on the route network varies according to the distance to the center of the typhoon (because the wind strength is related to that distance). Therefore, we need to make further modifications to the RSA of Hu, et al, (2016a), in order to achieve a realistic case study of path optimization under typhoon scenarios. Due to limited space, in this work, we only give a rough description of the major modifications which are as follows:

In real world, there exist many different route categories. In this paper, we only consider one origin to one destination path optimization. A node represents a city/town. We use an adjacency matrix to record routes between nodes as well as route categories. Let C_1, ..., C_n denote different route categories. Associated with different route categories C_1, ..., C_n are

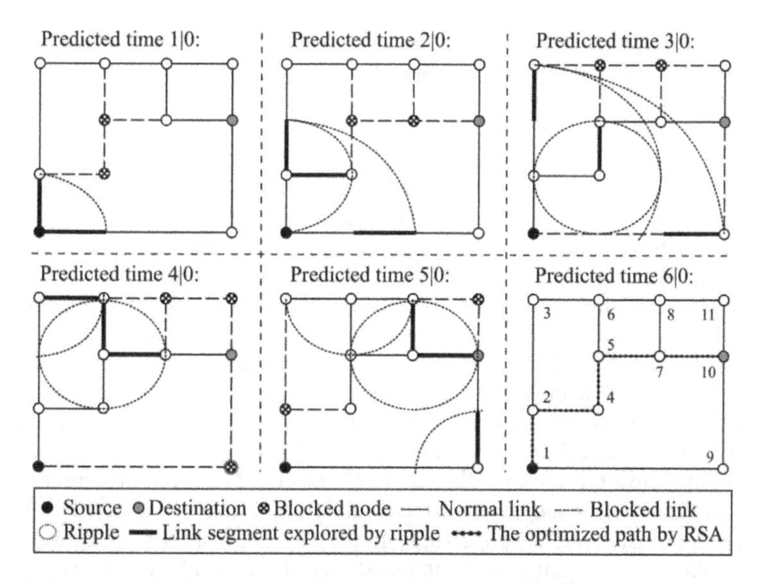

Figure 1. Illustration of the Ripple-Spreading Algorithm (RSA) for CEPO.

different speed limits S_1, ..., S_n under normal weather conditions. Therefore, ripples have different spreading speeds on routes of different categories. Let us assume that a ripple travels along a link (incoming link) to a node, and triggers the node to generate a new ripple to spread along another link (outgoing link). Consider that the incoming link and outgoing link have different route categories. And then, in the modified RSA, the incoming ripple and the outgoing ripple will spread at different speeds, which is determined by the associated route categories of the incoming link and the outgoing link.

The above modification is for ripples under normal weather conditions; in other words, for ripples, which are outside of the area covered by a typhoon. For those ripples, which are within the area covered by the typhoon, we need to introduce another modification to the RSA.

Within a typhoon-covered area, the wind strength varies depending on the distance to the center of the typhoon. For example, a typhoon-covered area may be divided into three rings in terms of wind strength. The first ring is around the typhoon center with the strongest wind and therefore, within this ring, all routes will be closed until the typhoon center moves away. The second ring encircles the first ring, and within this ring, the wind strength is weakened; and so, the routes in the second ring are not necessary to be closed, but the travelling speed must be largely reduced when compared with the associated normal speed limits. The third ring is the outer layer of the typhoon-covered area, and the wind strength is further weakened; and so, within the third ring, people may travel nearly at, but it is still slower than the normal speed limits. Only outside the typhoon-covered area (i.e., beyond the third ring) can people travel at the normal speed limits. Therefore, the spreading speed of ripples need to be adjusted according to both route categories (i.e., normal speed limits) and the distance to the typhoon center. In the modified RSA, we introduce a set of parameters α_i, related to the *i*th ring within the typhoon-covered area. For a route with a normal speed limit C_j, if it is within the *i*th ring, then, people can travel at a speed of $\alpha_i x C_j$.

With the above-mentioned modifications, the RSA can effectively deal with route category and wind strength in the real-world path optimization under typhoon scenarios.

3 SOME PRELIMINARY EXPERIMENTAL RESULTS

To test the modified method, we choose the Hainan Island of China as a case study. The route network in the Hainan Island has four categories of routes, a highway system, a national route system, a provincial route system, and county level routes, as shown in Figure 2. We use an adjacency matrix to record routes between nodes as well as speed limits for routes. It is assumed that a traveler may drive at a speed of 80 km/h on the highway system, 60 km/h on the national route system, 40 km/h on the provincial route system, and 30 km/h on county level routes under normal weather conditions.

The Hainan Island is a place that is hit by typhoons often (Lu, et al, 2015). We set up two typhoon scenarios according to historical data. In scenario 1, a typhoon lands at SanYa City, and in scenario 2, a typhoon lands at WanNing City. In each scenario, the typhoon-covered area has three rings with different wind strengths, i.e., high wind area, whole gale area, and hurricane area. Therefore, we have three parameters $\alpha_1 = 80$, $\alpha_2 = 60$, and $\alpha_3 = 0$. When a link is located within the high wind area, the traveling speed on the link will reduce to α_1% of the normal speed limit. When a link is located within the whole gale area, the traveling speed on the link will reduce to α_2% of the normal speed limit. When a link is located within the hurricane area, the link will be closed. It should be noted that, not every typhoon-covered area will have three rings as discussed above, but we can modify those relevant parameters in the RSA, in order to deal with the different number of typhoon rings.

Firstly, we simulate a typhoon landing at WanNing city. The typhoon center changes over time. Let us assume that we can forecast the direction and speed of the typhoon movement, as well as the size of the three typhoon rings. In the study of WanNing City, we set the direction of the typhoon movement as 45° northwest, at a speed of 42.4 km/h, the radius of high wind area is set as 50 km, the radius of the whole gale area is set as 40 km, and the radius

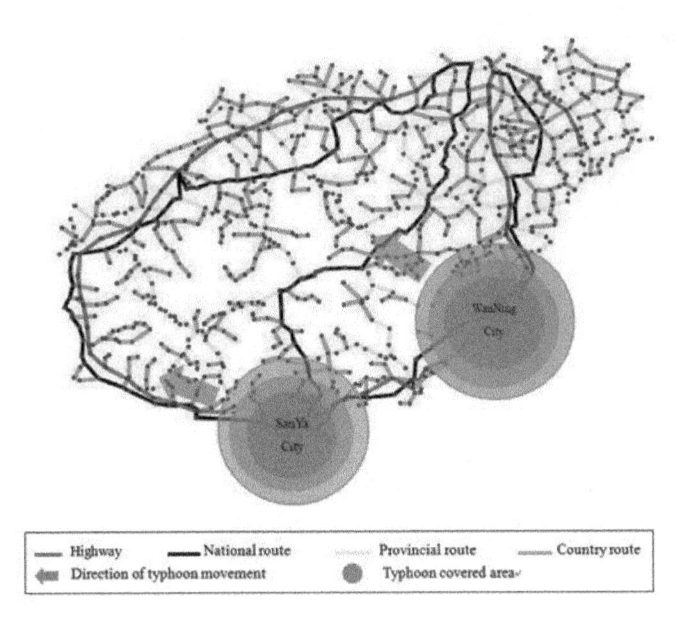

Figure 2. The route network of Hainan Island.

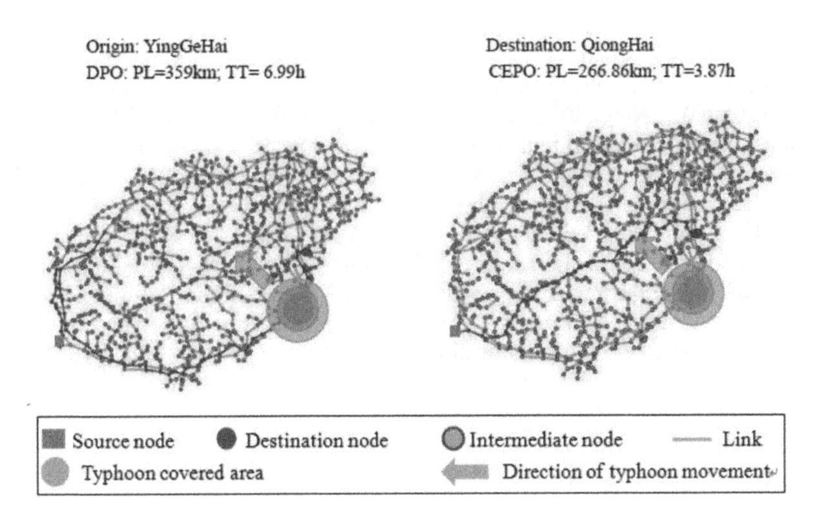

Figure 3. Examples of results of DPO and CEPO.

of the hurricane area is set as 30 km. Therefore, if the anode/link is less than 30 km away from the center of the typhoon, the node/link will be closed. If a link is within the range of [30 km, 40 km] from the center of the typhoon, the traveling speed on the link will reduce to α_2% of the normal speed limit. If a link is within the range of [40 km, 50 km] from the center of the typhoon, the traveling speed will reduce to α_1% of the normal speed limit. We randomly choose a pair of origin and destination cities/towns. To better evaluate the performance of the modified CEPO and RSA, a traditional DPO method is used for comparative purposes. The simulation result is shown in Figure 3. Secondly, we simulate a typhoon landing at SanYa city, and the moving direction is 37° northwest at a speed of 50 km/h. We randomly choose a pair of origin and destination cities/towns. The result of simulation is shown in Figure 4.

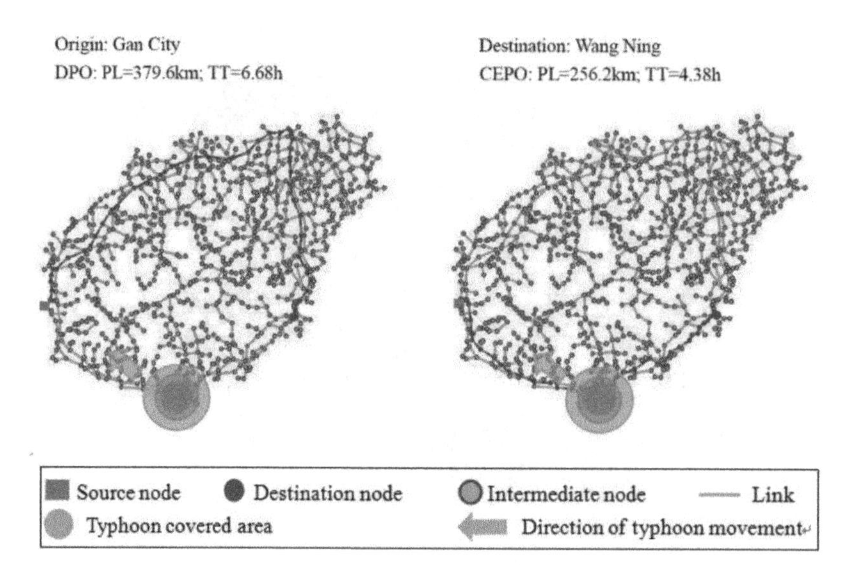

Figure 4. Examples of results of DPO and CEPO.

4 CONCLUSIONS

Based on the above-illustrated simulation results, we can find that the modified CEPO and RSA are clearly advantageous against the traditional DPO method in the simulated typhoon scenarios, in term of travelling time and path length. In the first experiment, where the typhoon lands at WanNing city, as the DPO method cannot make use of typhoon dynamics in its online optimization, it initially planned to go northeast, but with the typhoon moving, the initial plan became feasible. And then, the DPO re-conducted the path optimization, and turned back to travel southeast. Therefore, the DPO method resulted in a long traveling time and a large path length. In contrast, the modified CEPO and RSA achieved an actual traveling trajectory with a much shorter traveling time and much smaller path length, by using a single run of offline optimization. In fact, the RSA achieved the optimal actual traveling trajectory. In the study of SanYa city, we can arrive at the same conclusion that the CEPO and RSA can achieve better travelling times under the path length than the DPO method. Therefore, the modified CEPO and RSA in this paper have a good potential of applying to path optimization under typhoon scenarios.

ACKNOWLEDGMENTS

This work was supported in part by the National Natural Science Foundation of China (Grant No. 61472041) and the National Key Research and Development Programme (Grant No. 2016YFA0602404).

REFERENCES

Frigioni, D., Marchetti-Spaccamela, A., and Nanni, U. (2000). "Fully dynamic algorithms for maintaining shortest paths trees", *Journal of Algorithms*, 34: 251–281.

Hu, X.B. and Liao, J.Q. (2016a), "Co-Evolutionary Path Optimization by Ripple-Spreading Algorithm", *The 2016 IEEE World Congress on Computer Intelligence (WCCI2016), Congress on Evolutionary Computation (CEC2016)*, 24–29 July 2016, Vancouver, Canada.

Hu, X.B., Wang, M., Leeson, M.S., Hines, E.L., Paolo, E. Di. (2016b). "A Deterministic Agent-Based Path Optimization Method by Mimicking the Spreading of Ripples", *Evolutionary Computation*, 24(2): 319–346.

Hu, X.B., Sun. Q., Wang. M., Leeson, M.S., Paolo, E. Di. (2013). "A Ripple-Spreading Algorithm to Calculate the k Best Solutions to the Project Time Management Problem", *2013 IEEE Symposium Series on Computational Intelligence* (IEEE SSCI 2013), 16–19 April 2013, Singapore.

Keay, K., Simmonds, I. (2005). "The association of rainfall and other weather variables with road traffic volume in Melbourne, Australia", *Accident Analysis & Prevention*, 37: 109–124.

Keay, K., Simmonds, I. (2006). "Road accidents and rainfall in a large Australia city", *Accident Analysis &Prevention*, 38: 445–459.

Krozel, J., Lee, C., Mitchell, J. (1999). "Simulating time of arrival in heavy weather conditions", *Guidance, Navigation, and Control Conference and Exhibit, Guidance, Navigation, and Control and Co-located Conferences*, http://dx.doi.org/10.2514/6.1999–4232.

Lam, W.H.K., Shao, H. and Sumalee, A. (2008). "Modeling impacts of adverse weather conditions on a road network with uncertainties in demand and supply", *Transportation Research Part B: Methodological*, 42: 890–910.

Lu, G.R., Wang, W., Zheng, M.Q., Cai, Q.B. (2015). "Spatial and temporal distribution characteristics of typhoon precipitation in Hainan", *Trans Atmos Sci*, 38(5): 710–715.

Maarten K. Van Aalst (2005). "The impacts of climate change on the risk of natural disasters", *Disaster*, 30: 5–18.

Ramalingam, G., Reps, T. (1996). "On the computational complexity of dynamic graph problems", *Theoretical Computer Science*, 158: 233–277.

Trani, C.Q.A., Srinivas, S. "Modeling the Economic Impact of Adverse Weather into Route Flights," *Journal of the Transportation Research Board*, doi: 10.3141/1788–10.

Weisser, W.W., Volkl, W., Hassell, M.P. (1977). "Importance of Adverse Weather Conditions for behaviour and population Ecology of an Aphid Parasitoid", *Journal of Animal Ecology*, 66: 386–400.

Wu, J.D., Yu, F., Zhang, J., Li, N (2014). "Meteorological Disaster Trend Analysis in China: 1949–2013", *Journal of Natural Resources*, 29(9): 1521–1530.

Zhang, M.K., Hu, X.B., Liao, J.Q. (2016). "A New Path Optimization Method in Dynamic Adverse Weathers", *2016 12th International Conference on Natural Computation, Fuzzy Systems and Knowledge Discovery*, 13–15 August 2016.

Risk Analysis and Management – Trends, Challenges and Emerging Issues – Bernatik, Huang & Salvi (Eds)
 ISBN 978-1-138-03359-7

Recognizing the relationship between weather and yield anomalies and yield risk simulation of crops

Sijian Zhao & Qiao Zhang
Agricultural Information Institute, Chinese Academy of Agricultural Sciences, China
Key Laboratory of Digital Agricultural Early-warning Technology, MOA, Beijing China

ABSTRACT: In developing countries like China, weather-based index insurances of crops play an important role in the field of agricultural insurance. However, a big challenge in designing such an insurance product is to identify the key meteorological variables and to measure their impacts on the risk of crop yield loss. In the present study, the quantitative relationship between weather and crop yields, as well as yield risks were modeled using meteorological variables and Double Cropping Rice (DCR) from Hunan Province as a case study. In general, a weather-related disaster such as a drought, flood, typhoon, or cold damage is an extreme weather event type. More specifically, it is an anomalous weather event type. As a consequence, crop yields are expected to appear anomalous too when anomalous weather events appear. Therefore, it is investigated what would be the best mathematical relationship between anomalies of DCR yield and meteorological variables per county in Hunan Province. The meteorological variables were further ranked according to their contributions to the DCR yield loss risk. Furthermore, the DCR yield loss risk was simulated to generate a 4-dimensional yield risk chart, which can illustrate the DCR yield anomalies under various return period scenarios for the first-ranked meteorological factor. Such a chart has great potential for assisting the design of weather-based index insurances.

Keywords: Weather-based index insurance; Weather and yield anomaly relationships; Yield risk simulations; Hunan Province; Double Cropping Rice (DCR)

1 INTRODUCTION

China is one of the countries in the world most affected by natural disasters. Over the past decade, it experienced five of the world's top ten deadliest natural disasters and the top three of this list occurred in China (Guo Shujun 2012, Sijian Zhao and Qiao Zhang 2012, Wikipedia 2013). Among all the industries, agriculture is the one most affected by natural disasters because its production is highly dependent on natural conditions such as the weather, prevailing climate, and water availability. China's overall crop yields are increasing, but annual fluctuations in yields still occur. Such fluctuations are generally attributed to social, weather-related, and technological factors. The influence of weather, especially weather-related disasters, is thought to be the most important. According to statistics, outbreaks of major weather-related disasters have occurred every year in China since the 1990s. These disasters include droughts, floods, cold damage, hailstorms, and typhoons. The average annual affected area of crops is estimated to be 46,966 kha^2, which accounts for 30.57% of the total agriculturally used area and consequently has a great impact on crop yields (MOA, 2015). To reduce crop losses and protect crop yields, assessing and managing weather-related risks of crops is essential.

Weather-based index insurance is an efficient weather-related risk management tool for crops and can serve as an alternative to traditional multi-peril crop insurances. Morocco (Skees et al. 2001), South Africa (Geyser 2004), India (Giné et al. 2008), and Mexico (Barry and Olivier 2007) have implemented weather-based index insurance schemes for droughts. Canada uses weather-based index insurances to reduce and disperse the risk of yield reductions due to rain in dairy production in the country, and to decrease the risk of loss from corn and forage grass due to high temperatures (Denga et al. 2007). In Mexico, the agricultural industry uses weather-based index insurances (Agroasemex 2008), and Argentina adopts weather-based index insurances to disperse financial weather-related risks for fertilizer loans (Varangis et al. 2005). Moreover, weather-based index insurances seem more suitable to the developing countries. A rapidly growing body of literature has investigated the impact of weather-based index insurances on crops in developing countries, such as Berg et al. (2009) in Burkina Faso, De Bock (2010) in Mali, Chantarat et al. (2008) in Kenya, Molini et al. (2010) in Ghana, and Zant (2008) in India.

Traditional multi-peril crop insurances widely used in developed countries like the USA and Canada have been more detailed than modest in terms of regionalization and their high level of collating and stipulating. Furthermore, these countries have reduced the occurrence of adverse selection and moral hazards, which still existed at least several years ago in China (Huang et al. 2007). Weather-based index insurance uses a weather index (e.g., precipitation, temperature, or sunshine) as the critical factor and the insurer is responsible for payment of agricultural insurance mode when the index exceeds a trigger point. With this insurance, there is no relationship with a real crop loss due to a disaster, and there is no need to examine and determine individual holder-based losses. As a product based on the weather, weather-based index insurance is a financial tool for weather-related disaster risk management. It creates a community fund, which is a new approach for transferring weather-related risks and reduces the risks to the insurance company or its management in China (Cao and Wei 2004).

In designing weather-based index insurance, a key step is selecting one or two weather indices that relate best with crop yield losses. More specifically, recognizing which meteorological variable influences crop yield loss and assessing the yield risk caused by this meteorological variable is the principal part of this step. Therefore, the aim of this study is to find the optimal relationship between anomalies of both meteorological variables and crop yields. As a case study for our weather-based index insurance design, we used counties in the Hunan Province as the spatial unit and Double Cropping Rice (DCR) growing there as the representative crop.

2 DATASET

This study used three types of datasets, which are described below:

1. Annual DCR yields from 1984 to 2013 from 101 counties of Hunan Province (Figure 1), which were retrieved from the county-level rural economy statistics database of the Chinese Ministry of Agriculture.
2. Monthly observations of meteorological variables, including precipitation and temperature, from 25 weather stations located in Hunan Province from 1951 to 2013 (Figure 1). These data were obtained from the China Meteorological Data Sharing Service System.
3. A Digital Elevation Model (DEM) of Hunan Province with a 30 m spatial resolution (Figure 1) was provided by the Resource Environment Science Data Centre, Institute of Geographic Sciences and Natural Resources Research, Chinese Academy of Sciences.

3 METHODS

3.1 *Selection of key meteorological variables*

A first step was to select key meteorological variables that indicate weather-related disasters that potentially affected DCR yields. These meteorological variables can be obtained through an integrative analysis of the DCR phenological calendar, as well as a calendar of reoccurring weather-related disasters (Figure 2).

Figure 1. Map of Hunan Province.

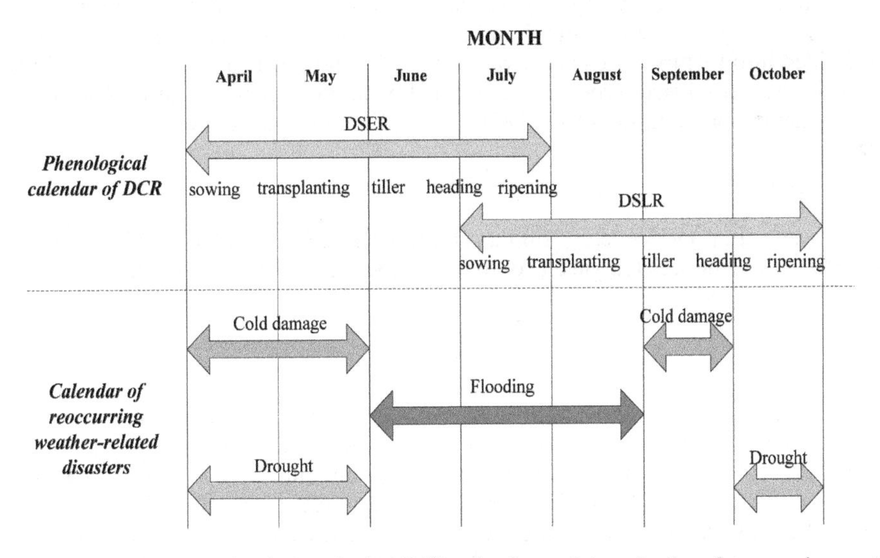

Figure 2. Integrated analysis of phenological DCR calendar and the calendar of reoccurring weather-related disasters in Hunan Province.

The DCR phenological calendar of Hunan Province was obtained by statistically analyzing local long-term planting logs. The annual DSER planting and seedling period was usually between 25 March and 15 April; the transplanting between 22 April and 8 May; booting and heading happened on an average between 27 May and 4 July; and ripening and harvesting happened on an average between 6 July and 3 August. In addition, the DSLR planting and seedling period was between 25 June and 9 July; transplanting between 16 July and 8 August; booting and heading between 26 August and 24 September; and ripening and harvesting happened on an average between 1 October and 29 October. Furthermore, a calendar of reoccurring weather-related disasters was obtained by analyzing local long-term historical disaster records. Three types of weather-related disasters were identified that affected overall DCR growth: cold damage, floods, and droughts. Cold damage happens when abnormally low temperatures cause a delay in DCR growth or damage to its physiological functions. In Hunan, cold damage frequently occurs in April, May, and September. Furthermore, floods generally result from storms with abnormally heavy precipitation. Rivers subsequently rise and go over their banks leading to the inundation and drowning of the DCR. Over the whole study period, flooding in Hunan frequently happened in June, July, and August. Finally, droughts occur when a region consistently receives below average precipitation for a long time. During the drought, crops cannot survive due to long-term water deficiencies. In Hunan, droughts concentrated in April, May, and October.

According to the analysis described above, nine key meteorological variables were selected as below.

$$W_g = \{T_4^-, T_5^-, T_9^-, P_6^+, P_7^+, P_8^+, P_4^-, P_5^-, P_{10}^-\}, g = 1, \dots 9 \tag{1}$$

where T_4^-, T_5^- and T_9^- are low temperatures of April, May, and September, which are indicative of cold damage happening; P_6^+, P_7^+ and P_8^+ are high precipitation rates of June, July, and August, which are indicative of flooding of the DCR fields; P_4^-, P_5^- and P_{10}^- reflect the low precipitation rates of April, May, and October, which indicate drought occurrences.

3.2 *Calculating relative weather-related yields*

DCR yield is affected by many factors such as weather conditions, local productivity, production inputs, and cropping technologies. However, in general, there are two types of factors influencing the yields: natural and unnatural factors. Therefore, annual DCR yields were divided into two parts. The first was the yield trend determined by unnatural factors, whereas in the second yield type was determined by natural, mainly weather-related factors. Thus, a general representation of the DCR's annual yield was as follows:

$$Y = Y_{td} + Yw \tag{2}$$

where Y was actual DCR yield (Ton/ha), Y_{td} was the DCR trend yield (Ton/ha), and Y_w was the DCR weather-related yield (Ton/ha). Weather-related yield Y_w could be easily calculated as:

$$Yw = Y - Y_{td} \tag{3}$$

As all counties in Hunan Province have nearly the same economic level and planting policy, it was assumed that the DCR yield per county had similar trends. Nonetheless, since there were still small differences at planting scale, planting technology, and labor inputs among the counties, the county-specific DCR trend yield needed to be adjusted from the overall trend yield.

Given the above situation, the trend yield could be divided into 2 parts: the province-level trend yield and the county-level trend yield. The province-level trend yield could be fitted with a logarithmic regression method using a time-series of average province-level yields according to the following equations:

$$y_{t,td}^P = a^* \ln(t - b) + c \tag{4}$$

$$y_t^P = \frac{1}{n}\sum_{i=1}^{n} y_t^i \tag{5}$$

where $y_{t,td}^P$ was the province-level trend yield (Ton/ha) for the t-th year; y_t^P the yield (Ton/ha) of the province as an average from the county-level yields for the t-th year; y_t^i was the yield

(Ton/ha) for the i-th county for the t-th year and a, b and c represented the logarithmic regression parameters. Based on the province-level trend, the county-level yield trends were fitted with a linear regression method between the actual county yields and the province-level trend yield, as described below:

$$y_{t,td}^{i} = A^{*} y_{t,td}^{P} + B^{i} \quad (6)$$

where $y_{t,td}^{i}$ was the trend yield (Ton/ha) of the i-th county for the t-th year; A^{i} and B^{i} are the linear regression parameters for the i-th county. In order to reflect the weather-related yield fluctuations relative to the trend yield, the relative weather-related yield $\Delta y_{t,rw}^{i}$ was calculated as follows:

$$\Delta y_{t,w}^{i} = \frac{y_{t,w}^{i}}{y_{t,td}^{i}} = \frac{y_{t}^{i} - y_{t,td}^{i}}{y_{t,td}^{i}} \quad (7)$$

with positive values of $\Delta y_{t,rw}^{i}$ meaning yields increased, whereas negative values meant a yield reduction.

3.3 *Spatial interpolation of meteorological variables*

Overall, 25 weather stations could not cover to represent all the 101 counties of Hunan Province (Figure 1). As some counties had no weather station, it was unreasonable to directly use the meteorological parameters from other stations for them. Therefore, the ANUSPLIN model was employed to interpolate meteorological parameters over the terrain of Hunan using data from 25 weather stations.

ANUSPLIN is a suite of FORTRAN programs developed at the Australian National University. It calculates and optimizes thin plate smoothing splines fitted to data sets distributed across an unlimited number of weather stations (Hutchinson, 1991). It has been applied to many regions such as Australia, Europe, South America, Africa, China, and parts of Southeast Asia. Details of the mathematical theory behind the thin plate smoothing splines can be found in Wahba (1990). A general representation of a thin-plate smoothing spline Z_i fitted to n data values at position x_i is given by (Hutchinson, 1995):

$$Z_i = f(x_i) + \varepsilon_i \quad (i = 1, \ldots, n) \quad (8)$$

where f is an unknown smooth function to be estimated and x_i represents longitude, latitude, and scaled elevation. The ε_i represents the zero mean random errors, which account for measurement errors as well as deficiencies in the spline model, such as local effects below the resolution of the data network.

ANUSPLIN performs very well in interpolations of meteorological parameters in regions with a complex topography. As a result, grid-level values of meteorological variables were generated for Hunan. To obtain county-level values of these variables, grid-level values were summed up and averaged according to county boundaries.

3.4 *Principal Component Analysis (PCA)*

The Principal Component Analysis (PCA) was invented in 1901 by Karl Pearson (Pearson 1901) as an analogue of the principal axis theorem in mechanics. It was later independently developed and named by Harold Hotelling in the 1930s (Hotelling 1933, 1936). It is a statistical procedure that uses an orthogonal transformation to convert a set of observations of possibly correlated variables into a set of values of linearly uncorrelated variables. The latter are called principal components. Considering the inherent correlation among nine meteorological variables, the PCA was applied to obtain meteorological principal components. For each county, the nine meteorological variables were orthogonally transformed to a set of principal components as follows:

$$(F_t^i)_k = (a_1^i)_k W_{1,t}^i + \ldots + (a_c^i)_k W_{c,t}^i + \ldots + (a_9^i)_k W_{9,t}^i \tag{9}$$

where $(F_t^i)_k$ is the k-th meteorological principal component for the i-th county and the t-th year; $W_{c,t}^i$ is the value of the c-th meteorological parameter for the i-th county and the t-th year; $(a_c^i)_k$ is the transform parameter from the c-th meteorological variable $W_{c,t}^i$ to the k-th principal component $(F_t^i)_k$ for the i-th county and the t-th year. As the first three principal components can contain all information of nine meteorological elements, k is set as 3.

3.5 *Optimal matching of anomalies*

As discussed above, weather-related disasters are generally anomalous weather events. For example, floods occur after precipitation events beyond the normal standard for the region. Consequently, the DCR yield would appear anomalous after such weather-related disasters happen. Therefore, it is investigated about the impacts of weather-related disasters on DCR production through the relationship between the anomalies in DCR yield and the anomalies in meteorological principal components.

Overall, an anomaly is a degree of deviation from a normal or average value, and the degree of deviation can vary among different individuals. DCR yield can be assumed as anomalous if its relative weather-related yield is below a pre-defined threshold (this threshold is flexible). It was expressed as follows:

$$y_{t,a}^i = \begin{cases} \Delta y_{t,rw}^i, & \Delta y_{t,a}^i < \Delta y_v^i \\ 0, & \text{else} \end{cases} \tag{10}$$

where $\Delta y_{t,rw}^i$ was the relative weather-related yield for the i-th county and the t-th year; $\Delta y_{t,a}^i$ was the anomaly of relative weather-related yield for the i-th county and the t-th year; Δy_v^i was the weather-related yield anomaly threshold for the i-th county, with $-1 \le \Delta y_v^i \le 1$.

A similar function was assumed for the meteorological principal components if the difference with its historical average level went beyond a certain (flexible) threshold. This could be expressed as follows:

$$(F_{t,a}^i)_k = \begin{cases} \left\| (F_t^i)_k - (\overline{F^i})_k \right| - (F_v^i)_k \right|, & \left| (F_t^i)_k - (\overline{F^i})_k \right| > (F_v^i)_k \\ 0, & \text{else} \end{cases} \tag{11}$$

where $(F_t^i)_k$ was the k-th meteorological principal component for the i-th county and the t-th year; $(\overline{F^i})_k$ was the average of the k-th meteorological principal component for the i-th county; $(F_{t,a}^i)_k$ was the anomaly of the k-th meteorological principal component for the i-th county and the t-th year; $(F_v^i)_k$ was the anomaly threshold for the k-th meteorological principal component for the i-th county.

A linear regression was used to fit the relationship between both anomaly types. To obtain an optimal mathematical relationship, the linear regression method was applied repeatedly for two types of thresholds, Δy_v^i and $(F_v^i)_k$, which were justified until the largest adjusted R^2 was obtained. This was expressed as follows:

$$f_{op}\left(y_a^i, (F_a^i)_k\right) = \begin{cases} y_a^i = g_1^i (F_a^i)_1 + \ldots + g_k^i (F_a^i)_k + g_0^i \\ \Delta y_v^i = \Delta y_v^i + \delta \cdot \sigma_{y_v}^i; \quad (F_v^i)_k = (F_v^i)_k + \beta_k \cdot \sigma_{F_k}^i \\ R_{adjusted}^2 \to \text{largest} \end{cases} \tag{12}$$

where f_{op} was the optimal mathematical relationship between the relative weather-related yield anomaly y_a^i and the anomalies of $(F_a^i)_k$ for the first to the k-th meteorological principal components of the i-th county; g_k^i was the optimal linear regressive parameter for the k-th meteorological principal component of the i-th county; $\sigma_{y_v}^i$ and $\sigma_{F_k}^i$ referred to the standard deviations of the relative weather-related yield and the k-th meteorological principal component for the i-th county; δ and β were the adjustable coefficients of $\sigma_{y_v}^i$ and $\sigma_{F_k}^i$, respectively; $R^2_{adjusted}$ was the adjusted R^2.

3.6 *Yield risk simulation*

For the i-th county, based on the optimal mathematical relationship of both anomalies, the nine key meteorological variables were ranked by their sensitivities to DCR yield anomalies. The sensitivity of a meteorological variable represents the extent to which the anomalies affected DCR yield when other variables were kept normal. The higher this degree was, the greater was the sensitivity of this meteorological variable. Combined with equation (12), the sensitivity $S_{w_c}^i$ of any one meteorological variable W_c^i was expressed as:

$$S_{w_c}^i = \frac{\partial y_a^i}{\partial W_c^i} = g_1^i \frac{\partial (F_a^i)_1}{\partial W_c^i} + \ldots + g_k^i \frac{\partial (F_a^i)_k}{\partial W_c^i} = g_1^i (a_c^i)_1 + \ldots + g_k^i (a_c^i)_k \tag{13}$$

With all meteorological variables together, the sensitivity $S_{w_c}^i$ of any one meteorological variable w_c was further expressed as the contribution $P_{w_c}^i$:

$$P_{w_c}^i = \frac{\left|S_{w_c}^i\right|}{\left|S_{w_1}^i\right| + \ldots + \left|S_{w_c}^i\right| + \ldots + \left|S_{w_9}^i\right|} \tag{14}$$

According to their contributions, all nine meteorological variables were ranked with the top ranking variable having the greatest impact on DCR yield anomalies. When this variable was defined as W_c^i, its probability density function could be estimated with a Gaussian Kernel Density estimation method (Barry et al. 1998, Alan et al. 2000) and its long-time serial historical observations $\{w_{c,1951}, \ldots, w_{c,t}, \ldots, w_{c,2013}\}$ from 1951 to 2013 as given below:

$$\tilde{f}(w_c^i) = \frac{1}{nh} \sum_{i=1}^{n} K\left(\frac{w_c^i - w_{c,t}^i}{h} \right) \tag{15}$$

where $\tilde{f}(w_c^i)$ was the probability density function of W_c^i; h is the bandwidth; n is the sample size; K is the Gaussian Kernel Density function as expressed below:

$$K(u) = \frac{1}{\sqrt{2\pi}} \exp\left(-\frac{u^2}{2} \right) \tag{16}$$

Using the return period T_0 to simulate the anomaly scenario of W_c^i, the amount W_{c,T_0}^i under T_0 was determined by using the inverse probability distribution function W_c^i as given below:

$$w_{c,T_0}^i = F^{-1}\left(1 - \frac{1}{T_0} \right) \tag{17}$$

Assuming that other variables were normal, the yield anomaly y_a^i of -impacted DCR under the return period T_0 was calculated by substituting W_{c,T_0}^i to equation (12) as given below:

$$y_a^i\left(w_{c,T_0}^i\right) = g_1^i \cdot (F_a^i)_1 + \ldots + g_k^i \cdot (F_a^i)_k + g_0^i \tag{18}$$

where

$$(F_a^i)_k = \begin{cases} \left|(a_c^i)_k \cdot \left(w_{c,T_0}^i - \overline{w_c^i}\right)\right|, & \left|(a_c^i)_k \cdot \left(w_{c,T_0}^i - \overline{w_c^i}\right)\right| > (F_v^i)_k \\ 0, & \text{else} \end{cases} \tag{19}$$

Combined with the trend yield, the W_c^i-impacted DCR yield under the return period T_0 could also be predicted by substituting $y_a^i(w_{c,T_0}^i)$ to equation (7) as expressed below:

$$y_{t_0}^i\left(w_{c,T_0}^i\right) = y_{t_0,td}^i + y_{t_0,td}^i \cdot y_a^i\left(w_{c,T_0}^i\right) \tag{20}$$

where t_0 was the predicted time, which could be the next future year. Given a group of return periods $\{T_0, T_1,\ldots,T_m\}$, a group of 4-dimensional DCR yield risk information could be worked out by W_c^i, i.e., $\left\{\left(T_0, w_{c,T_0}^i, y_a^i(w_{c,T_0}^i), y_{t_0}^i(w_{c,T_0}^i)\right),\ldots,\left(T_m, w_{c,T_m}^i, y_a^i(w_{c,T_m}^i), y_{t_0}^i(w_{c,T_m}^i)\right)\right\}$. By plotting all points $< T_m, w_{c,T_m}^i, y_a^i(w_{c,T_m}^i), y_{t_0}^i(w_{c,T_m}^i) >$ of a group in a chart, a 4-dimensional yield risk chart of a single meteorological variable W_c^i was created as a final result.

4 RESULTS AND DISCUSSION

The analyses and simulations generated a large amount of possible results, which are too numerous to present collectively. Therefore, this section is illustrated mostly with the (intermediate) results of the Nanxian County as an example.

4.1 *Modeling of relationships*

4.1.1 *Calculations of relative weather-related yield*

Based on the annual DRC yield of 101 counties in the Hunan Province from 1984 to 2010, the annual average DCR yield of the whole province from 1984 to 2010 could be calculated easily. The province-level DCR trend yield was subsequently fitted with a logarithmic regression-like equation (4), which is expressed as given below:

$$y_{t,td}^P = 0.7502 \times \ln(t - 1982.4) + 4.0401 \tag{21}$$

After obtaining the province-level DCR trend yield, the county-level DCR trend yield per county was fitted with linear regressions such as equation (6). Using the Nanxian County as an example, the DCR trend yield of Nanxian is expressed as follows:

$$y_{t,td}^{NX} = 0.775 \times y_{t,td}^P + 1.407 \tag{22}$$

Furthermore, the relative weather-related DCR yield for Nanxian was calculated with equation (7) as follows:

$$\Delta y_{t,w}^{NX} = \frac{y_t^{NX} - y_{t,td}^{NX}}{y_{t,td}^{NX}} \tag{23}$$

4.1.2 *Spatial interpolation and PCA of meteorological variables*

The ANUSPLIN algorithm was used to interpolate monthly average temperatures and precipitation of 25 weather stations from 1951 to 2013. Since the complex terrain of Hunan Province has a great effect on the weather, a DEM of the province with a 30-meter resolution was used as a covariate in the algorithm. Grid-based maps showing the annual meteorological variables could be generated for the entire province such as the monthly precipitation in 2001 for Hunan Province (Figure 3).

Figure 3. Interpolation of the 2001 monthly precipitation in Hunan Province.

Furthermore, using Nanxian as an example, after the PCA transformation of the nine meteorological variables, the first three meteorological principal components could be expressed as given below:

$$\begin{bmatrix} F_1^{NX} \\ F_2^{NX} \\ F_3^{NX} \end{bmatrix} = \begin{bmatrix} -0.0007,-0.0030,0.0036,0.5743,0.6208,0.3400,0.2223,0.3285,0.1086 \\ -0.0031,0.0009,0.0016,0.0914,-0.3566,-0.1874,0.8939,0.0476,0.1676 \\ -0.0019,-0.0018,-0.0045,-0.5218,-0.1598,0.7390,0.1307,0.3714,-0.0319 \end{bmatrix} \cdot \begin{bmatrix} T_4^- \\ T_5^- \\ T_9^- \\ P_6^+ \\ P_7^+ \\ P_8^+ \\ P_4^- \\ P_5^- \\ P_{10}^- \end{bmatrix} \quad (24)$$

4.1.3 *Relationships between weather and yield anomalies*

The optimal relationship between meteorological principal component anomalies and those of the yields was established with optimal data matching. For example, the Nanxian optimal relationship was as follows:

$$y_a^{NX} = 0.01114 + 0.000126 \times \left(F_a^{NX}\right)_1 - 0.000205 \times \left(F_a^{NX}\right)_2 + 0 \times \left(F_a^{NX}\right)_3 \quad (25)$$

The standard deviation of the relative weather-related yield in Nanxian σ_{y_a} was 0.065 and the responding adjustable coefficient δ was 0.08. The standard deviation of the first meteorological principal component σ_{F_1} was 102.58 and the responding adjustable coefficient β_1 was 1.52. The standard deviation of the second meteorological principal component σ_{F_2} was 86.34 and the responding adjustable coefficient β_2 was 0.04. Finally, the standard deviation of the third meteorological principal component σ_{F_3} was 78.69 and the responding adjustable coefficient β_3 was 0.0.

4.2 *Yield risk simulation*

Using the optimal relationship, the sensitivity of meteorological variables could be ranked and the contribution per meteorological variable to the yield anomaly calculated (Table 1).

In Nanxian, the high precipitation in July and August had the largest effect on DCR yield, followed by the low precipitation in October and May, the high precipitation in June, and the low precipitation in April, respectively (Table 1). The low temperatures in September, May, and April had almost no effect on DCR yield. In terms of weather disasters, the flooding that occurred in July and August may have been the main cause of yield reduction followed by the drought in October and May. In contrast, the cold damage had no significant effect in this county.

When selecting the most influential variable on DCR yields, historical observations from 1951 to 2013 could be used to create a probability distribution through a kernel density estimation. For example, this was done for the precipitation in July of Nanxian (Figure 4).

Assuming that the other eight meteorological variables are all normal, the yield risk of DCR could consequently be estimated for anomalies in the July precipitation. Based on the July precipitation probability distribution, the extreme precipitations of different return periods were calculated. Furthermore, with the optimal relationship, the relative weather-related yields and predicted yields corresponding to different return periods could be obtained. By graphing four types of information together (i.e., the return period, extreme precipitation, relative weather-related yield, and predicted yield), a 4-dimensional yield risk chart of a single meteorological variable (precipitation in July) was generated (Figure 5).

Table 1. Ranking of the contribution of the nine meteorological variables to crop yield anomalies in Nanxian County.

Meteorological variables	P_7^+	P_8^+	P_{10}^-	P_5^-	P_6^+	P_4^-	T_9^-	T_4^-	T_5^-
Contribution (%)	52.4	18.7	13.6	8.0	6.7	0.4	0.1	0.0	0.0

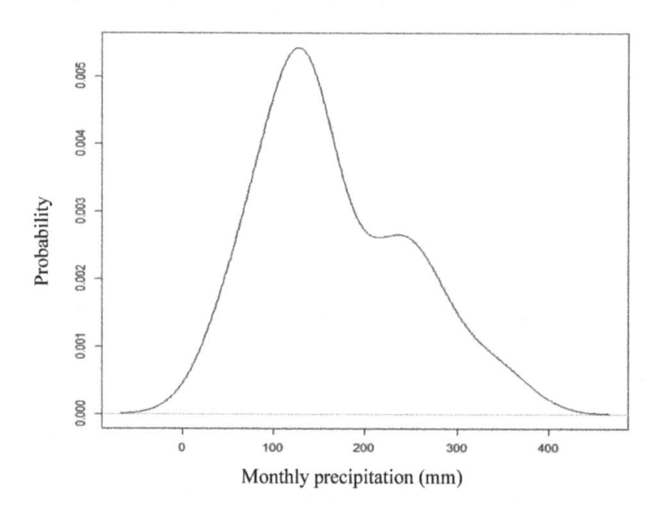

Figure 4. Probability distribution of July precipitation in Nanxian County.

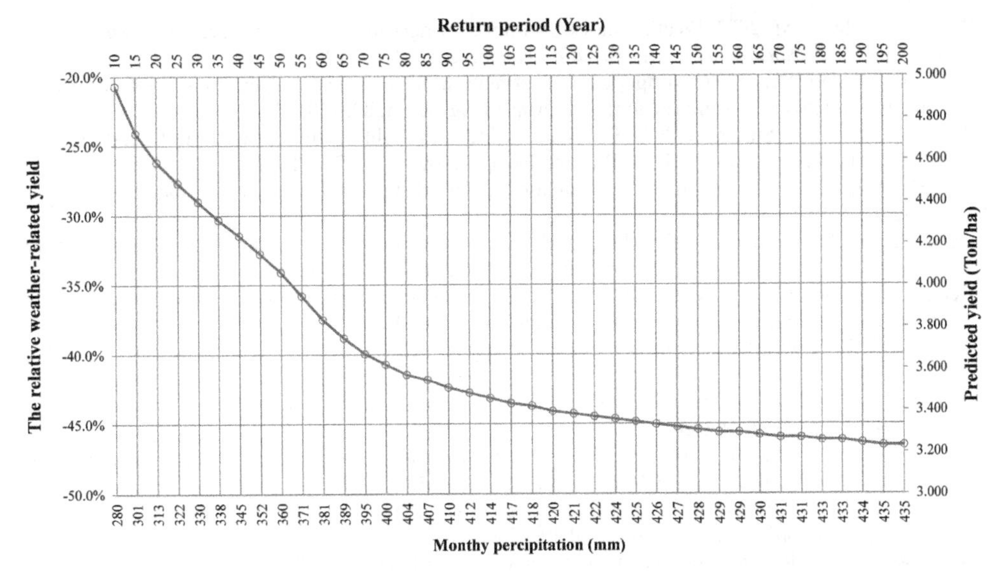

Figure 5. A 4-dimensional yield risk chart of a single meteorological variable (July precipitation) in Nanxian.

5 SUMMARY

In this study, county-level weather-and-yield anomaly relationships of DCR were determined using DCR in Hunan Province as a case study. Furthermore, the weather-related yield risks of single meteorological variables were simulated. Our conclusions were as follows:

1. According to the generalization that weather-related disasters are caused by weather anomalies, and weather anomalies further lead to crop yield anomalies, the relationship between the anomalies of meteorological variables and crop yield can be identified through the optimal matching of data. This optimal matching subsequently provides a new method for determining the weather-related yield risks of crops.
2. Crop yield anomalies comprise the effects of various anomalies from meteorological variables throughout the growing seasons. However, the influence of some variables is larger than from others. By recognizing the weather-and-yield anomaly relationship, it becomes possible to identify the most influential variables on crop yields. This is a key characteristic for supporting crop monitoring in weather-related disasters.
3. A 4-dimensional yield risk chart can provide a new quantitative expression of weather-related yield risks. These charts illustrate the crop yield anomalies under the various return period scenarios of a meteorological variable. Such charts can, therefore, play an important role in designing weather-based index insurances.

ACKNOWLEDGMENTS

This study was financed by the Natural Science Foundation of China (41471426).

REFERENCES

Agroasemex 2008. The experience of Mexico in the development and operation of parametric insurances applied to agriculture. [online]: http://www.agroasemex.gob.mx.

Alan, P.K., Barry, K.G. 2000. Nonparametric estimation of crop insurance rates revisited. Amer J Agr Econ, 83, 463–478.

Barry, J.B., Olivier, M. 2007. Weather index insurance for agriculture and rural areas in lower-income countries. *American Journal of Agricultural Economics*, 89, 1241–1247.
Barry, K.G., Alan, P.K. 1998. Nonparametric estimation of crop yield distributions: implications for rating group-risk crop insurance contracts. Amer J Agr Econ, 80, 139–153.
Berg, A., Quirion, P., Sultan, B. 2009. Can weather index drought insurance benefit to least developed countries' farmers? A case study on Burkina Faso. Weather Clim Soc, 1, 7184.
Cao, M., Wei, J. 2004. Weather derivatives valuation and market price of weather risk. *Journal of Futures Markets*, 24, 1065–1089.
De Bock, O. 2010. Etude de faisabilité: Quels mécanismes de micro-assurance privilégier pour les producteurs de coton au Mali ? Discussion paper, CRED, PlaNet Guarantee.
Denga, X.H., Barnettb, B.J., Vedenovb, D.V., et al. 2007. Hedging dairy production losses using weather-based index insurance. Agricultural Economics, 36, 271–280.
Geyser, J.M. 2004. Weather derivatives: concept & application for their use in South Africa. Agrekon, 43, 444–464.
Giné, X, Tounsend, R, Vickery, J. 2008. Patterns of rainfall insurance participation in rural India. The World Bank Economic Review, 22, 539–566.
Guo, Shujun 2012. The Meteorological Disaster Risk Assessment Based on the Diffusion Mechanism. *Journal of Risk Analysis and Crisis Response*, 2(2): 124–130.
Hotelling, H. 1933. Analysis of a complex of statistical variables into principal components. *Journal of Educational Psychology*, 24, 417–441, and 498–520.
Hotelling, H. 1936. Relations between two sets of variates. Biometrika, 27, 321–77.
Huang, Y., Li, X., Zhang, G.S. 2007. International innovative products of agricultural insurance and their useability in China. *Journal of Shenyang Agricultural University* (Social Sciences Edition), 9: 848–850. (in Chinese).
Hutchinson, M.F., 1991. The application of thin-plate smoothing splines to continent-wide data assimilation. In: Jasper, J.D. (Ed.), Data assimilation systems. BMRC Res. Report No. 27, Bureau of Meteorology, Melbourne, 104–113.
Hutchinson, M.F., 1995. Interpolating mean rainfall using thin plate smoothing splines. Int. J. GIS 9: 385–403.
Hutchinson, M.F., Gessler, P.E., 1994. Splines-more than just a smooth interpolator. Geoderma 62: 45–67.
Chantarat, S., Turvey, C.G., Mude, A.G., Barrett, C.B. 2008. Improving humanitarian response to slow-onset disasters using famine-indexed weather derivatives. Agric Financ Rev 68(1): 169–195.
MOA, 2015. Database of natural disasters of crops. [Online] available: http://www.zzys.moa.gov.cn/.
Molini, V., Keyzer, M., Zant, W., et al. 2010. Safety nets and index-based insurance: historical assessment and semiparametric simulation for Northern Ghana. Econ Dev Cult Change 58(4): 671–712.
Pearson, K. 1901. On Lines and Planes of Closest Fit to Systems of Points in Space. Philosophical Magazine, 2 (11): 559–572.
Sijian, Zhao, Qiao, Zhang, 2012. Risk Assessment of Crops Induced by Flood in the Three Northeastern Provinces of China on Small Space-and-Time Scales. *Journal of Risk Analysis and Crisis Response*, 2(3): 201–208.
Skees, J.S., Gober, P., Varangis, R., et al. 2001. Developing Rainfall-Based Index Insurance in Morocco. The World Bank, Policy Research Working. p. 2577.
Varangis, P., Skees, J., Barnett, B. 2005. Weather Indexes for Developing Countries. World Bank, University of Kentucky and University of Georgia.
Wahba, G., 1990. Spline models for observational data. CBMS-NSF Regional Conf. Ser. Appl. Math., Philadalphia Soc. Ind. Appl. Math.: 169.
Wikipedia 2013. Natural disasters in China. [Online] available: http://en.wikipedia. org/wiki/Natural_disasters_in_China.
Zant, W. 2008. Hot Stuff: index insurance for Indian small holder pepper growers. World Dev 36(9): 1585–1606, 0305-750X.

Risk Analysis and Management – Trends, Challenges and Emerging Issues – Bernatik, Huang & Salvi (Eds)

CBRN remote sensing using unmanned aerial vehicles: Challenges addressed in the scope of the GammaEx project regarding hazardous materials and environments

Mario Monteiro Marques, V. Lobo, Rodolfo Santos Carapau & Alexandre Valério Rodrigues
Centro de Investigação Naval, Portugal

Júlio Gouveia-Carvalho
Portuguese Army, Portugal

Alfredo José Martins Nogueira Baptista
Instituto Superior Técnico, Lisbon, Portugal

Jorge Almeida
I-SKYEX, Portugal

Cristina Matos
Instituto de Soldadura e Qualidade, Portugal

ABSTRACT: In this paper, the GammaEX project, which focuses on the use of Unmanned Aerial Systems (UASs) as a response to Chemical, Biological, Radiological, and Nuclear (CBRN) threats or incidents, is presented. The GammaEX project is a product of a collaborated effort between various entities, and its motivation is based on the creation of a CBRN threat emergency response system that uses unmanned systems in order to minimize the risk of endangering human lives. In order to meet the requirements to operate in hazardous CBRN environments, a Remotely Piloted Aircraft System (RPAS) was created, according to the ATEX (ATmosphères EXplosibles) legal and standard framework directives for the certification of electric equipment for use in potentially explosive atmospheres and its application to UAS, the ATEX certification being one of the primary goals of the GammaEX project. The development of the project involved studies to determine and customize the appropriate sensors to detect CBRN agents (built in the RPAS), the construction of RPAS prototypes, the creation of navigation and communications modules, and the creation of validation scenarios based on real life CBRN incidents. The final product of the GammaEx project aims to be a tested and functional system, with the capability of being deployed in hazardous areas by specialized teams of the Navy, Army, or any other entity, in order to detect, identify, and pin-point through GPS (Global Positioning System) coordinates, radiation sources and chemical agents, as well as trapped victims, with minimal risk of endangering the human operators' lives.

Keywords: Unmanned Systems, ATEX, CBRN, Robotics

1 INTRODUCTION AND MOTIVATION

Throughout history, various conflicts have motivated the search for a better weapon, in order to achieve victory, since swords, bows and arrows, firearms, to bombs, missiles, and even unmanned systems. Even though these were effective tools in their own day and time, nowadays, there is a wider roster of weapons and threats. Some special kinds of weapons are related to Chemical, Biologic, Radiologic, and Nuclear threats (CBRN). CBRN threats can be traced back to the days of World War I, through the use of chlorine, mustard, and phosgene chemical gas to decimate waves of trench soldiers, to the detonation of the Hiroshima and Nagasaki Atom bombs during World War II, the radiation leak in the Chernobyl and Fukushima power plant, the outbreaks of the Black Plague and Ebola, terrorist attacks which employed lethal neurotoxic chemical agents in a Tokyo subway and, nowadays, throughout Syria, among other unfortunate events (Hendricks et al., 2012). As such, CBRN threats can be of intentional and non-intentional nature; the intentional kind leads to the dawning of a harsh reality when it comes to its use by radical and terrorist's groups. Adding to the fact that this is a growing threat, with the current development of weapons of mass destruction and bio-terrorism technology, and that terrorist attacks are of great unpredictability, it only serves to enhance the need to develop and create CBRN threats emergency response systems (Shea, 2013; Unal, et al. 2016).

CRBN incidents' response relies on the following four main pillars: information gathering assessment and dissemination, scene management, saving and protecting lives, and additional/specialist support. Information gathering assessment and dissemination is paramount, since it comprehends the initial contact to the victim, the detection of the contaminating agent, threat information dissemination, and recognition and labeling of the contaminated area. In order to contain the hazardous agent area, scene management is crucial. It is based on procedures of isolating the contaminated scene, access control, and victim extraction. An all-out priority, saving and protecting lives prioritizes an initial examination of the victim, followed by a decontamination process, and further medical treatment, having in mind the reduction of exposure to other people. Further and additional/specialist support is required to deal with the outcome of a CRBN event, which involves the identification or confirmation of the contaminating agent, the verification of the levels of contamination, as well as the further medical support, victim extraction means, and posterior emergency service resources (NATO, 2014).

All of these procedures involve a considerable amount of risk of exposure for the individuals providing support, especially at the initial phase of detection and identification of the hazardous agent, which presents an obstacle to the full CBRN incident emergency intervention process. When the human element of this equation is removed, the risk of developing exposure is greatly reduced, which can be made possible through the use of Unmanned Systems (UxSs). UxSs can operate either as Unmanned Aerial Vehicles (UAVs), Unmanned Ground Vehicles (UGVs), Unmanned Underwater Vehicles (UUVs), or as Unmanned Surface Vehicles (USV). UxSs provide a flexible and versatile response, since they vary in size, type of material, shape (allowing access to confined space), and in functions and purpose (which defines its sensors and conducted actions). A UxS communicates with a control station, which performs the vehicle's command and control; it also carries navigation systems for controlled and autonomous movement and sensors that can be built on to the UxS structure, or these are carried as payload (Austin, 2010). The main advantage of using a UxS is focused on considerable reduction of endangering human lives and also, it allows for integration of various types of sensors that can detect CBRN agents, medical material delivery, and communication relaying for information dissemination.

Regarding this technology, the GammaEX project has the objective of developing a Remotely Piloted Aircraft System (RPAS) to be deployed in CBRN incidents, thereby providing the adequate emergency response and reducing the risks of further human hazardous exposure to a minimum. The GammaEx project involves five partners: *Centro de Investigação Naval* (CINAV), *Unidade Militar Laboratorial de Defesa Biológica e Química/Centro de Investigação da Academia Militar* (CINAMIL), *Instituto Superior Técnico* (IST), *Instituto de Soldadura e Qualidade* (ISQ), and I-SKYEX.

This RPAS, named as "M6", is built based in a three module architecture: sensors, navigation and control, and communications modules. Implementing the M6 involves a process of identifying initial challenges and challenges presented during testing, followed by a process of overcoming such challenges. Some challenges, are the M6's sense-and-avoid capabilities, compatibility and implementation of modules, and validation of CBRN detection sensors and scenarios. The GammaEX project's validation is based on scenarios of CBRN incidents on board of ships and ashore, and this is supported by the Damage Control Department of the Naval Technology School of the Portuguese Navy.

The M6 is aimed to operate in a maritime environment as a tool to examine potentially hazardous atmospheres at the exterior areas of ships and off-shore platforms, and to do so according to an ATEX (*ATmosphères EXplosibles*) certification.

2 CHALLENGES

In order to evaluate the success of the project, there were certain requirements established for the RPAS, chemical detector, and radiation detector. These requirements can differ by type, since some are essential when it comes to functionality, operation, and safety, whereas other requirements are only desirable. For the RPAS, some of the essential requirements are as follows: action radius – 2 km; maximum weight at take-off – 20 kg; autonomy – 15 minutes; maximum velocity – 30 km/h; maximum size – 2 m; and maximum vertical velocity – 3 m/second (lift-off).

An RPAS system designed for the detection of radioactive sources on the soil surface from an aircraft normally senses gamma rays emitted by the source. Gamma rays have the longest path length (least attenuation) through the air of any of the common radioactive emissions and will thus permit source detection at large distances. A secondary benefit from gamma ray detection is that, nearly all radioactive isotopes can be identified by the spectrum of gammas emitted. Major gaseous emissions from fuel reprocessing plants emit gammas that may be detected and identified. Some types of Special Nuclear Material (SNM) also emit neutrons that are also useful for detection at a distance.

A gamma ray detection system must be sensitive enough to allow rapid source location from a reasonable altitude with the expectation that sources of small activity will be found. Once a source is found, the system should be able to identify the source isotope by its characteristic gamma emission spectrum. The twin goals of high sensitivity (necessary to find sources) and very good energy resolution (for source identification) are difficult to achieve simultaneously with current detector technology.

For example, one of the most important overall descriptive parameters of a detection system is the Minimum Detectable Activity (MDA). If the system cannot differentiate a source from the background, then other operations with the system are meaningless.

The ATEX requirements prevent the RPAS from the formation of explosive atmospheres as well as potential sources of ignition, such as electric sparks, arcs and flashes, electrostatic discharges, electromagnetic waves, ionizing radiation, hot surfaces, and mechanically generated sparks.

The ATEX directives are a product of the directives 94/9/EC and 1999/92/EC of the European Union (which became mandatory since July 1st 2003 and recently replaced by the 2014/34/EC, mandatory since April 20th 2016), which establish the minimum requirements for health and safety, which should be applied in all equipment intended for use in places in which explosive atmospheres caused by gases, vapors, mists, or air/dust mixtures are unlikely to occur. According to the 2014/34/EC directive, an 'explosive atmosphere' indicates the formation of a mixture of air, under atmospheric conditions, and flammable substances in the form of gases, vapors, mists, or dusts in which, after ignition has occurred, combustion spreads to the entire unburned mixture.

The ATEX directive states that equipment used in explosive atmospheres belong to one of two different groups, which are divided into the following sub-groups according to Table 1: 1 – equipment used inside and over mines and 2 – equipment used in areas of probable risk of becoming an explosive atmosphere.

To classify dangerous areas, the ATEX directives separate gases from dusts and differentiate the duration of the occurrence of an explosive atmosphere using the numbers 0, 1, and 2 and 20, 21, and 22, respectively.

Table 1. ATEX directive equipment classification. Source: http://www.dosatron.com/sites/default/files/1994-9-CE_en.jpg.

Equipment group	Equipment categories	Flammable substances	Protection levels	Protection, faults
I (Mines)	M1	Methane dusts	Very high	2 means of protection or 2 independent faults
	M2	Methane dusts	High	1 means of protection Normal operation
II (Surface industry)	1	Gases, vapours, mists, dusts	Very high	2 means of protection or 2 independent faults
	2	Gases, vapours, mists, dusts	High	1 means of protection Ordinary and frequent failure
	3	Gases, vapours, mists, dusts	Normal	Required level of protection

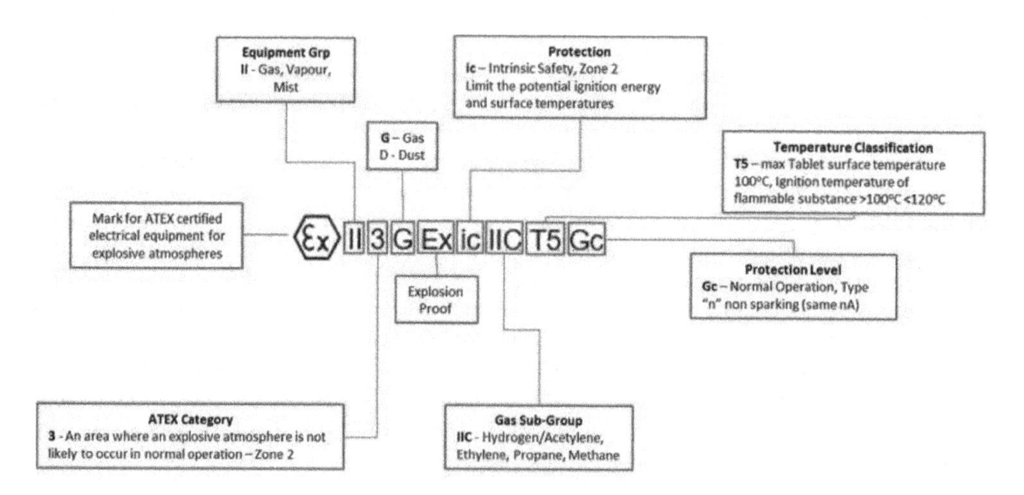

Figure 1. ATEX directives for marking certified equipment. Source: http://www.spiritdatacapture.co.uk/data-capture/atex-chart.jpg.

Electronic equipment that comply with the ATEX directives must be marked, as shown in Figure 1.

The GammaEx project intends to have in the RPAS the ATEX mark: – **CE ⓔ II 3 G EX n IIC T6.**

3 ARCHITECTURE

The GammaEx project uses two different RPASs: one is a "M6 CBRN" UAV, a hexacopter with retractable arms and landing tray, which is made out of carbon fiber reinforced with Kevlar, with a contained encapsulation made to meet the ATEX requirements, payload up to 2 kg, autonomy of 30 minutes, a weight of 8 kg, flight camera, and is able to withstand a maximum wind speed of 40 km/h. The second RPAS is a tricopter with a fixed landing tray and made out of carbon fiber reinforced with Kevlar, flight camera, and with contained encapsulation made to meet the ATEX requirements (Figure 2).

The GammaEx RPAS system architecture is based on a structure composed of a sensors module, a navigation and control module, a communications module, and a Ground Control Station (GCS).

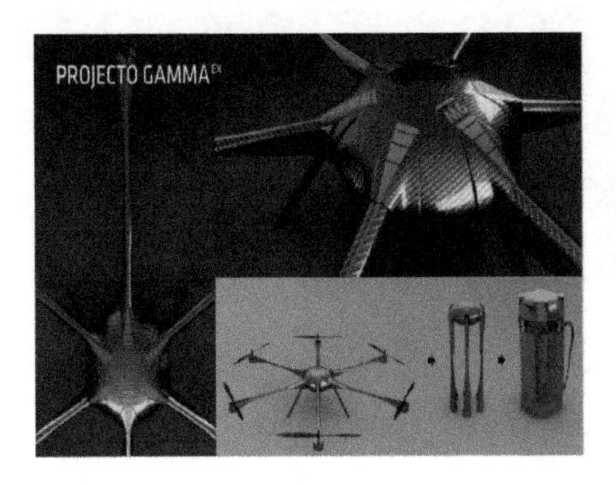

Figure 2. Pictures of GammaEX RPASs: an M6 hexacopter on the left and a tricopter on the right.

3.1 *Sensors module*

This module receives information from its sensors and it transmits it to the communication module. For the GammaEx project, radiation, chemical, and thermal cameras are the primary sensors of this module, in order to achieve the GammaEx goals and requirements. These sensors allow for the detection of chemicals agents, radiation, and victims of a hazardous area or confined space with reduced visibility. The choice of the chemicals and radiation was prioritized, due to the nature of the project. The initial search for chemical sensors was based on the following parameters: time of response, cost, autonomy, and weight. It was a challenge to acquire the right product, since the majority of companies did not meet the correct requirements of a communication interface and aerodynamics of a chemical sensor, which is to be integrated into a UAV, since the access to the communication protocols of the sensors was denied and the sensors were mostly of manual use. As such, there was the need to build custom sensors for the RPAS, a task that was fulfilled by SGX SensorTech. The sensor was built in an electronic board with the capability of supporting three electrochemical sensors, with signal conversion and transmitting capabilities, and power supply elements. The chemical sensors used are as follows: SGX-4DT Gas Sensor; EC410 Electrochemical Sensor Oxygen; and VQ548ZD Gas Sensing Element (Figure 3).

The choice of the radiation sensor was based on a device that could perform as well as other portable gamma spectrometry devices, but with reduced size and volume, so that it could be integrated into the RPAS. Through a market study, the best results were delivered by using the RadEye SPRD, from Thermo Scientific, since it presented a wide gamma radiation measurement, an easily RPAS-integrated interface, and signal processing that does not require further adjustments. The RadEye SPRD is a light device, with an autonomy of 170 hours, can be connected via USB or wireless, and the function of activating the radionuclide identification mode when it detects radiation levels over a certain threshold (Figure 3).

3.2 *Navigation and control module*

The RPAS navigation functionality is made possible through the use of an integrated Global Positioning System (GPS), an Inertial Measurement Unit (IMU), and a LIght Detection And Ranging (LIDAR) system. This module processes information related geographical coordinates in order to present sites of interest, waypoints, platform stability, object detection, and sense-and-avoid functions.

3.3 *Communications module*

The communication module regulates the flux of information through datalinks, sending downlinks to transmit sensor collected data to the GCS, and receiving uplinks from the GCS

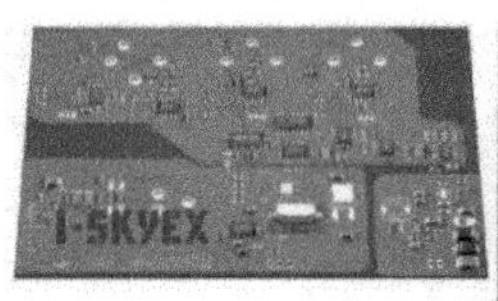

Figure 3. Pictures of SGX Sensortech Chemical sensors (left) and RadEye SPRD radiation sensor (right).

with control information to the RPAS platform. It allows for low latency and high bandwidth connection with the GCS, which enhances the quality of the communication process.

4 VALIDATION

According to the operational requirements of the GammaEx project, validation scenarios were created with the purpose of testing and validating the RPAS prototype's operation inserted in Chemical, Biological, and Radiologic environments (CBR), and to present and obtain the approval of the prototype from the Portuguese Ministry of Defence. These tests serve, as well, to detect weaknesses of the RPAS prototype in order to further develop and enhance it.

The tests take place during two different phases: the RPAS prototype testing phase and the operational demonstration RPAS phase. The first phase is focused on achieving the functional and operational requirements goals set for this project, according to each scenario, and retrieving information from the tests for post-tests adjustments. The second phase has the objective of demonstrating the RPAS's capabilities in the operational environments, which are associated to each scenario.

These tests will take place at the Lisbon Naval Base (BNL, *Base Naval de Lisboa*) in the Portuguese Navy's Damage Control School (*Escola de Limitação de Avarias*).

This location allows for the following: joint military training; simulation of maritime and land environments; safety conditions to operate with CBR; safety conditions to operate the RPAS; and cost reduction associated to the mobilization of operational means.

The proposed scenarios are focused on two different environments: maritime and land. As such, the maritime scenarios involve Navy intervention, whereas the land environment is focused on Army intervention. For each one of the scenarios, there are different threats associated; the chemical threat is employed in the Army scenario and the radiation threat is employed in the Navy scenario.

The Army scenario is used to test an unintentional chemical threat and an intentional chemical threat in the Alfeite's Port. The unintentional scenario is based on an explosion that took place in the port of Tianjin in an industrial compound of the Ruihiai Logistics Company that included storages of hazardous material, in August 2015. As such, the test scenario involves an industrial chemical leakage caused by a fire that started in one of the storages. In order to detect the hazardous and dangerous chemicals, the Army was called to deploy an Unmanned Aerial Vehicle (UAV) to perform reconnaissance in the affected area, in order to determine and identify the types of chemicals that leaked.

The intentional scenario is based on the terrorist attack that took place in a Tokyo subway. In the sequence of a terrorist group getaway, there was a hazardous chemical spill by one of the evading terrorists, thereby resulting in a number of non-lethal victims. This spill created a dangerous environments filled with unknown dangerous chemicals, and as such, the Army was called upon to deploy an UAV in order to perform area reconnaissance of the chemicals and, if possible, chemical agents sampling and identification to confirm the types of chemicals.

These scenarios are to be played according to the following sequence: the RPAS takes-off from the ground control station that is located 100 to 500 meters of the Hot Zone, aerial visual reconnaissance (in real time) of the hazardous area is performed in 15 minutes, the RPAS identifies its targets and sends the GPS coordinates of the targets to a CBR team on stand-by, the RPAS concludes reconnaissance and returns to the base that is to be decontaminated. The tasks to be completed are as follows: real-time image capturing, identification, and geo-referencing of targets; user remote controlled navigation; autonomous navigation enabled by the user or by loss of communications, with the capability of returning to the decontamination point; navigational sense-and-avoid; and chemical agent measurement.

The Navy scenario is used to test an unintentional radiation threat in interior and exterior areas at Alfeite's Port. The scenario is based on an accidental radiological accident that involved radiation leakage to water, air, and soil. It involves a nuclear propulsion submarine that has visited the Alfeite Port, and during its stay, it was activated by using an Environment Radiologic Surveillance plan, which had an objective to measure the amount of radiation through sampling analysis and area monitoring. This monitoring process found uncommon radiation level variations in the analyzed samples, which could be a pointer of a radiologic or nuclear incident. This scenario is to be played according to the following sequence: the RPAS takes-off from the Ground Control Station (GCS) located 100 to 500 meters from the Hot Zone, aerial visual reconnaissance (in real time) of the hazardous area is performed in 15 minutes, the RPAS identifies its targets and sends the GPS (Global Positioning System) coordinates of the targets to a Lab of Radiological Safety and Protection team on stand-by, and the RPAS concludes reconnaissance and returns to base to be decontaminated.

The tasks to be completed are as follows: potential "hot spots" identification through imagery; identification/measurement of radionuclides; creating a safety perimeter, according to the amount of radiation in the area; and according to the wind state, predict dispersion areas of the radioactive material.

5 CONCLUSIONS

The GammaEX project has the objective of developing UAS technology to be deployed in CBRN incidents; the consortium involves the following five partners: *Centro de Investigação Naval* (CINAV), *Unidade Militar Laboratorial de Defesa Biológica e Química/Centro de Investigação da Academia Militar* (CINAMIL), *Instituto Superior Técnico* (IST), *Instituto de Soldadura e Qualidade* (ISQ), and I-SKYEX.

For this project, the most concerning challenges are as follows: interference between sensors, flight system and communication links, system integration and compatibility, validating the accuracy and sensitivity of the radiological sensors to detect 10 mCi radioactive sources at a 5-meter distance, collision avoidance through sense-and-avoid capability implementation, and compliance with the ATEX directives in order to obtain an ATEX certification of the project, which increases the difficulty and the level of complexity in the development of the RPAS.

The validation phase constitutes an important part of this project. The validation scenarios were created with the main purpose of testing and validating the RPAS prototype's operation inserted in Chemical, Biological, and Radiologic (CBR) environments. In this particular case, the ATEX capabilities can contribute to a safer and secure operation in a multitude of approaches with inherent applications in security or industry configuring a broader and thorough spectrum of RPAS use in CBR operations. In this case, it was considered as a joint demonstration concept with both land and maritime components of the proposed scenarios for validation.

REFERENCES

Austin, R. Unmanned Aircraft Systems: UAVS Design, Development and Deployment, Wiltshire: Wiley Publications, 2010.

Hendricks, G.C., Hall, M.G. "The History and Science of CBRNE Agents, Part 1," 2012.

NATO, "Guidelines For First Responders To a Cabin Incident," NATO Civil Emergency Planning Civil Protection Group, 2014.

Shea, D.A. "Chemical Weapons: A Summary Report of Characteristics and Effects," Congressional Research Service, 2013.

Unal, B., Aghlani, S. "Use of Chemical, Biological, Radiological and Nuclear Weapons by Non-State Actors—Emergency trends and risk factors," Lloyd's Emerging Risk Report 2016, Innovation Series, 2016.

Risk Analysis and Management – Trends, Challenges and Emerging Issues – Bernatik, Huang & Salvi (Eds)
 ISBN 978-1-138-03359-7

A framework for uncertainty analysis of flood risk assessment in the near-response phase

Toni Kekez, Snježana Knezić & Roko Andričević
Faculty of Civil Engineering, Architecture and Geodesy, University of Split, Split, Croatia

ABSTRACT: Flood risk assessment methods are usually based on measured water level values and vulnerability estimation of an affected area. Choosing adequate structural or non-structural measures for reducing the negative impacts of flood is often difficult because of the influence of different uncertainty sources. The aim of this paper is to provide a framework for the uncertainty analysis of flood risk assessment in the near-response phase, which mostly considers epistemic uncertainties resulting from ambiguity or lack of information. Decision-making in disaster response is complex and includes choosing between different risk-mitigation options. Uncertainty analysis helps to identify flood risk management options that are likely to be robust to uncertainty. This study helps to identify the main sources of uncertainties, which can affect risk value at the observed point or section of the river, and to minimize the information gap between the location where the water level is measured and the endangered area. Sources of uncertainties are identified and determined using the systems analysis approach, and relevant uncertainties are included in the risk assessment model. This approach enables decision-makers, especially at tactical level, to plan the disaster response phase more effectively. Adequate information gives support to decision-makers in order to make optimal decisions, which can only be expected when all relevant uncertainties are taken into consideration. Both reliable information and more comprehensive risk analysis are necessary to enhance safety of citizens and improve risk reduction.

1 INTRODUCTION

Flood risk relates to the combination of the probability of a flood event and the potential adverse consequences for human health, the environment, cultural heritage, and economic activity associated with such an event (DIRECTIVE 2007/60/EC). In general, risk consists of hazard and vulnerability, which in different functional relationships define risk values (Pistrika and Tsakiris, 2007). The state-of-the-art flood risk management not only depends on flood defense structures (structural measures) but also considers several nonstructural measures that may be used to reduce the severity of flood or reduce its consequences (Petry, 2002). Choosing adequate measures for reducing the negative impacts of flood is often difficult because of the influence of different uncertainty sources, especially in a response phase. Emergency response plans are usually based on predefined flood risk maps, which define the most vulnerable areas and prioritize evacuation and rescue actions. A better understanding of uncertainties and their impact on decision-making helps to identify flood risk management options that are likely robust to uncertainty. Results of MATRIX project (EU funded FP7) point out the importance of uncertainty characterization and propagation in risk assessments as a support for decision-making, which can be under significant risk and can lead to inadequate emergency response.

2 UNCERTAINTY ANALYSIS

Uncertainty analysis is required for a better understanding of complex systems, and it can support risk-based decision-making (Hall and Solomatine, 2008). In general, uncertainties

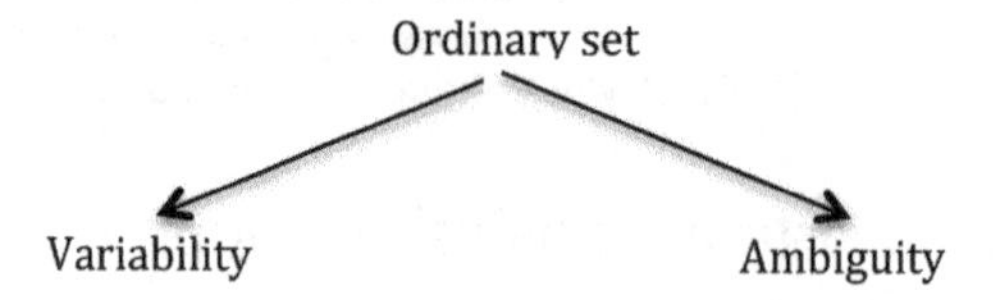

Figure 1. General uncertainty classification (Ling and Simonović).

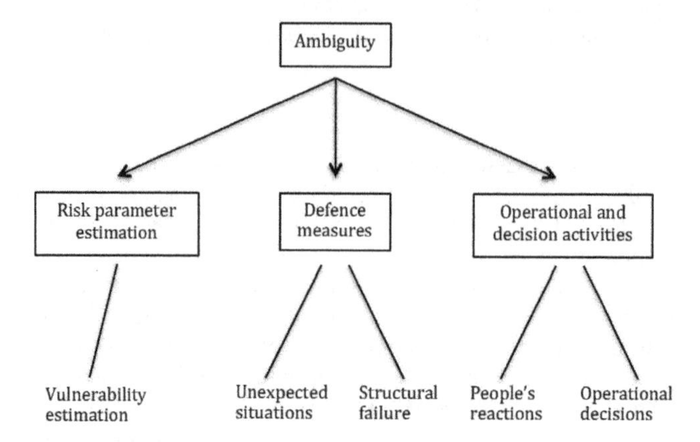

Figure 2. Uncertainty classification in near-response flood risk management.

can be divided into two main groups: aleatory uncertainties and epistemic uncertainties. Natural uncertainty arises from variability of the underlying stochastic process, and epistemic uncertainty results from incomplete knowledge about the process under study (Merz and Thieken, 2005). Ling (1993) and Simonović (2011) proposed uncertainty classification in two major groups: variability and ambiguity (Figure 1).

Apel et al. (2004) analyzed uncertainties of different parameters in the preliminary phase of flood risk assessment, but still many unknowns are yet to be discovered and analyzed. In case of floods, variability due to natural randomness is represented through river discharge values or flood magnitudes, and they are analyzed with hydrological or hydraulic analyses. Natural variability of floods in the near-response disaster phase can be neglected because the upstream information about flood magnitude is considered reliable.

Uncertainty due to ambiguity is the main consideration, which includes other sources of uncertainty that can directly affect the risk value. Ambiguity arises from lack of knowledge or unexpected situations that can affect security of people. Ling (1993) and Simonović (2011) defined ambiguity in three basic parts: model, parameter, and decision. Model ambiguity refers to variable and model imperfections that represent the system; parameter ambiguity refers to data errors; and decision ambiguity quantifies the social impact of risk.

In case of floods, the concept proposed by Ling (1993) and Simonović (2011) is modified in a way that it represents major parts of flood risk management and that it can be applicable in any area endangered by floods. In case of natural disasters, such as floods, it is important to assume that uncertainties will arise no matter if we consider them or not. Sources of uncertainty due to ambiguity in case of floods can be defined in three major groups (Figure 2): risk parameter estimation, defense measures, and decision and operational activities.

3 METHODOLOGY FOR UNCERTAINTY IDENTIFICATION

Risk analysis can be represented as a model of system with possible consequences estimated. Risk parameter estimation (or risk estimation) represents the basis for risk management options. In the near-response phase, uncertainty may arise from lack of knowledge about system vulnerability; when the flood magnitude is known, flood hazard and exposure are also considered familiar. Defense measures play an important role in flood risk management, and they are associated with risk reduction and calibration. Defense measures depend on possible unexpected situations in defense reliability, which can lead to a possible increase of risk. Operational and decision activities consider human activities, which may have a direct influence on the risk value and negative consequences. Human operational activities can lead to a major increase of risk because of their unpredictability. Decision activities of affected people and their reactions can significantly complicate the activities of emergency responders.

Identified uncertainties in the near-response flood risk management are listed in Table 1. They cover different parameters and values that can be ambiguous during response planning and implementation.

The goal of this study is to develop a robust methodology for uncertainty identification in the near-response phase of flood risk management that can be applicable to any area that is under possible threat from floods. Using the systems analysis approach, major uncertainties are identified, and their connection with risk parameters is presented. Uncertainties have a major influence on hazard, vulnerability, and risk. The aim of this paper is to establish a robust framework for incorporating the main aspects of uncertainty in flood risk management decision-making, which refers to near-response time, and to identify system elements that contribute to insecurity and risk. Uncertainties are defined as objects, and through their relationships, it is possible to determine which uncertainties contribute to risk parameters and/or on each other (Figure 3). For example, an early warning system failure can directly

Table 1. Identified uncertainties in near-response flood risk management.

Risk parameter estimation	Defense measures	Operational and decision activities
Population age	Probability of dike overtopping	Evacuation decision
Number of swimmers	Probability of dike failure	Floodgate or dam human operational error
Population density	Probability of floodgate or dam failure	
	Early warning system reliability	

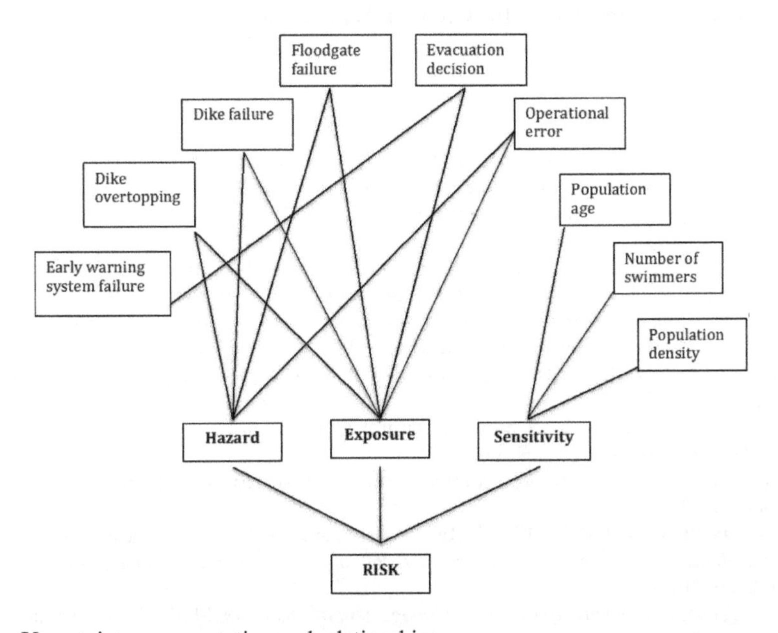

Figure 3. Uncertainty representation and relationships.

influence evacuation decision, which leads to an increase of exposure and risk. Uncertainties can be described using mathematical functions and can be propagated with different methods, such as Bayesian analysis or event trees, resulting in a comprehensive risk analysis.

The major goals in flood management are optimal decisions, which provide security of people and accelerate the recovery phase after flood disaster. Information uncertainty can generate risk in decision-making, which can lead to an inadequate system response. Inadequate response (under-reaction or over-reaction) can lead to multiple expenses and, in addition, different reactions of other stakeholders can increase the negative consequences and the expected damage. With this approach, it is possible to achieve an integrated flood risk management, which involves the integration of residual risks (risk that may remain unmanaged) and epistemic uncertainties that refer to lack of knowledge about human reactions during flood (evacuation decision) and operational errors (humans operating spillways on dams or floodgates) into a comprehensive risk management. Taking uncertainties into account, risk management can provide a less impact on system, that is, a shorter recovery time and an increase of system resilience.

4 CONCLUSIONS

Uncertainty analysis is required for a more efficient flood risk management through different flood mitigation measures. Understanding uncertainty in flood risk management can support risk-based decision-making in emergency response, whose dominant role has epistemic uncertainties resulting from ambiguity or lack of knowledge about different processes that may occur. Using the systems analysis approach, an uncertainty identification framework has been developed. The identified uncertainties are associated with risk parameters in order to determine their possible influence on risk values and on each other. The integration of residual risks and epistemic uncertainties can lead to integrated flood risk management. Taking uncertainties into account, the impact on the system can be reduced, thereby enhancing security of people and shortening the recovery time of the system.

ACKNOWLEDGMENTS

This work was fully supported by the Croatian Science Foundation.

REFERENCES

Apel, H., Thieken, A.H., Merz, B., Bloschl, G. (2004). Flood risk assessment and associated uncertainty, Natural Hazards and Earth System Sciences, 4: 295–308.

DIRECTIVE 2007/60/EC OF THE EUROPEAN PARLIAMENT AND OF THE COUNCIL of 23 October 2007 on the assessment and management of flood risks.

Hall, J., Solomatine, D. (2008). A framework for uncertainty analysis in flood risk management decisions, *International Journal of River Basin Management*, Vol. 6, No. 2, pp. 85–98.

Ling, C.W. (1993). *Characterising Uncertainty: A Taxonomy and an Analysis of Extreme Events*, MSc Thesis, School of Engineering and Applied Science, University of Virginia, VA.

MATRIX (New methodologies for multi-hazard and multi-risk assessment methods for Europe), EU funded FP7 project.

Merz, B., Thieken, A.H. (2005). Separating natural and epistemic uncertainty in flood frequency analysis, *Journal of Hydrology*, 309, 114–132.

Petry, B. (2002). Coping with floods: complementarity of structural and non-structural measures, *Flood Defence*, 60–70.

Pistrika, A., Tsakiris, G. (2007). Flood Risk Assessment: A Methodological Framework, *Water Resources Management: New Approaches and Technologies*, European Water Resources Association, Chania, Crete-Greece.

Simonović, S.P. (2011). *Systems Approach to Management of Disasters*, Methods and Applications, John Wiley & Sons, Inc.

Risk Analysis and Management – Trends, Challenges and Emerging Issues – Bernatik, Huang & Salvi (Eds)
© 2017 Taylor & Francis Group, London, ISBN 978-1-138-03359-7

Reaching out: Demonstration of effective large-scale threat and crisis management outside the EU

Tomasz Grzegory
Astri Polska Sp. z o.o., Tamka, Warszawa, Poland

ABSTRACT: The recent catastrophes involving huge earthquakes in many places (e.g., Italy, Nepal) around the world; nuclear disaster in Fukushima; large evacuation and humanitarian support in Tunisia; and the Ebola epidemic in West Africa require strict international cooperation in both the prevention and response phase. Effective EU support to a large external crisis requires new approaches. The topic is wide and cuts across multiple knowledge domains and functions, e.g., risk modeling, responders CBRNe protection, healthcare equipment, deployable facilities, identification systems, surveillance systems, location and tracking systems, communications, simulation, and training. In response to this challenge and to identify the user and market needs from previous projects, the Reaching Out initiative proposes an innovative multi-disciplinary approach that will optimize efforts, address a wide spectrum of users and maximize market innovation success. The Reaching Out consortium is responding to the Horizon 2020 security topic "Crisis management topic 3: Demonstration activity on large-scale disasters and crisis management and resilience of EU external assets against major identified threats or causes of crisis" where the 3-year project will demonstrate the deployable EU disaster and crisis management capabilities applied in real situations outside the EU. The project 5 demonstration outputs will aim to allow capabilities to be shared among multinational and intercontinental stakeholders, which is paramount in cross-border incident/crises management and over time will allow for a build-up of common capability not only across European boundaries but international boundaries as well. Reaching Out is supported by different *Associated Partners* from the demonstration countries outside the EU. The project will provide *innovative solutions* to improve crisis resilience and allow enhanced interoperability and effectiveness between the EU and other crisis operators from different continents. This project will increase the technical maturity and performances of distributed situation awareness, monitoring, and decision support systems.

ACKNOWLEDGMENTS

Project name: Reaching Out; *Funding*: European Union, European Commission, Horizon 2020 (H2020); *Topic*: Crisis management Topic 3 (DRS-03-2015); *Budget*: close to 19M€; *Duration*: started October 2016 (duration 38 months); *Consortium Partners*: 27 partners representing a mix of end users, industry, SMEs, academia and RTOs coordinated by Airbus Defence and Space SAS.

Risk Analysis and Management – Trends, Challenges and Emerging Issues – Bernatik, Huang & Salvi (Eds)

European standardization in nanotechnologies and its relation to international work: How can standardization help the industry and regulators to develop safe products?

Patrice Conner
AFNOR Standardization, Management and Consumer Services Department, La Plaine Saint-Denis, France

ABSTRACT: Nanotechnology has enormous potential to contribute to human flourishing in responsible and sustainable ways. It is rapidly developing the fields of science, technology, and innovation. As enabling technologies, their full scope of applications is potentially very wide. Major implications are expected in many areas, e.g. healthcare, information and communication technologies, energy production and storage, materials science/chemical engineering, manufacturing, environmental protection, consumer products, etc. However, nanotechnology is unlikely to realize its full potential unless its associated societal and ethical issues are adequately attended. Namely, nanotechnology and nanoparticles may expose humans and the environment to new health risks, possibly involving quite different mechanisms of interference with the physiology of human beings and environmental species. One of the building blocks of the "safe, integrated and responsible" approach is standardization. Both the Economic and Social Committee and the European Parliament have highlighted the importance to be attached to standardization as a means to accompany the introduction on the market of nanotechnology and nanomaterials, and a means to facilitate the implementation of regulation. ISO and CEN have respectively started in 2005 and 2006 to deal with selected topics related to this emerging and enabling technology. In the beginning of 2010, EC DG "Enterprise and Industry" addressed the mandate M/461 to CEN, CENELEC, and ETSI for standardization activities regarding nanotechnology and nanomaterials. Thus, CEN/TC 352 "Nanotechnologies" has been asked to take the leadership for the coordination in the execution of M/461 (46 topics to be standardized) and to contact relevant European and International Technical committees and interested stakeholders as appropriate (56 structures have been identified). Prior requests from M/461 deal with characterization and exposure of nanomaterials and any matters related to health, safety, and environment. The following questions will be answered: what are the structures and how they work? Where are we right now and how work is going from now onwards? How CEN's work and targets deal with and interact with global matters in this field?

Keywords: nanotechnology, responsible and sustainable ways, enabling technologies, healthcare, information, communication technologies, energy production, storage, materials science, chemical engineering, manufacturing, environmental protection, consumer products, nanoparticles, health risks, nanomaterials, regulation, ISO, CEN/TC 352 "Nanotechnologies", characterization, exposure, health, safety

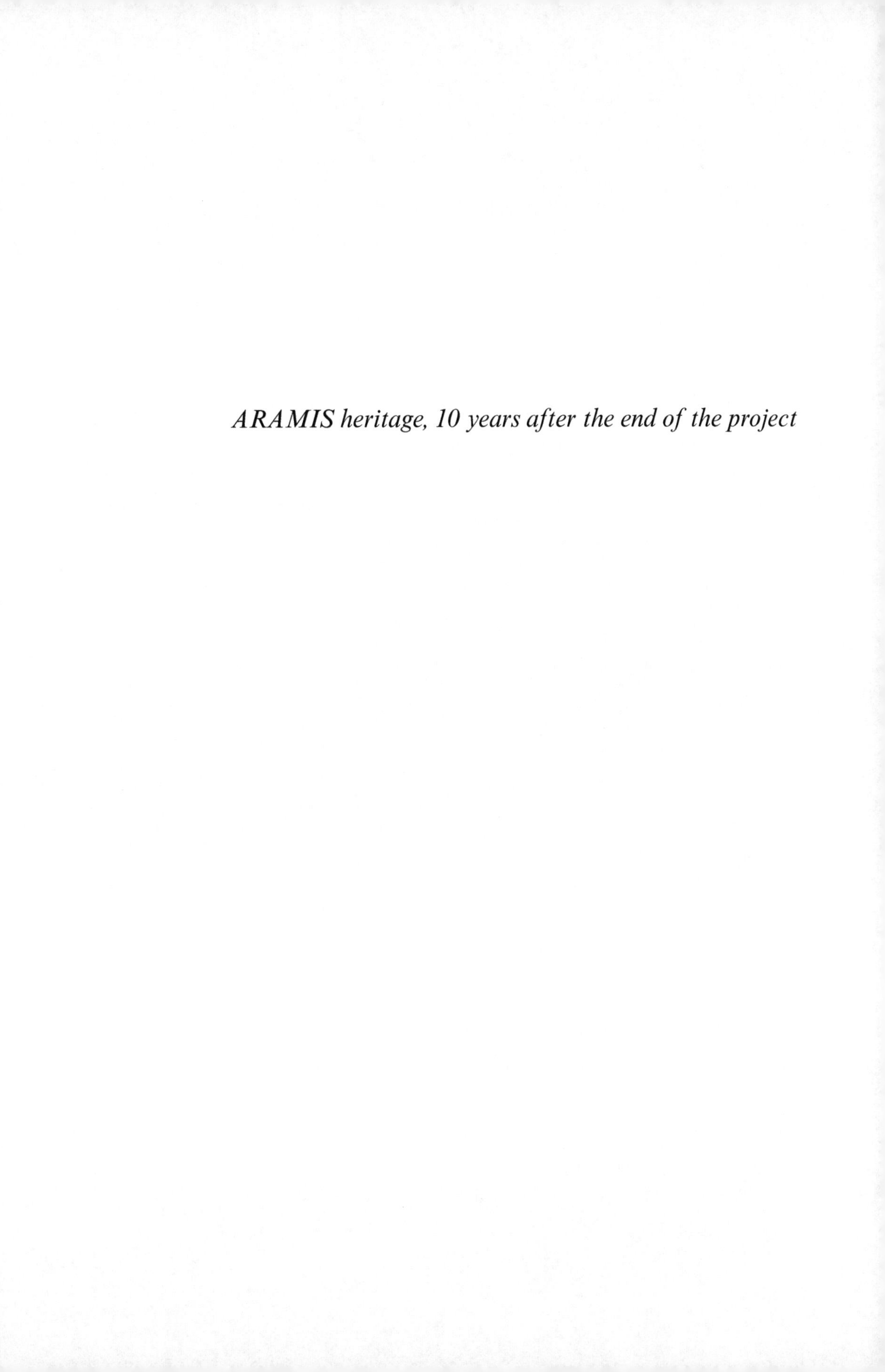

ARAMIS heritage, 10 years after the end of the project

Risk Analysis and Management – Trends, Challenges and Emerging Issues – Bernatik, Huang & Salvi (Eds)
ISBN 978-1-138-03359-7

Improving the safety management systems at small Seveso establishments through the bow-tie approach

Paolo Angelo Bragatto, Silvia Maria Ansaldi, Annalisa Pirone & Patrizia Agnello
Department of Technological Innovations, INAIL Italian Workers' Compensation Authority, Monteporzio Catone, Rome, Italy

ABSTRACT: Many Small and Medium-sized Enterprises (SMEs) are obliged to adopt a Safety Management System (SMS) by the Directive 2012/18/EU for the control of major accident hazards (Seveso III). The difficulties of implementing an SMS within a small-sized work organization are well known by practitioners, including inspectors and auditors. In this paper, how the bowtie model, introduced 10 years ago by the ARAMIS project, may be exploited to have an easy and effective SMS, suitable also for small establishments, is discussed. In the audit procedure adopted by Italian competent authorities, the bow-tie approach is already present, even though in a complementary position with respect to the classic check list. The proposed method stresses the use of the bow-tie approach. The bow-ties connect equipment, operating instructions, procedures, and safety documents to the top events, as identified in the risk analysis. Near misses' discussion exploits the potential of a net representation to find the weaknesses of safety barriers and the distance from potential top events. In the proposed method, the audit too is based on the operating experience, rather than on checklists. The method has been implemented by using a web-based application, named as AGILE-G, which has been tested in a few sample sites. It has been demonstrated as being suitable to involve workers in an effective prevention of major accidents, especially in SMEs.

1 BACKGROUND AND OBJECTIVES

1.1 *Background*

Before 20 years, the Seveso II Directive (1996) had required the Competent Authorities to organize a system of ongoing inspections that are adequate to a planned and systematic examination of the technical, organizational, or managerial systems being adopted at the establishment. The new Seveso III Directive (2012) has confirmed definitely the importance of inspections. The Seveso Inspection system is the most powerful and effective tool available to authorities for enforcing the major accident hazard legislation, as described by Basso et al. (2004) and Wood (2005). The Seveso Inspection is basically an audit on the Safety Management System (SMS). The most common auditing method is carried out by using a categorical check list, which is aimed at scrutinizing the main SMS chapters, including the following: policy, organization and personnel, risk assessment, operating control, management of changes, emergency planning, performances monitoring and accident analysis, and control and revision.

A study by De Bruin & Swuste (2008) suggests that the traditional categorical audits should be replaced by risk-based audits. Audits should take a hazard as a starting point to ensure that the SMS that is being assessed is capable of handling the actual hazards. By auditing from the perspective of a possible incident scenario, the approach is much more practical and vivid. Risk-based audits can be based on bow-tie diagrams. The audit should assess whether the barriers in between are adequate in such a scenario. Bow-tie-based audits are a more risk-based approach to auditing. Instead of focusing on abstract categories, the scenario focuses on the barriers that reduce the unique risks of an organization. The direct link to scenarios and barriers makes it easier to make concrete recommendations for improvement. The seed of the idea was initially placed by AVRIM2 and TRAM methods. Both TRAM and AVRIM2 were developed to assist inspectors. TRAM auditing and the inspection tool by the UK Health and Safety Executive (HSE) for application at sites falling within the scope of the Seveso Directive is based on scenarios defined and the associated protective measures, which is termed as "Lines Of Defense" (Naylor et al. 2000). One of the principal features of AVRIM2 is the explicit link that has been constructed between Lines of Defense and aspects of the Safety Management System (Bellamy et al. 1999).

Thus, both methods were not so far from the idea of preventive and protective barriers, which the bow-tie method is based on. A further and essential step was then taken by the ARAMIS project, which developed a comprehensive methodology for the identification of reference accident scenarios in process industries. As described by Salvi and Debray (2006), the concept was developed and applied in the framework of the EU Seveso II Directive by using the ARAMIS project. The bow-tie method is basically a combination of fault trees and event trees. Fault trees were drawn on the right-hand side and captured the threats, while event trees showing the mitigation systems were drawn on the right-hand side.

The central top event is usually a loss of containment. In the ARAMIS model, the bow-ties guide the dialogue between the company and audit team. The result of this dialogue is a mapping of the company's choice of barriers onto the generic bow-ties (Delvosalle et al. 2006). The bow-tie approach is the most important legacy of the ARAMIS project, which influenced the management of the major accident hazard to a great extent.

1.2 *Objectives*

The audit of safety systems at the Italian Seveso establishments has been one of the tasks of our Department for almost 20 years. The task is performed with other control bodies, including National Fire Brigades and Regional Environment Protection Agencies. Such an experience enables us to discuss the issue from the regulatory point of view. Our Department, first of all, is a research institute; therefore, the operating experience is exploited to provide ideas and inspiration to develop innovative tools and which, at the end, is targeted to meet user needs. Therefore, the scope of the paper embraces both the regulatory point of view and the operating one.

This paper has two basic objectives:

1. to discuss the actual effects of the bow-tie approach on improving the practice of Seveso Inspections in the last decade in Italy;
2. to discuss the potential effect of bowtie approach for the improvement of SMS.

For objective 2, the author discusses the experience of an innovative tool for SMS, which was developed starting from a few basic ARAMIS concepts. This tool is suitable for the improvement of SMSs at small Seveso establishments, where a systematic approach for an SMS could be too heavy. Sections 2 and 3 deal with objectives 1 and 2, respectively. A few more general thoughts are provided in Section 4.

2 SEVESO INSPECTIONS IN ITALY

2.1 *Seveso II*

2.1.1 *Seveso inspection guideline, 1st release*

Just 1 year after the Seveso II (Directive 1999), the Italian Competent Authority adopted the very first guidelines for the inspections at Seveso establishments. It was based on a detailed check list, which was organized in a hierarchical way and featuring 100 questions about Safety Management Systems (SMSs). Before scrutinizing the procedures of the SMS, the inspectors were required to analyze the operating experience, including near misses, in order to prioritize the points in the check list.

2.1.2 *Seveso inspection guideline 2nd release*

In 2008, a revision of the guidelines was decided, which was aimed at answering a few remarks of the EU Commission. There was an interesting debate in the working group that was in-charge of preparing the new guideline. The ARAMIS project had been concluded recently, and some enthusiasts wanted to adopt the approach based on the bow-tie even in Seveso inspections.

According to the ARAMIS approach, the main purpose of the audit activities was to quantify those aspects of the management system that have a direct impact on the reliability and effectiveness of the barriers and hence, the probability of the scenarios involved (Guldemund et al. 2006).

These ideas encountered a very strong resistance in the Italian working group, because the check list was considered by the majority in a comprehensive and systematic manner. According to them, it was the only method that could ensure equitable treatment for all companies. They criticized the bow-tie approach for inspection as the results are supposed to be, despite the check list, dependent on the specific knowledge of the inspectors. At the end, enthusiasts had to settle for considering the bow-tie as a complementary method.

Plant operators were asked to identify the barriers (technical, procedural, and managerial) put in place to prevent accidents and mitigate the consequences for each top event. These were exactly bow-ties, though they had to be used only for examination of the technical systems, while for the rest of the inspection, it was mandatory to follow the checklist, which had since been further burdened.

2.1.3 *Five years of experience with the 2nd release*

The new Seveso inspection guideline was delivered in 2008 and since that year, our inspectors have gathered hundreds of bow-tie documents. In order to understand the impact of the bow-tie approach on the practice of Seveso inspections, Bragatto et al. (2014) published a survey, which analyzed the bow-ties of 26 establishments, which were inspected after the release of the new guideline. Three hundred and seventy-four top-events with their bow-ties were analyzed. The bow-ties included some 2500 preventive measures and 1500 protective measures. The sample was very diverse indeed, including a number of LPG depots, chemical warehouses, and galvanic plants, as well as one refinery.

2.2 *Seveso III*

In July 2012, a new Directive on major accident hazards was issued (Directive 2012) and it was implemented in the national legislation in July 2015. The 2008 guidelines were included, with moderate revision, in the Seveso III text. Thus, in Italy, we continue to use the double approach: the risk-based approach (i.e. bow-tie) and the systematic (i.e. check list) approach. In time, the application of the check list in the inspections became less and less formal, giving more and more space for the evaluation of preventive and protective barriers to the inspectors.

In the transition from Seveso II to Seveso III, plants involved were more or less the same; thus, in many cases, they have already received a number of inspections always with the same check list over the past 15 years, which is less and less effective. Thus, it is better to focus on the study of near misses and analysis of safety barriers.

In the inspectors' experience, when a risk-based approach (i.e. bow-tie + near miss) is adopted, may be a few trivial non-conformities are not detected; but for sure, all significant flaws are discovered. This is the exact the opposite of the check list, which looks very well at the very small defects, but does not capture the more serious ones.

There is also an economic reason for one to prefer, in Seveso Inspections, the bow-tie to the check list approach. Companies, in fact, pay to the competent authorities a fee for the inspection, which depends, basically, on the type of establishment. The political decision makers have fixed very low fees at the national level, which in a few regions may be even lower. There are, furthermore, discounts to reward complying companies. Thus, it is essential to optimize the time spent at establishments, thereby preventing uneconomic conditions.

For all these reasons, the study of the near misses and the analysis of the safety barriers represented through the bow-ties are now becoming the true pillars of the Seveso inspections, but for new establishments, where a first systematic screening of the SMS is irreplaceable indeed.

3 AGILE-G PROJECT

The AGILE-G project was originated by a project that was funded a few years ago by an Italian Ministry of Health, aiming at developing an SMS, suitable also for SMEs. The need of a more effective SMS for SMEs was recognized by both practitioners and scholars (McGuiness & et al. 2012). In particular, the gap between safety documents and real operation at the shop floor may be a major drawback for the successful implementation of the SMS (Agnello et al. 2012).

The basic idea of the AGILE-G project was born from the experience of the inspections. After the survey described in the previous chapter, we saw the potential of the bow-tie to focus all the efforts on the actual hazards and counter them. Through the bow-tie approach, a more effective and integrated management of technical systems, the operating instructions, safety procedures, safety training, and other resources that used to prevent accidents and mitigate their consequences was achieved. We understood also the difficulties of implementing such a focused system at small-sized companies and recognized the need of a software tool, which is preferably web-based and usable also via smartphones or tablets.

The core of the new product was the Safety Digital Model, based on which we developed a web-based application that is also usable on smartphones and tablets.

3.1 *Safety digital model*

The original idea of the Safety Digital Representation (SDS) was introduced by Bragatto et al. (2015). In this representation, as shown in Figure 2, the very center is the hazard and,

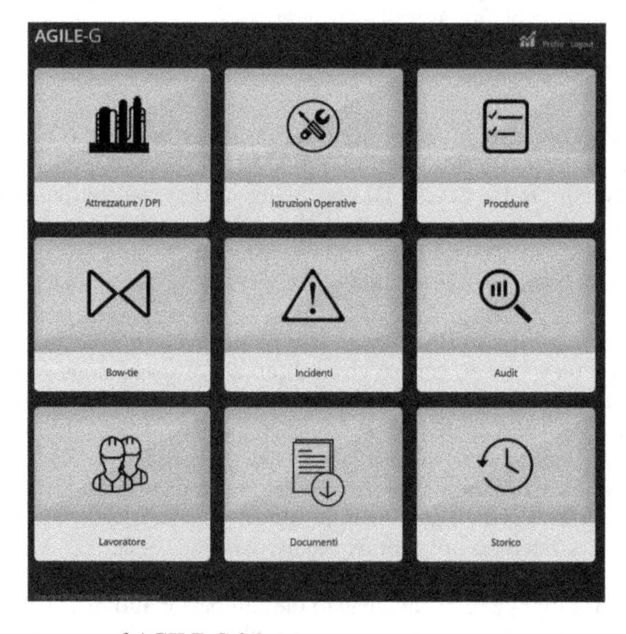

Figure 1. The home screen of AGILE-G 2.0.

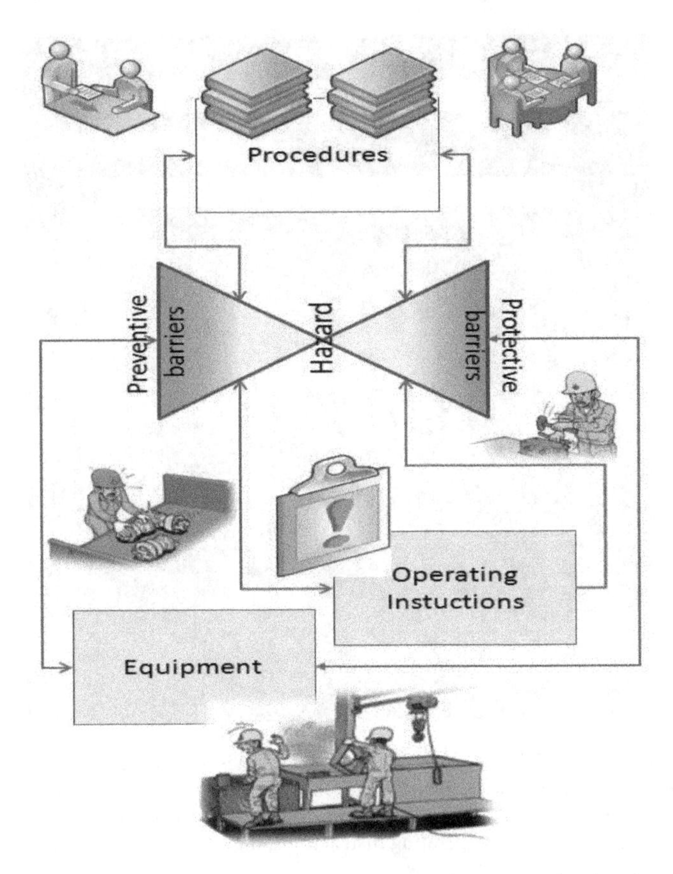

Figure 2. The "Safety Digital Representation" that connects hazards, safety barriers, and workers on the shop floor.

in particular, to the top events. SDS connects the procedures, the operating instructions, and the technical systems, which are both preventive and protective, to the potential hazards. SDS builds a digital network, which includes all resources that are relevant for accident prevention and protection, both physical and intangible. The items that are not related to a hazard are not included in the model. The very core of this logical model is the bow-tie representation, as shown in Figure 3.

For each top event, there is a bow-tie, which links to the technical and organizational systems that are adopted for prevention or mitigation purpose. Both technical systems and procedures and operating instructions are considered. In such a way, physical and intangible barriers are linked to each other. This digital model may be a backbone, around which everything revolves, as the safety documents should contain the knowledge of the establishment, which could improve safety management at every stage.

3.2 *The "AGILE-G" prototype*

The main scope of the AGILE-G prototype is to achieve control of major accident hazards, but occupational hazards may be included. There are different user profiles to access the software, including worker, supervisor, safety manager, and owner.

The AGILE-G prototype is able to manage the equipment, the operating instruction, and the procedures, linking them to the major hazards, according to the bow-tie schema. By means of AGILE-G for each hazard, it is possible to see together all the safety barriers. Green and red tones are used to discriminate preventive and protective barriers.

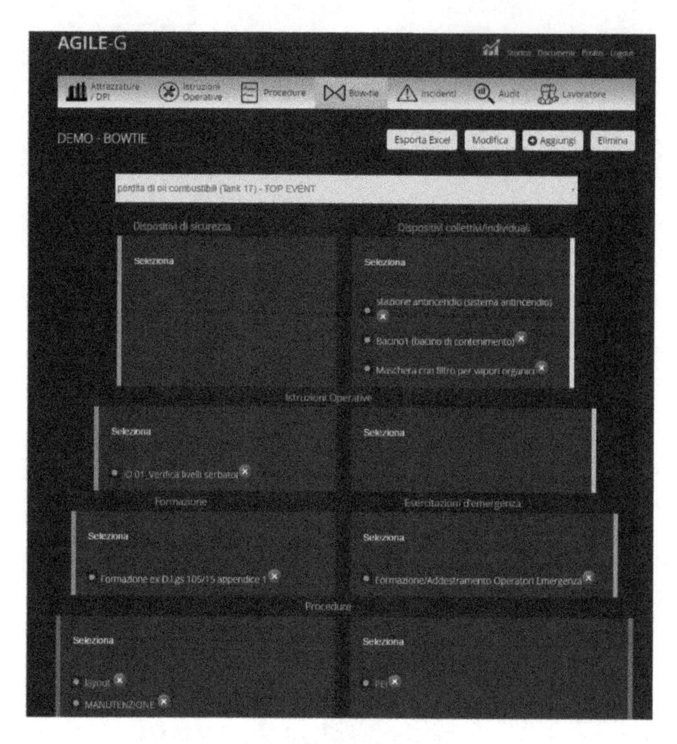

Figure 3. A screenshot of AGILE-G. A bow-tie combination is represented here. Under the menu, on the top, the centered white pane shows the hazards. On the left-hand side, the preventive barriers (shade of green) are present; on the right-hand side, the protective barriers (shade of red) are present. From top to bottom, the physical barriers, the operating instructions, the training resource, and the management procedures are present.

By means of the AGILE-G project, workers are involved in the improvement of the systems by reporting and discussing near misses and anomalies. The near miss discussion is facilitated by the bow-ties embedded in the SDS, as they allow to detect the event to which the near miss has approached and to understand the holes in the safety barriers, which may be repaired. The weaknesses found out in safety barriers during the discussions of near misses are memorized to address the audit of the system, according to the schema introduced 1 year ago by the ARAMIS project.

AGILE-G helps also Safety Managers to provide the workers with training packages and protective personal equipment considering the tasks and the related hazards. AGILE-G capabilities, furthermore, include the verification of the system consistency and the automatic monitoring of selected indicators.

3.3 *Experience with AGILE-G*

3.3.1 *Testing at SMEs*

Testing of AGILE-G initially was carried out in a simulated environment, by using real data collected from various plants. Out of these, three small Seveso plants were particularly significant: a galvanizing plant, a polyurethane adhesives factory, and a mineral oil depot. The tool was demonstrated as being very effective for the discussion of near misses. The procedures and instructions questioned during the near miss discussion are marked by using a color code. This was very useful for the periodical revision of the Safety Management Systems, as most suggestions coming from workers were implemented by the operator (Figure 4).

Figure 4. A worker is using the AGILE-G project to report an anomaly detected at the shop floor. This will activate a co-operative discussion, which will eventually lead to an improvement of safety barriers.

3.3.2 *Integration with occupational safety*

Further tests were carried out to test the applicability of AGILE-G in contexts where there was already a formal management system for occupational safety (e.g., 18001). These tests have revealed significant incompatibility of the product with the formal constraints of a certified system. We have made note of the highly dynamic approach to the AGILE project as well, which could clash with the formalism of the occupational Safety Management System. We have maintained our approach, but we had to accept some trade-offs with specific aspects required by the occupational Safety Management System. In particular, the new version has a lot more attention for individual workers and considers in particular, training requirements, certification, and qualification required by different tasks. We have also taken into account the development of ISO 45000, in anticipation that this standard will finally be approved.

4 CONCLUSIONS

The idea of the bow-tie, albeit introduced about 10 years ago, is more alive than ever; we expect further technological developments to lead us in this direction. The three points briefly discussed here are just examples of the huge potential of the bow-tie approach.

4.1 *Smart SMS and IoT*

The dynamic properties of the bow-tie approach are compatible with the revolution achieved by the Internet of Things (IoT) and the smart objects that are now available in many industrial plants. There are many proposals, both in research and in the market, for new Smart Systems for the Safety (SSSs). The proposals are often intriguing, but there is the risk to have just a number of nice "safety gadgets" that are not relevant for major hazards.

In order to avoid that, it is essential to have a clear and logical model in mind and use it to choose the SSS and to integrate them in the safety management.

Something like the AGILE-G project, based on bow-tie schema, could be the backbone of future "smart" applications for safety. In other words, any new "smart systems for the safety" should be placed within an ideal "bow-tie". It should be useful for the following:

1. *or to prevent directly an accident or mitigate the consequences*
2. *or to support the application of safety procedures and operating instructions*
3. *or to facilitate the workers to have a safer behavior, both in normal and in emergency situations.*

Data and information gathered by the smart sensors could be sent to the right position in the framework of barriers, thereby reducing effectively the likelihood or the severity of potential accident events.

In other words, the new smart systems for the safety are expected to become additional safety barriers (preventive or protective) to be included in the bow-tie approach. Otherwise, these are just "gimmicks". The focus of the SMS should move from the formal compliance to an effective management of the safety barriers, which is supported by both workers' involvement and smart systems.

4.2 *Smart audits*

The control of major accident hazards involves plant operators and regulatory bodies. Mandatory inspection required by the Seveso Directive should adopt audit procedures based on the bow-tie schema. This could improve the effectiveness of the inspections and could promote a more dynamic management of safety throughout the process industries. The time is definitely right for that.

4.3 *Dynamic risk assessment*

The benefits of a dynamic updating of the quantitative risk assessment are a new frontier in risk research, as discussed by Villa et al. (2016). For this purpose, there is a requirement for a huge potential of the bow-tie approach. There are a few SSSs, which have been able to gather and transmit data continuously on issues that are critical and useful for the control of major accident hazards (e.g. thinning rate or micro cracks). If the new safety systems are managed within the bow-tie approach, the gathered data may be exploited for updating the values used for likelihood of failures, in order to obtain a true "dynamic" risk assessment.

REFERENCES

Agnello, P., Ansaldi, S.M., Bragatto, P.A. 2012. Plugging the gap between safety documents and workers perception, to prevent accidents at Seveso establishments.. Chemical Engineering Transactions, 26, 291–296.

Basso, B., Carpegna, C., Dibitonto, C., Gaido, G., Robotto, A., Zonato, C. 2004. Reviewing the safetymanagement system by incident investigation and performance indicators. J. of Loss Prevention, 17 (3), 225–231.

Bellamy, L.J. & Brouwer W.G.J. 1999. 'AVRIM2, a Dutch major hazard assessment and inspection tool'. J. of Hazardous Materials 65 (1–2) 191–210.

Bragatto, P.A., Ansaldi, S., Agnello, P. 2015. Small enterprises and major hazards: How to develop an appropriate safety management system. J. of Loss Prevention 33, 232–244.

Bragatto, P.A. Ansaldi, S. Antonini, F. Agnello, P. 2014. Bowtie approach for improved auditing procedures at "Seveso" establishments Safety, Reliability and Risk Analysis: Beyond the Horizon—Taylor & Francis, London.

De Bruin, M., & Swuste, P. 2008. Analysis of hazard scenarios for a research environment in an oil and gas exploration and production company. Safety science, 46(2), 261–271.

De Dianous, V., Fiévez, C. 2006. ARAMIS project: A more explicit demonstration of risk control through the use of bow-tie diagrams and the evaluation of safety barrier performance. J. of Hazardous Materials, 130 (3 spec. iss.), 220–233.

Delvosalle, C., Fievez, C., Pipart, A., Debray, B. 2006. ARAMIS project: A comprehensive metodology for the identifi cation of reference accident scenarios in process industries. J. of Hazardous Materials, 130(3), 200–219.

Guldenmund, F., Hale, A., Goossens, L., Betten, J., Duijm, N.J. 2006. The development of an audit technice to assess the quality of safety barrier management. J. of Hazardous Materials 130, 234–241.
McGuinness, E., Utne, I.B., Kelly, M. 2012. Development of a safety management system for Small and Medium Enterprises (SME's) Advances in Safety, Reliability and Risk Management Taylor & Francis, London 1791–1799.
Naylor P.J, Maddison, T. and Stansfield, R. 2000. 'TRAM: Technical Risk Audit Methodology for COMAH Sites'. Hazards XV The Process, its Safety, and the Environment—Getting it Right, IChemE Symposium Series 147.
Salvi, O., & Debray, B. 2006. A global view on ARAMIS, a risk assessment methodology for industries in the framework of the SEVESO II directive. J. of Hazardous Materials, 130, 187e199.
Seveso II Council Directive 96/82/EC of 9 December 1996 on the control of major-accident hazards involving dangerous substances. Retrieved from http://eur-lex.europa.eu.
Seveso III Directive 2012/18/EU of the European Parliament and of the Council of 4 July 2012 on the kontrol of majoraccident hazards involving dangerous substances. Retrieved from http://eur-lex.europa.eu.
Wood, M. 2005. The Mutual Joint Visit Programme on Inspections Under Seveso II: Exchanging lessons learned on inspection best practices. Institution of Chemical Engineers Symposium Series, (150), 977–994.

Risk Analysis and Management – Trends, Challenges and Emerging Issues – Bernatik, Huang & Salvi (Eds)
 ISBN 978-1-138-03359-7

ARAMIS heritage and land use planning in the Walloon Region

Christian Delvosalle, Sylvain Brohez & Jérémy Delcourt
Major Risk Research Centre, University of Mons, Mons, Belgium

Damien Beaudoint, Nathaël Cornil & Fabian Tambour
Polyris, Gosselies, Belgium

ABSTRACT: The ARAMIS project was supported by a funding in the 5th Framework Program of the European Commission under the program "Environment and Sustainable Development" and more specifically in "The fight against major natural and technological hazards" in 2002. In this 3 years' project, the Major Risk Research Centre of the Faculté Polytechnique de Mons played a major role in the work package related to the identification of accident scenarios in the process industries. At the same time, the Ministry of the Walloon Region worked in collaboration with the Major Risk Research Centre in order to develop a consistent and transparent methodology to assure a sustainable Land Use Planning (LUP) around Seveso plants. The approach selected for the risk assessment is similar to a full probabilistic approach, which is called a "QRA". This type of study, for the quantification of the accident consequences, required to identify the potentially hazardous equipment in the plant and to select relevant hazardous equipment. The MIMAH methodology developed in Aramis and more specifically the event tree approach was followed in the LUP methodology. For the Identification of Reference Accident Scenarios (MIRAS methodology), a generic value is taken for the failure frequency in the LUP methodology. However, in some cases, a fault tree can be developed taking into account the safety systems. This paper proposed to point out how the results of the ARAMIS project were integrated in the methodology of Walloon Region to assure a sustainable Land Use Planning.

1 INTRODUCTION

ARAMIS, acronym for Accidental Risk Assessment Methodology for IndustrieS, was an European project aiming to build up a new integrated risk assessment method for being used as a supportive tool to speed up the harmonized implementation of Seveso II Directive (Salvi & Debray, 2016). This project was supported by a funding in the 5th Framework Program of the European Commission under the program "Environment and Sustainable Development" and more specifically in "The fight against major natural and technological hazards" in 2002.

One of the aims of the ARAMIS project was to develop a methodology able to focus on the influence of safety systems and safety management in the definition of accident scenarios. The methodology proposed by ARAMIS is divided into four major steps: the identification of major accident hazards (MIMAH) (Delvosalle et al., 2005–2007), the identification of reference accident scenarios taking into account the safety barriers and the assessment of their performances (MIRAS) (Debray et al., 2004), the assessment and mapping of the risk severity (Planas et al., 2004) of reference scenarios and of the vulnerability (Tixier et al., 2004) of the plant surroundings.

The Major Risk Research Centre of the Faculté Polytechnique de Mons played a major role in the work package related to the identification of major (MIMAH) and reference (MIRAS) accidents scenarios.

Let us recall that in the end of the nineties, the article 12 of the Seveso II Directive (1996) requested that member states assure that their land use policy takes account of the need, in the long term, to maintain appropriate distances between establishments covered by the Directive and public areas. In Belgium, land use planning falls within the competence of regional authorities. With the experience achieved within the framework of the ARAMIS project, the Ministry of Walloon Region asked the Major Risk Research Center to develop a consistent and transparent methodology to assure a long-term land use planning around Seveso companies.

It is clear that different aspects of the ARAMIS results have been integrated in the Walloon methodology for LUP around Seveso plants. This paper proposes to point out how the results of the ARAMIS project were implemented in the land use planning around Seveso sites in Walloon Region.

2 THE ARAMIS PROJECT

Establishments where dangerous substances are present in one or more installations are obliged by the Article 5 of Seveso Directive to take all necessary measures to prevent major accidents and to limit their consequences for human health and the environment. In order to do so, a risk analysis is certainly an indispensable tool for an effective management of safety. This requires an awareness of the hazards and an identification of accidents scenarios.

The methodology developed in ARAMIS is based on a definition of risk which is the probability (or frequency) that an element of the territory suffers a damage. To observe a damage, a dangerous phenomenon with a given intensity has to hit a target element. The combination of frequency and intensity of the dangerous phenomenon is called the severity. The level of expected damage is determined by the vulnerability of the element and the severity of the phenomenon. The method aims at assessing separately the severity and the vulnerability to provide to decision makers elements to assess the resulting risk.

For the assessment of the severity of the phenomenon and the vulnerability of the environment, as already mentioned the ARAMIS methodology is divided into four major steps:

1. Identification of the major accident hazards (MIMAH);
2. Identification of the reference accident scenarios (MIRAS) taking into account the safety systems and management efficiency;
3. Risk severity mapping from the set of Reference Accident Scenarios;
4. Vulnerability mapping representing the sensitivity of one plant‘s surrounding environment.

The objective of MIMAH is to predict which major accidents are likely to occur in a chemical plant. A first step aims to identify relevant hazardous equipment in the considered plant. The equipment containing potentially hazardous substances (flammable, explosive, oxidising, toxic, dangerous for the environment) are selected if the mass of the hazardous substance in the equipment is higher or equal to a mass threshold. Then, in a second step, bow-ties are built for each equipment studied which represent the major accident hazard which could occur on the selected equipment. MIMAH offers tools to construct the bow-ties, centred on a critical event and developed through a fault tree on the left-side (causes) and an event tree on the right-side (consequences).

The objective of MIRAS is to select Reference Accident Scenarios (RAS) among the Major Accident Hazards identified with MIMAH. These RAS give an acute estimation of the risk level, because they take into account the safety systems implemented on the equipment. They will have to be modelled in order to calculate the Severity, which in turn will be compared with the Vulnerability of the surroundings of the plant.

3 LUP APPROACH IN WALLOON REGION

3.1 *Introduction*

In the Walloon region, the chosen approach for LUP is similar to a full probabilistic approach, called "QRA" (Quantitative Risk Assessment) in the Netherlands and United Kingdom, with some different assumptions from a classic QRA method on several points. The successive steps of the LUP Walloon methodology are summarized in Annex 1 (Delvosalle et al., 2006; Delvosalle et al., 2008; Servranckx, 2010; Tambour, 2010).

3.2 *The different steps of the approach chosen in Walloon Region*

3.2.1 *Gathering the information*

In a first step of the walloon methodology, it is necessary as in step 1 of MIMAH to collect information about the Seveso plant around which vulnerable zones have to be drawn:

list of substances;
list of equipment;
process data;
weather conditions;
...

These data are taken from the analysis of safety reports and completed by visits of the plant.

3.2.2 *Choosing the equipment contributing to the risk*

In a second step, it is necessary to select equipment which will be regarded for the risk assessment as it is done in steps 2 and 3 of MIMAH. An equipment is retained if it contains a quantity of hazardous substance higher or equal to a threshold-quantity depending on the

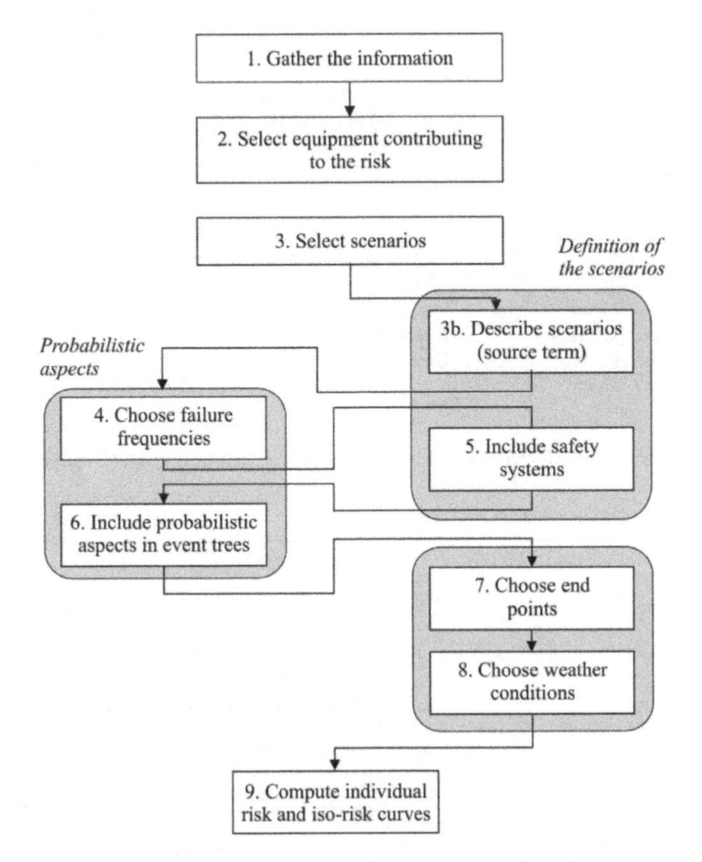

Figure 1. Summary of the steps followed for the determination of the consultation zones (Delvosalle et al., 2011).

property of the substance and its physical state. Thus it is possible to considerer only equipment contributing to a risk beyond the fence of the plant. In the frame of the method, a hazardous substance is defined as a substance belonging to one of the Seveso III categories. This method is presented in the Vade-Mecum of the Service public de Wallonie (Service public de Wallonie, 2016). Initially based on the Seveso II Directive, an update has been provided to take account of the CLP Regulation (Beaudoint et al., 2012). The threshold is then further adjusted according to the possibility of vaporisation and domino effects.

3.2.3 *Selecting appropriate scenarios*

In the Walloon Region, some scenarios are considered in a systematic way for each selected equipment. In general, the full rupture of the equipment is considered (catastrophic rupture of a tank, a pipe, a pump,...) and also breaches or leaks of various equivalent diameters (breaches of 35 and 100 mm diameter on a vessel, leaks of 22 and 44% of the nominal diameter for a pipe). These scenarios are summarized in Figure 2. These scenarios come partly from MIMAH methodology.

3.2.4 *Choosing tailure frequencies for the selected scenarios*

The main tool on which MIMAH is based, is the bow-tie (see Figure 3) which represents a major accident scenario centred on a critical event (defined as a loss of containment or a loss of physical integrity). The left part of the bow-tie (fault tree) identifies the possible causes of

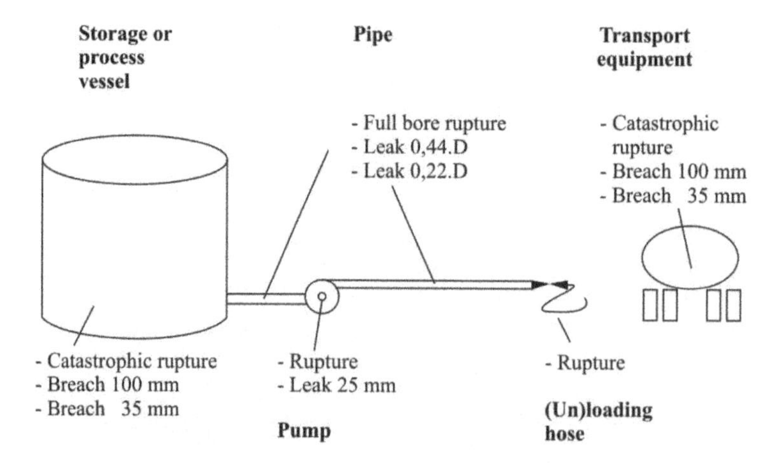

Figure 2. Example of pre-selected scenarios in the Walloon metodology (Delvosalle et al., 2006).

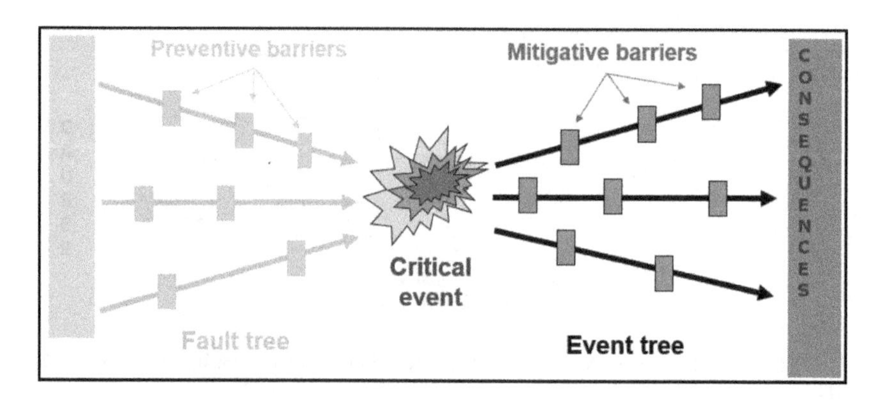

Figure 3. Bow-tie in the Walloon methodology.

a critical event. The right part (event tree) identifies the possible consequences of a critical event.

Based on generic frequencies (Fiévez et al., 2008), the walloon methodology does not take into account the fault tree. Indeed, it has been decided to use the same frequency failure rates as those used in the Flemish Region of Belgium, published in the "Handbook failure frequencies" (Handbook failure frequencies, 2009). A generic approach is used so there is not a failure frequency specific to the equipment studied with all the particularities of the plant. The bow-tie of ARAMIS is thus simplified without taking into account the fault tree, as presented in Figure 3. However, when the consequences are not acceptable outside the plant, the Seveso company can develop a specific fault tree taking into account the safety systems.

3.2.5 *Taking into account the influence of safety systems*

As in MIRAS methodology, safety systems are taken into account in order to perform an actual estimation of the risk around a Seveso site. We have to distinguish preventive safety systems (left part of the bow-tie; fault tree) from the mitigating ones (right part of the bow-tie; event tree). In the approach retained, it is not possible to take into account the devices acting upstream of the loss of containment (maintenance program, corrosion prevention,...). The generic frequencies of the critical event are independent of the quality of the design and operation of the installation. In some cases, it is possible to consider preventive systems, such as bunds and remote bunds, equipment in ventilated building, equipment with a scrubbing system,... Furthermore, generic failure frequencies are sometimes available including intrinsic safety measures (distinction between a regular and double wall atmospheric tank).

In the walloon methodology, only automatic mitigation safety systems are considered (see Figure 3). These safety systems have a significant impact on the consequences of the dangerous phenomenom (for example in terms of release duration) and on the frequencies of dangerous phenomana such as fire, explosion, dispersion of toxic cloud... They are characterized by their probability of failure on demand. There will be two or more scenarios resulting from the loss of containment, according to the failure or the success of the safety system.

3.2.6 *Choosing endpoint values*

Once all the scenarios are characterised, their consequences have to be estimated. Effect thresholds are related to the exposition of people to first irreversible damage, rather than endpoint values related to fatalities, more common in QRA approaches. Endpoint values are summarised in Table 1.

3.2.7 *Results*

As in the step 3 of the ARAMIS methodology, the severity of the phenomenon has to be estimated (combination of the frequency and intensity of the major effect of the dangerous

Table 1. Endpoint values used in Walloon Region for land use planning purposes.

Effects	Endpoint values	Dangerous phenomenon	Related on
Radiation	6,4 kW/m^2	Poolfire, Jetfire, Flashfire	2nd degree burns after 20 seconds
Radiation	237 $(kW/m^2)^{1.33}.s$ (Dose)	BLEVE	2nd degree burns
Overpressure	50 mbar	Explosion	Indirect effects, NATO value
Toxic	$ERPG^n$ x 60 min (Dose)	Toxic dispersion	Maximum airborne concentration below which it is believed that nearly all individuals could be exposed for up to 1 hour without experiencing or developing life-threatening health effects

phenomenon). The calculations are carried out by the software used in land use planning (PHAST RISK 6.6 developed by DNV Software). Results are provided in the shape of iso-risk curves superimposed on an aerial view of the area and the sector plan. It is possible to know the value of the risk for each precise location. The area delimited by the 10^{-6} per year iso-risk curve is called the "consultation zone", inside which Walloon authorities must give an advice for every project concerning land use.

4 CONCLUSIONS

ARAMIS was an European project aiming to build up a new integrated risk assessment method for being used as a supportive tool to speed up the harmonized implementation of Seveso II Directive. A methodology was proposed to stakeholders interested by risk assessment for land use or emergency planning, enforcement or, more generally, public decision-making. It is evident that the benefits of ARAMIS results were immediate in Belgium. Indeed, with the experience achieved within the framework of ARAMIS, the Ministry of Walloon Region asked the Major Risk Research Center to develop a consistent and transparent methodology to assure a long-term land use planning around Seveso companies. The LUP methodology for the Walloon Region has been developed relying on the ARAMIS results.

REFERENCES

Beaudoint, D., Delvosalle, C., Delcourt, J., Nourry, J., Rosmorduc, A., Benjelloun, F. 2012. Updating Selection of hazardous Equipment for the Land use planning around Seveso Sites in Walloon Region after the CLP Regulation, Land Use Planning and Risk-Informed Decision Making, Proceedings of the 43rd ESReDA Seminar, Rouen, October 21–23, 2012.

Debray, B., Delvosalle, C., Fiévez, C., Pipart, A., Londiche, H., Hubert, E. 2004. Defining safety functions and safety barriers from fault andevent trees analysis of major industrial hazards, PSAM7-ESREL2004 conference, Berlin, June 2004.

Delvosalle, C., Benjelloun, F., Brohez, S., Cornil, N., Fiévez, C., Servranckx, L., Tambour, F. 2007. The influence of safety systems in the land use planning methodology used in Walloon region (Belgium), IChemE Symposium Series n° 153, Edinburgh (2007).

Delvosalle, C., Benjelloun, F., Brohez, S., Fiévez, C., Niemirowski, N., Tambour, F. 2006. Overview of Current practices regarding Land Use Planning around Seveso sites in Walloon Region. *5th European Meeting on Chemical Indus-try and Environment, EMChIE, Vienna, Austria, 3rd - 5th May 2006.*

Delvosalle, C., Cornil, N., Fiévez, C., Servranckx, L., Tambour, F., Yannart, B., Benjelloun, F. 2008. Land use planning methodology used in Walloon Region (Belgium) for tank farms of gasoline and diesel oil; - Proceedings 17th European Safety and Relability Conference (ESREL 2008), 22–25 September 2008, Valencia, Spain.

Delvosalle, C., Fiévez, C., Cornil, N., Nourry, J., Servranckx, L., Tambour, F. 2011. Influence of new generic frequencies on the QRA calculations for land use planning purposes in Walloon region (Belgium), Journal of Loss Prevention in the Process Industries 24 (2011) 214–218.

Delvosalle, C., Fiévez, C., Pipart, A. 2006. Aramis project: reference accident scenarios definition in Seveso Establishment, Journal of Risk research, 9(5) (2006): 583–600.

Delvosalle, C., Fiévez, C., Pipart, A., Casal, Fabrega, J., Planas, E., Christou, M., Mushtaq, F. 2005. Identification of reference accident scenarios in Seveso establishments, Reliability Engineering & System Safety, 90 (2005): 238–246.

Delvosalle, C., Fiévez, C., Pipart, A., Debray, B. 2006. Aramis project: a comprehensive methodology for the identification of reference accident scenarios in process industries, Journal of Hazardous Materials, 130 (2006): 200–219.

Fiévez, C., Delvosalle, C., Cornil, N., Servranckx, L., Tambour, F., Yannart, B., Benjelloun, F. 2008. Influence of safety systems on land use planning around Seveso sites; example of measures chosen for a fertiliser company located close to a village; Proceedings 17th European Safety and Relability Conference (ESREL 2008), 22–25 September 2008, Valencia, Spain.

Handbook failure frequencies 2009 for drawing up a safety report, AMINAL, Afdeling Algemeen Milieu—en Natuurbeleid, Cel Veiligheidsrapportering—Ministerie van Vlaamse Gemeenschap.

Planas, E., Ronza, A., Casal, J. 2004. ARAMIS project: the risk severity index, in: H.J. Pasman, J. Skarka, F. Babinec (Eds.), Proceedings of the 11th International Symposium on Loss Prevention and Safety Promotion in the Process Industries, Petro Chem Eng, 31 May-3 June 2004, Praha, Czech Republic, 2004, pp. 1126–1132.
Salvi, O., Debray, B. 2016. A global view on ARAMIS, a risk assessment methodology for industries in the framework of the SEVESO II directive, Journal of Hazardous Materials 130(3) (2015):187–99.
Service public de Wallonie (SPW) 2016. VADE-MECUM: Spécifications techniques relatives au contenu et à la présentation des études de sûreté, des notices d'identifications des dangers et des rapports de sécurité. Cellule Risques d'accidents majeurs (2016) 01–83.
Servranckx, L., Benjelloun, F., Cornil, N., Fiévez, C., Lheureux, E., Katz, T., Tambour, F., Delvosalle, C. 2010. Overview of results after six years of risk assessment for Land Use Planning in Walloon Region (Belgium), 13th International Symposium on Loss Prevention, 6th - 9th June 2010, Brugge, Belgium.
Tambour, F., Benjelloun, F., Cornil, N., Fiévez, C., Lheureux, E., Katz, T., Servranckx, L., Delvosalle, C. 2010. Methodology of risk assessment of Seveso explosives storages sites for land use planning in Walloon Region (Belgium), 13th International Symposium on Loss Prevention, 6th - 9th June 2010, Brugge, Belgium.
Tixier, J., Dandrieux, A., Dusserre, G., Bubbico, R., Luccone, L.G., Mazzarotta, B., Silvetti, B., Hubert, E., Salvi, O., Gaston, D. 2004. Vulnerability of the environment in the proximity of an industrial site, in: Pasman, H.J., Skarka, J., Babinec, F. (Eds.), Proceedings of the 11th International Symposium on Loss Prevention and Safety Promotion in the Process Industries, PetroChemEng, 31 May-3 June 2004, Praha, Czech Republic, 2004, pp. 1260–1267.

ANNEX 1. Table issued from Vade-Mecum for the selection for hazardous equipment (Service public de Wallonie, 2016).

	Categories		Threshold mass Ma (kg)		
	Seveso	Additional	Solid	Liquid	Gas
Health hazards	H1: Acute Toxic Cat 1 All exposure routes		1000	100	10
	H2: Acute Toxic Cat 2 All exposure routes H3: SPECIFIC TARGET ORGAN TOXICITY (STOT)- Single exposure STOT SE cat 1	Acute Toxic Cat 3 (ingestion route and dermal route) Germ celle mutagenic, Cat 1 Carcenogenicity Cat 1 and 2 Serious eye damage/ irritation Cat 1 Reproductive toxicity Cat 1 and 2 Reproductive toxicity, additional category: effects on or via lactation	10 000	1000	100
		Acute Toxic Cat 4 Serious eye damage/ irritation Cat 2 STOT- Single exposure STOT SE Cat 2 and 3	100 000	10 000	1000
Physical hazards	P1a: Unstable explosives P1a: Explosives div 1.1 - 1.2 - 1.3 - 1.5 - 1.6 P1a: Explosives: explosive according method A.14		250	250	---
	P1b: Explosives div 1.4	EUH006: Explosive with or without contact with air	500	500	---
	P2: Flammable gases: Cat 1 or 2		---	---	1000
	P3a: Flammable aerosols Cat 1 or 2, flammable containing Cat 1 or 2 or flammable liquids Cat 1		10 000		
	P3b: Flammable aerosols Cat 1 or 2, flammable not containing Cat 1 or 2 or flammable liquids Cat 1		100 000		
	P4: Oxidizing Gases Cat 1		---	---	10 000
	P5a: Flammable liquids Cat 1		---	10 000	1000
	P5a: Flammable liquids: - Cat 2 or 3 maintained at a temperature above thier boling point or - with a flash point ≤ 60 °C, maintained at a temperature above their boiling point				
	P5b: Flammable liquids: - Cat 2 or 3 where particular processing conditions, such as high pressure or high temperature, may create major-accident hazards, or - With a flash point ≤ 60 °C where particular processing conditions, such as high pressure or high temperature, may create major-accident hazards		---	10 000	1000
	P5c: Flammable liquids Cat 2 ou 3 not covered by P5a and P5b		---	10 000	1000
	P6a: Self-reactive and organic peroxides, type A or B		250	250	---
	P6b: Self-reactive and organic peroxides, type C, D, E or F		500	500	---
	P7: Pyrophoric liquids and solids Cat 1	EUH209: Can become highly flammable is use	1000	1000	---
	P8: Oxidizing liquids and solids Cat 1, 2 or 3		10 000	10 000	---
		Flammable solids Cat 1 and 2	10 000	---	---
Environment hazards	E1: Hazardous to the aquatic environment in cat. acute 1 or chronic 1		If LC50 96h for fishes (in mg/l) > = 1, so 1000. Otherwise, 1000*LC50 96h (in mg/l)		
	E2: Hazardous to the aquatic environment in cat. chronic 2	Hazardous to the aquatic environment in cat. chronic 3 and 4	10 000	10 000	10 000
Other hazards	O1: Substances or mixtures with hazard statement EUH014		10 000	10 000	10 000
	O2: Substances and mixtures which in contact with water emit flammable gases, Cat 1		10 000	10 000	10 000
	O3: Substances or mixtures with hzard statement EUH029		10 000	1000	100
		Substances and mixtures which in contact with water emit flammable gases, Cat 2	10 000	10 000	---

Risk Analysis and Management – Trends, Challenges and Emerging Issues – Bernatik, Huang & Salvi (Eds)
© 2017 Taylor & Francis Group, London, ISBN 978-1-138-03359-7

SafetyBarrierManager, a software tool to perform risk analysis using ARAMIS's principles

Nijs Jan Duijm
Department of Management Engineering, Technical University of Denmark DTU, Kgs. Lyngby, Denmark
Nicestsolution, Jyllinge, Denmark

ABSTRACT: The ARAMIS project resulted in a number of methodologies, dealing with among others: the development of standard fault trees and "bowties"; the identification and classification of safety barriers; and including the quality of safety management into the quantified risk assessment. After conclusion of the ARAMIS project, Risø National Laboratory started developing a tool that could implement these methodologies, leading to SafetyBarrierManager. The tool is based on the principles of "safety-barrier diagrams", which are very similar to "bowties", with the possibility of performing quantitative analysis. The tool allows constructing comprehensive fault trees, event trees and safety-barrier diagrams. The tool implements the ARAMIS idea of a set of safety barrier types, to which a number of safety management issues can be linked. By rating the quality of these management issues, the operational probability of failure on demand of the safety barriers can be calculated. The paper will give a short description of the features of the tool, with emphasis on the methodologies that originate from the ARAMIS project. The paper will also address developments and experiences over the last years, which have inspired additional features. This includes a discussion of the use of generic management issues as opposed to concrete safety measures targeted at specific safety barriers, which includes a discussion of the basic philosophy in the ARAMIS methodology of dealing with safety management. The adjustments to the barrier typology is also discussed.

1 INTRODUCTION

The ARAMIS methodology is based, among others, on a set of standardized fault trees and the concept of safety barriers, which can be presented in a bow-tie diagram. At the end of the ARAMIS project, Risø National Laboratory concluded that they did not have access to a tool that in a user-friendly way could present these fault trees or bowties, nor implement safety barriers in a logic and consistent way. At the time, Danish process industry used the concept of "safety-barrier diagrams" widely, but these diagrams were always constructed manually, or by using the drawing capabilities of spreadsheet tools like Excel®.

2 SAFETY-BARRIER PRINCIPLES

Since the ARAMIS project, the concept of safety barriers has become widely known. There is a plethora of definitions around. In many contexts, the term "safety barrier" is used for any safeguard or risk-reducing measure that decreases the probability of consequence of an accident. In our methodology we limit the notion of "safety barrier" to those risk-reducing measures that acutely and autonomously respond to a potentially dangerous situation, i.e. safety barriers perform unplanned actions. In contrast, many other risk-reducing measures consists of actions that are planned or scheduled in advance, and where there is no direct relationship in time between the carrying out of the action, and a potentially dangerous situation being pertinent. Such safety actions are in our framework considered to be "safety-management measures", or in short: "management measures". Of course, there is a relation between the management measures and the safety barriers. The chance of successful deployment of the safety barrier depends on the proper design, installation, inspection, maintenance and replacement of the safety barrier, and, in so far the safety barrier comprises human action, in providing the proper education, competence, training, procedures, etc. to be able to take the right decisions at the right time.

Useful definitions make use of the concept of a "barrier function". This means that there may be different ways of implementing a barrier function: a specific safety barrier is just one such possibility. A often referred definition of a safety barrier is (Sklet, 2006):

- A *barrier function* is a function planned to prevent, control, or mitigate undesired events or accidents;
- A *barrier system* is a system that has been designed and implemented to perform one or more barrier functions.

3 SAFETY-BARRIER DIAGRAMS

The "SafetyBarrierManager" tool uses safety-barrier diagrams to show accident sequences. Following the definition, the function of a safety barrier is to abort a potentially dangerous situation. Such functioning can be showed graphically as a response to a potentially dangerous development, the *Demand Condition*, avoiding the dangerous outcome, the *Condition on Failure*, which would occur if the barrier was not present, or is not working properly. A successful deployment of the barrier may lead to the *Condition on Success*, see Figure 1. Figure 1 can be interpreted as a minimal example of a safety-barrier diagram. Depending on the complexity of the system, one can add more conditions or events, and more safety barriers.

A safety-barrier diagram is a simplified cause-consequence diagram (Nielsen, 1971). It shows the sequence of possible events as an accident scenario. As the diagram shows a sequence of (potential) events, the order of the events, and thus the order of the safety barriers, has meaning, and the risk analyst shall consider the order of the barriers and events.

The safety-barrier diagram is very similar to the "bowtie" diagrams, which the ARAMIS project mentioned. Although the aspect of sequence was not addressed in the ARAMIS project, it is noted that common bowtie diagrams, including tools to graph bowtie diagrams, do not represent sequences of events, and indeed, sequential ordering of barriers is often not applied. Consequently, these bowties do not distinguish between the barriers that act in immediate response to potentially hazardous conditions and the risk reducing measures or safety management measures that are scheduled. Both types of risk reducing measures, the "real" barriers and the "safety management measures" may appear side by side in these bowtie diagrams.

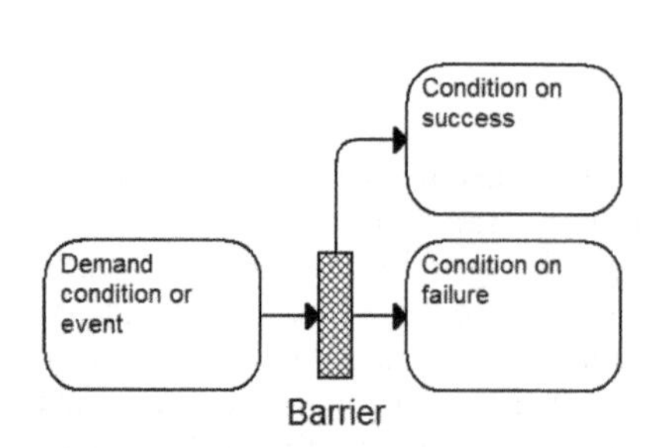

Figure 1. Definition of a safety barrier.

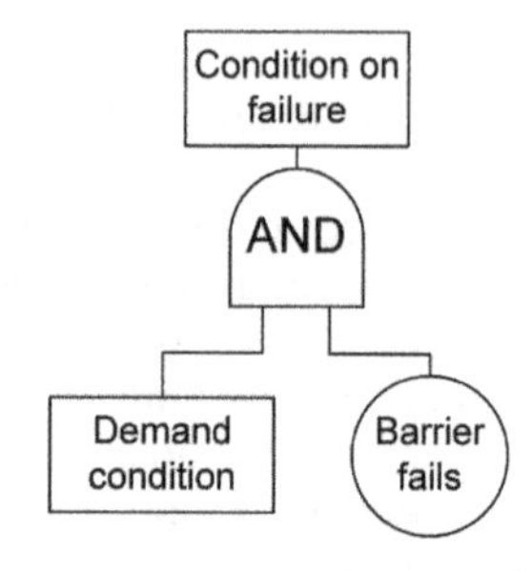

Figure 2. Fault tree representation of the safety barrier.

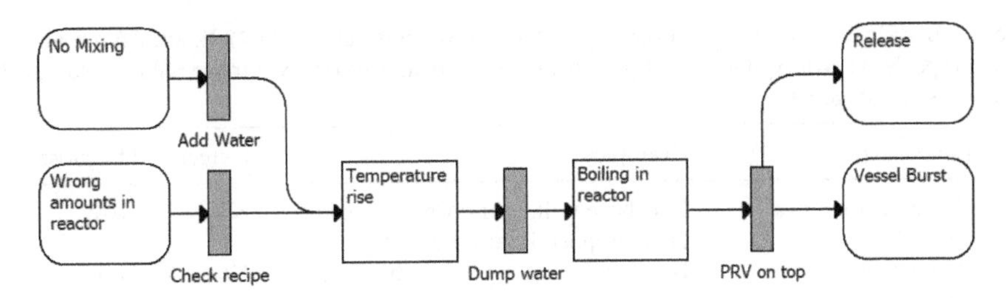

Figure 3. Example of a safety-barrier diagram.

As an example, Figure 3 shows a barrier diagram for a run-away reaction in a batch reactor. The "SafetyBarrierManager" tool includes the logic to ensure that safety barriers are properly placed between the Demand Condition and a Condition on Failure or a Condition on Success. Each safety barrier represents a simplified fault tree as shown in Figure 2. The logic is discussed in more detail in (N J Duijm, 2009). Due to the strict implementation of the logic, for each condition in the diagram it is possible to derive a fault tree that has this condition as top event. This is used to calculate the probability or expected frequency of the conditions. The Probability of Failure on Demand (PFD) of the safety barriers and the probability or expected frequency of the initial conditions are the inputs for that calculation.

4 SAFETY MANAGEMENT: HOW IT AFFECTS SAFETY-BARRIER QUALITY

The "SafetyBarrierManager" tool forces the risk analyst to assign a "safety-barrier type" to each safety barrier that is introduced in the safety-barrier diagram. The safety-barrier types belong to a predefined set of types. ARAMIS developed such a set of safety-barrier types, see Table 1. Barrier types 1–4 are normally considered to be "passive" barriers (there is no detection function), while types 5–11 are considered "active" barriers. By forcing the risk analyst to assign a barrier type, and thus compare the barrier with a description of a barrier type, we hope to minimize the use of incomplete barriers (incomplete barriers are typically barriers where either detection (e.g. a "shut-down valve") or action (e.g. an alarm) is missing.

The other function of the barrier type is to enable to link the safety barrier to a set of management functions. Depending on the type of barrier, different actions will determine the quality of the barrier: barriers that depend on hardware require inspection and maintenance; barriers that depend on human behaviour require training, procedures and resource planning. Table 2 shows the list of "management issues" as proposed by ARAMIS. This table also shows the "weight factors" that describe how important the management issue is for ensuring the integrity of the safety barrier of the given type.

Based on an audit of the safety management system according to the ARAMIS methodology as described in (Guldenmund, Hale, Goossens, Betten, & Duijm, 2006), for each of the safety management issues a quality rating can be obtained: 100% for a perfectly functioning management issue and 0% if the management issue fails to meet any of its requirements. Given a design PFD of each safety barrier, i.e. the best (lowest) obtainable PFD assuming that all conditions and support functions are according to specifications, it is now possible to derive the operational PFD given the quality of the safety management according to the audit (Nijs Jan Duijm & Goossens, 2006):

$$\log(PFD_{operational.k}) = \left(1 - \sum_{i=0}^{7}(1 - S_i) \cdot B_{i,k}\right) . \log(PFD_{design,k}) \tag{1}$$

Here S_i is the rating (between 0 and 1, with 1 for perfect rating) of management issue i, and $B_{i,k}$ the weight factor of management issue i for barrier type k, as presented in Table 2.

Table 1. Barrier typology as developed in ARAMIS (Andersen et al., 2004). Examples are mentioned for each type. N/A: Not Applicable; H: Hardware; S: Software; B: Behavioural, S: Skill-based; R: Rule-based; K: Knowledge-based.

No.	Barrier type	Examples	Detect	Diagnose	Act
1	Permanent passive control	Pipe/hose wall, anti-corrosion paint, tank support, floating tank lid	N/A	N/A	H
2	Permanent passive barrier	Bund, dyke, railing, fence, blast wall, lightning conductor	N/A	N/A	H
3	Temporary passive	Barriers round repair work, blind flange, helmet/gloves/ goggles	N/A	(B)	H
4	Permanent active	Active corrosion protection, heating/cooling system, ventilation, explosion venting, inerting system	N/A	(B)	H
5	Activated/on demand	Pressure relief valve, interlock with "hard" logic, sprinkler installation, pressure/temperature/level control	H	H	H
6	Activated—automated	Programmable automated device, control system or shutdown system	H	S	H
7	Activated—manual	Manual shutdown in response to instrument reading or alarm, donning breathing apparatus or calling fire brigade on alarm	H	B.SRK	B/H
8	Activated—warned	Donning personal protection equipment in danger area, refraining from smoking, keeping within white lines, opening labelled pipe,	H	B.R	B
9	Activated—assisted	Using an expert system	H	S/B.RK	B/H
10	Activated—procedural	Follow start up/shutdown procedure, adjust setting of hardware, warn others to evacuate, empty & purge line before opening, lay down water curtain	B	B.SR	B/H
11	Activated—emergency	Response to unexpected emergency, improvised jury-rig during maintenance, fight fire	B	B.K	B/H

Table 2. Management issues and weight factors for adjusting safety-barrier performance as proposed by (Nijs Jan Duijm & Goossens, 2006).

	ARAMIS weight factor $B_{i,k}$			
	Barrier types (Table 1)			
ARAMIS management issue	1, 2, 4, 5, 6 (Hardware)	3, 8 (Temporary)	7, 9, 10 (Behaviour—R/S)	11 (Behaviour—K)
0 Safety Culture	0%	8%	15%	25%
1 Manpower planning & availability	0%	29%	58%	87%
2 Competence & suitability	0%	36%	72%	100%
3 Commitment, compliance & conflict resolution	0%	10%	20%	33%
4 Communication & coordination	0%	25%	50%	83%
5 Procedures, rules & goals	0%	9%	18%	40%
6 Hard/software purchase, build, interface, install	43%	22%	0%	0%
7 Hard/software inspect, maintain, replace	17%	8%	0%	0%

Table 3. Example of ratings for safety management issues.

Safety management issue	Rating
0 Safety Culture	75% (default)
1 Manpower planning & Availability	100%
2 Competence & suitability	86%
3 Commitment, compliance & conflict resolution	80%
4 Communication & coordination	85%
5 Procedures, rules & goals	80%
6 Hard/software purchase, build, interface, install	100%
7 Hard/software inspect, maintain, replace	80%

Table 4. The management weights and ratings as listed in Table 2 and Table 3 applied to the barrier diagram in Figure 3.

Type	Name	Design PFD	Operational PFD	Frequency using design PFD (per year)	Frequency using operational PFD (per year)
Initiating event	No mixing			0.021	(idem)
Barrier	Add water	0.02	0.0639		
Initiating event	Wrong amounts in reactor			0.01	(idem)
Barrier	Check recipe	0.1	0.1963		
Intermediate event	Temperature rise			0.00142	0.003305
Barrier	Dump water	0.02	0.0248		
Intermediate event	Boiling in reactor			$8.649 \cdot 10^{-5}$	$13.61 \cdot 10^{-5}$
Barrier	PRV on top	0.001	0.001265		
Consequence	Release			$8.641 \cdot 10^{-5}$	$13.59 \cdot 10^{-5}$
Consequence	Vessel Burst			$8.649 \cdot 10^{-8}$	$17.21 \cdot 10^{-8}$

Table 3 and Table 4 demonstrate the effect of reduced management quality when applied to the barrier diagram in Figure 3. The ratings in Table 3 have been have been taken from the example in the ARAMIS User Guide (Andersen et al., 2004). As can be seen, even modest reductions in management efficiency from 100% down to 80% may double the expected frequency of the worst consequence.

5 NEW DEVELOPMENTS—AN ALTERNATIVE SAFETY-BARRIER CLASSIFICATION

During a number of studies, it was found that the barrier-type classification originally introduced by ARAMIS could be improved. This leads to the classification in Figure 4.

- The type "Excessively conservative design" replaces the ARAMIS passive "control" barrier. "Controls" in the sense of the equipment and actions necessary to perform the primary process should not be considered as safety barriers. Failure of a primary process control is an initiating event, not a barrier failure. In order to allow for primary process controls with "built-in" aspects that are specifically added for a better safety performance, the "excessively conservative design" is included as a "barrier" instead. This barrier covers e.g. extra wall thickness and mechanical redundancy such as double steering rods.
- An extra group "preventive barriers" is introduced. This group is a kind of grey area between safety management measures and "real" barriers. They are not "real" barriers because they are invoked before the threatening situation occurs, so they can be considered "planned" actions. On the other hand, they represent actions closely linked to the daily

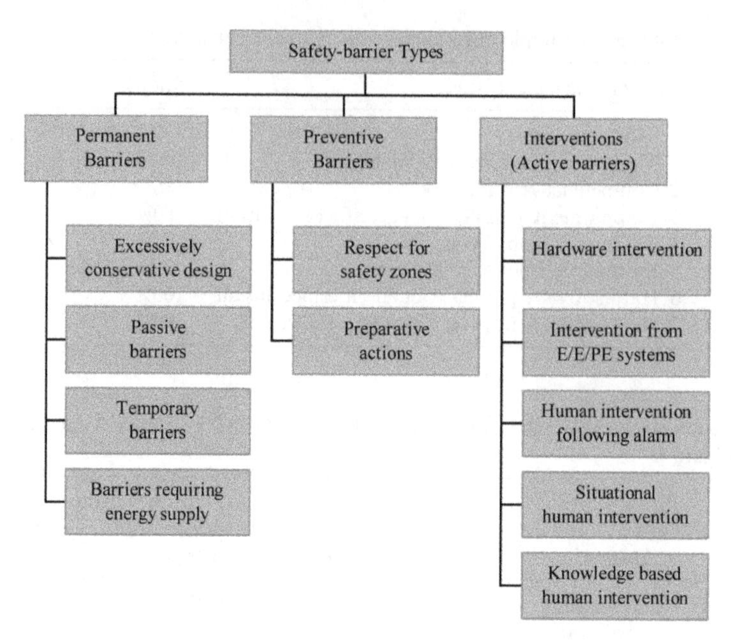

Figure 4. A new safety-barrier classification.

operations and performed by the operators at the sharp end. After a number of studies, it turned out that these actions best could be introduced in the barrier diagram as safety barriers. The group includes the behavior of "respecting safety zones", e.g. refrain from entering danger zones. It also includes "preparative actions", e.g. venting tanks before entering (which was barrier type No. 8 in ARAMIS's typology).

- On the other hand, the five different barrier types covering human intervention are reduced to three (type No.9 is removed while type No. 8 is moved into the "Preparative actions", see above).

6 NEW DEVELOPMENTS—RISK REDUCING MEASURES

The philosophy of the ARAMIS project was to deal with safety management through an abstract, top-level approach. This was justified by the consideration that an evaluation of a management system by means of a management audit always would be based on selective spot checks of the management system. Therefore, efforts were aimed at ensuring how spot checks of the management system could be used to provide general statements on the effectiveness of all safety barriers. This is established through a direct link between top-level management issues and safety barriers as expressed in formula. (1)

However, an industrial risk analyst drawing up barrier diagrams has access and knowledge of the detailed safety measures that link to specific safety barriers. Such safety measures can be e.g. inspection and maintenance plans, competence requirements for specific functions and tasks, and detailed procedures how to handle alarms and deviations. For an industrial risk analyst it is straightforward to link the detailed safety measures to the safety barrier. It is of importance to demonstrate (e.g. to the authorities) that such measures have been taken to ensure the integrity of the barrier Each measure fits within a management issue, but that link is of lesser importance to the risk analyst in this context.

Figure 5 shows how the SafetyBarrierManager tool displays the detailed management measures linked to the barriers, but also to the initiating events in the example in Figure 3: Avoiding initiating events from happening, by ensuring the integrity of the primary process controls, is equally important as ensuring the integrity of the safety barriers.

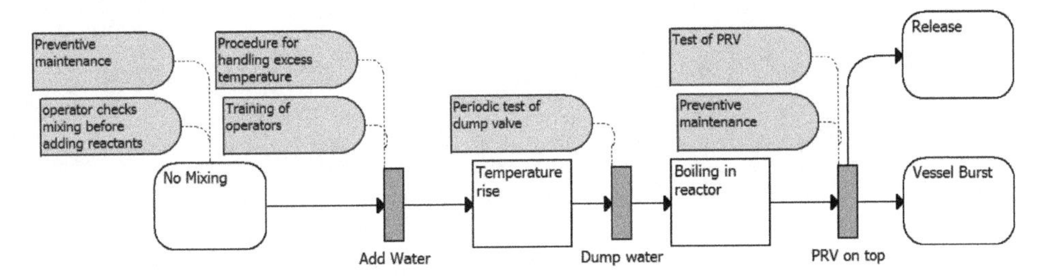

Figure 5. Safety-barrier diagram showing safety-management measures that a) prevent the initiating events or b) ensure the integrity of the safety barriers.

7 DISCUSSION

The "SafetyBarrierManager" tool implements a number of the methodologies developed during the ARAMIS project. The tool is helpful in both qualitative assessments (showing implemented risk reducing measures) and quantitative assessments (by calculating the expected frequency of consequences). Experiences gained by using this tool has initiated some changes and provoked the need for additional features. There is still a need for better understanding of the relation between safety management and the quality of safety barriers, and the role safety management may have as a common cause factor for the simultaneous failure of barriers, as has been suggested in (Markert, Duijm & Thommesen, 2013).

REFERENCES

Andersen, H., Casal, J., Dandrieux, A., Debray, B., de Dianous, V., Duijm, N. J., … Tixier, J. (2004). *ARAMIS User Guide*. Retrieved from http://safetybarriermanager.com/files/aramis/ARAMIS_FINAL_ USER_GUIDE.pdf.

Duijm, N.J. (2009). Safety-barrier diagrams as a safety management tool. *Reliability Engineering and System Safety*, 94(2), 332–341.

Duijm, N.J., & Goossens, L.H.J. (2006). Quantifying the influence of safety management on the reliability of safety barriers. *J.Haz.Mat.*, 130(3), 284–292.

Guldenmund, F.W., Hale, A.R., Goossens, L.H.J., Betten, J.M., & Duijm, N.J. (2006). The Development of an Audit Technique to Assess the Quality of Safety Barrier Management. *J.Haz.Mat., 130*(3), 234–241.

Markert, F., Duijm, N.J., & Thommesen, J. (2013). Modelling of safety barriers including human and organisational factors to improve process safety. In *Proc. 4th Int. Symp. on Loss prevention and safety promotion in the process industries* (pp. 283–288). Florence, Italy.

Nielsen, D.S. (1971). *The Cause/Consequence Diagram Method as a Basis for Quantitative Accident Analysis* (Vol. Risø-M-137). Roskilde: Danish Atomic Energy Commision, Risø.

Sklet, S. (2006). Safety barriers: Definition, classification, and performance. *Journal of Loss Prevention in the Process Industries*, 19(5), 494–506.

Risk Analysis and Management – Trends, Challenges and Emerging Issues – Bernatik, Huang & Salvi (Eds)
© 2017 Taylor & Francis Group, London, ISBN 978-1-138-03359-7

Post-ARAMIS project use of the safety climate questionnaire for the process industry

Marko Gerbec
Jozef Stefan Institute, Ljubljana, Slovenia

ABSTRACT: The ARAMIS project (Accidental Risk Assessment Methodology in IndustrieS) proposed some novel approaches on how to consider quality of the risk management within the risk assessment in the process industries. As a part of that, the safety climate questionnaire survey was developed. After the project, the author applied the safety climate survey in three industrial organizations. The results obtained allowed to follow the trends in time on the individual safety climate dispositions, when compared with results among the organizations, as well as to propose the management actions addressing the identified weak points. This work is concluded with the note that guidance is needed on how to consider identified deficiencies in the selection of the management responses, followed by brief guidance on how to approach transposition of the questionnaire in the local national and organizational context.

1 INTRODUCTION

The ARAMIS project (Accidental Risk Assessment Methodology in IndustrieS) was EU's 5th FP project that proposed a novel approach to consider the quality of management in relation to the safety measures (safety barriers) in the process of safety risk assessment (Salvi and Debray, 2006). The ARAMIS project brought to a foreground a number of methods and tools to the risk assessment and risk management, just as the concepts of safety barriers, levels of confidence, bow–tie diagrams, risk matrices for risk decisions, as well as evaluation of the effectiveness of the organizational risk management—related to safety barriers and their delivery systems, including the safety climate evaluation approach (Guldenmund et al., 2006; Duijm and Goossens, 2006). Considering that the risk assessment should not be static (e.g., likelihoods of the basic failure events are not invariant in time and safety performance) added to the "standard" risk assessment at that time and likely contributed to the current understanding of the dynamic risk assessment (e.g., Villa et al., 2016).

The concept of safety climate among the organization's personnel contributing to actual perception, attitude, and behaviour was developed in the context of the process industry and linked to the model of performance of safety barriers within the risk model (Duijm and Goossens, 2006). The snapshots of the safety culture among the personnel are to be determined through specific questionnaire surveys, thereby providing an insight into the concept of safety climate (Guldenmund, 2007). For that purpose, an analysis of the relevant safety climate dispositions in relation to the specific safety barriers and their delivery systems (organizational activities are oriented towards assurance of the safety barriers and their trust into their delivery systems) was performed and resulted in 11 dispositions/topics evaluated using specific questions (ARAMIS, 2006; Andersen et al., 2004):

1. Reporting of accidents;
2. If and when incidents and accidents (all types) do not become reported, this is because ...
3. Safety instructions and attitudes;
4. If and when incidents and accidents happen (all types) this is generally because ...

5. Prioritisation of safety at work;
6. Employee involvement in decisions about safety;
7. Who do you think *should* be taking responsibility for safety?
8. Who do you think is, *in fact*, taking responsibility for safety?
9. Commitment by management and leaders to safety;
10. Trust and fairness; and
11. Work and social relations.

An additional 12th topic is related to demographic data.

The author was involved in a number of risk management consultations with the process industries in the context of risk assessments in the licensing procedures. While formally the management performance assessments and safety climate surveys were not mandatory, the management was usually interested in obtaining information about the actual situation.

While the aim was not to consider the safety climate results within the risk assessment, the motivation was to look for possible weak points in the safety culture among the personnel and to build on the measures for improvements. For that purpose, the author developed a quantification approach to obtain numerical values from the individual responses to the questions in the questionnaire (Gerbec, 2013): statements (answers) to the questions are collected by following a five-point Likert-type scale (strongly agree, agree, neutral, disagree, and strongly disagree). The meaning of the statement depends on the exact question. For example, "strongly agree" with topic-question 1–1 "I am personally willing to report any minor injury incident" reflects a very positive behaviour. Thus, the statements were correspondingly processed to the scale as very positive, positive, neutral, negative, and very negative, with numerical values (grades) from 5 to 1, respectively, thereby enabling a calculation of meaningful averages for questions and topics.

This paper will present application and results from three case applications in process industry organizations in Section 2, discuss the usefulness of the results in Section 3 and conclude with summary of lessons and limitations of the approach in Section 4.

2 EXAMPLES OF THE USE

In this section, we will present a summary of the results of the application of the ARAMIS questionnaire in three anonymous organizations from the oil and gas industry. The organizations are from different countries; case 1 and case 2 are close to the typical medium size organization (100–200 employees), while case 3 is a very large organization.

2.1 *Case 1*

The questionnaire surveys were done in 2010, 2013, and 2016, involving 58, 76, and 89 attendees, respectively, being located on two separate sites. Almost all employees participated (from plant operations, line managers, office personnel, and mid and top managers).

The meaningful average grades per year are presented in Figure 1.

It is evident from the figures that, changes in the number of persons attending the surveys are far larger than changes in average scores per year and per topic. For example, topic no. 11 (work and social relations) appears with a remarkably constant score (2.94 to 3.06) and is consistently the worst evaluated topic. The next worst evaluated topic is no. 4 (if and when incidents and accidents happen …) and also is quite consistent in score (3.09 to 2.92). On the other hand, topics no. 7, 8 and 1 do reflect the above-mentioned positive score about declared and actual responsibility for safety, as well as to report accidents.

2.2 *Case 2*

The questionnaire surveys were carried out on a single site in 2010, 2013, and 2016, involving 13, 19, and 21 attendees, respectively. All employees who were present at the time on the production site (operators and line management) participated. The meaningful average grades per year are presented in Figure 2.

Results are in general similar to case 1, but the scores are higher. Interestingly, again, topics no. 4 and no. 11 are among the worst evaluated, but no. 11 has about a 0.4 higher score. On the other hand, topics no. 1, 7, and 8 have very high and consistent scores, which probably reflects actual differences among the organizations. However, the team in this case seems to

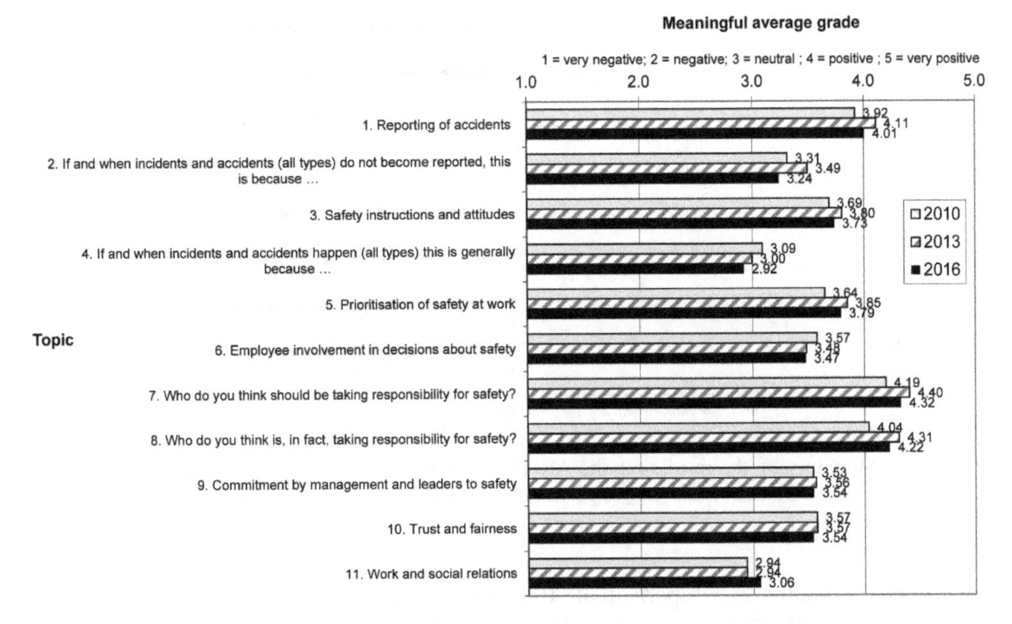

Figure 1. Summary of the results for the ARAMIS safety climate questionnaire survey for case 1. Meaningful average values of the grades per topic are shown (1 = worst and 5 = best).

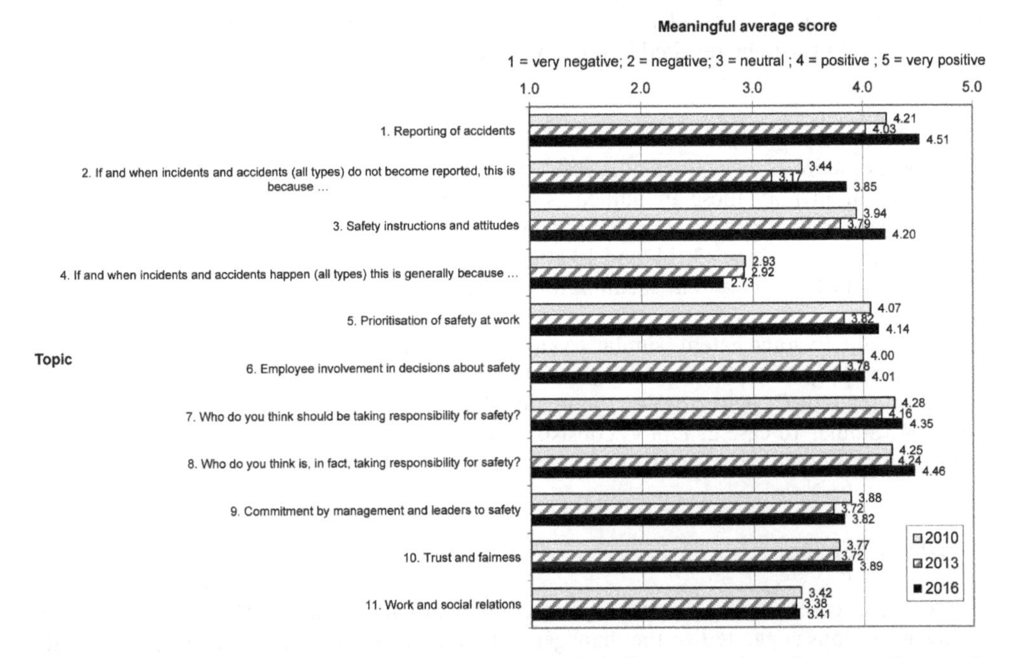

Figure 2. Summary of the results for the ARAMIS safety climate questionnaire survey for case 2. Meaningful average values of the grades per topic are shown (1 = worst and 5 = best).

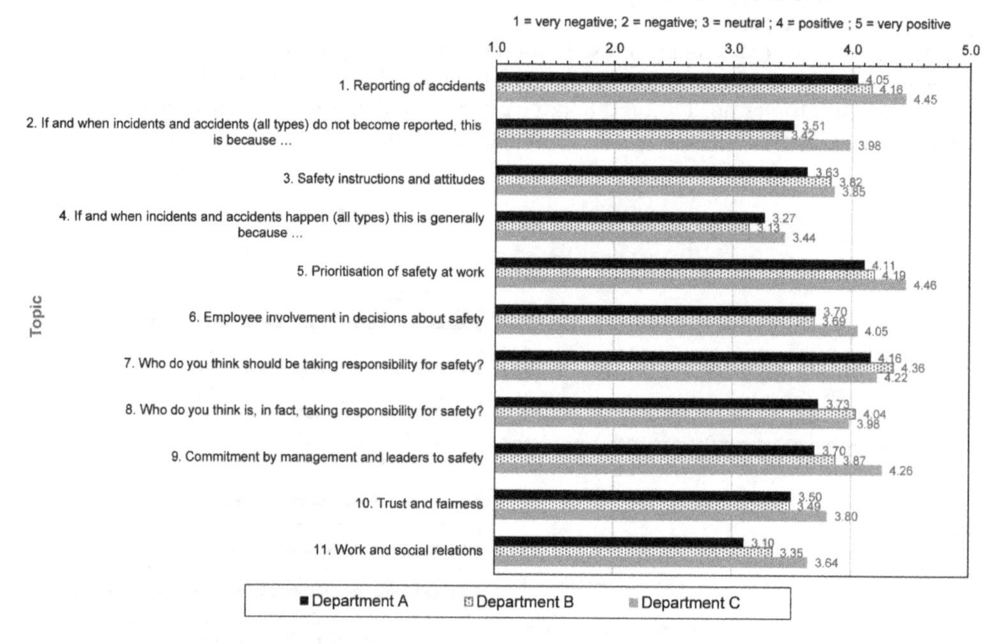

Figure 3. Summary of the results for the ARAMIS safety climate questionnaire survey for case 3. Meaningful average values of the grades per topic are shown (meaning of the grades: 1 = worst, and 5 = best).

be very closely related. A comparison of 2010 to 2013 scores reveals a decrease; this attracted the attention of the management, and one-on-one interviews were organized in consultation with the author about exact understanding and rationale for the responses to the selected survey questions. The finding was that, at the time of the survey (2013), they were under the season's highest work load and additional time was needed for safety training and survey that added stress and thereby resulted in the lower scores.

2.3 *Case 3*

The questionnaire surveys were organized in 2016 at three organization's departments (for the sake of anonymity, these are named as A, B, and C) as a part of a larger safety management implementation project. For that purpose, "test" surveys were conducted involving very low number of attendees: 43, 14, and 10 attendees per departments A, B, and C, respectively participated in the survey. The meaningful average grades per department are presented in Figure 3.

Results are, to some extent, similar to cases 1 and 2 (again topics no. 11 and 4 are worst evaluated and topics no. 7, 8, and 1 are among the best); however, generally scores are higher, especially for Dpt C. Nevertheless, the reader should note that the number of attendees is very low. Similar to case 2, Dpt C consists of a geographically separated plant, where the team members could be closely related.

3 DISCUSSION

The trend in case 1 safety climate topics scores seems to be stable over time, despite the following actions suggested to the management: enhance management visibility on the site floor, raise awareness on the hazards present, and foster responsibility of the leaders for supervision, as well as to assure compliance with the rules (see Gerbec, 2013). The trend in case 2 is similar, but some improvements (however, statistically hard to prove) can be seen. The management there is concerned and seeks additional measures in addition to those mentioned under case 1. Related to case 3, the full-scale implementation is envisaged in 2017 and more reliable results that should allow comparison and trending among departments are required. However, how to derive effective improvement actions from the safety climate results is the next topic for the management and consultants in all three cases.

4 CONCLUSIONS

The ARAMIS safety climate survey is a useful and effective tool for monitoring safety climate among process industry personnel. While additional guidance is needed in terms of orienting the management on how to practically improve safety climate, some guidance for transposition and translation into the local organizational context should be given:

1. Obtain management commitment about the safety culture;
2. Pay attention to correct translation and technical phrases used (match local terminology, work positions, duties, etc.);
3. Involve experienced local experts/managers;
4. Assure and demonstrate full anonymity;
5. Plan suitable place and space for each time and duration for surveys;
6. Provide feedback to employees on how the results were used for improvements.

REFERENCES

Andersen, H.B., Nielsen, K.J., Cartensen, O., Dyreborg, J., Guldenmund, F., Hansen, O.N., et al. 2004. Identifying safety culture factors in the process industry. In: *11th International symposium on loss prevention, proceedings of CSCHI and EFCE event 635*, Praha, June 2004.

ARAMIS Project 2006. Workpackage 3—Prevention management efficiency, deliverable D3B—Annex B.

Duijm, N. J. and Goossens, L. H. J., 2006. Quantifying the influence of safety management on the reliability of safety barriers. *Journal of Hazardous Materials*, 130(3), 284–292.

Gerbec, M. 2013. Supporting organizational learning by comparing activities and outcomes of the safety-management system. *Journal of Loss Prevention in the Process Industries*, 26, 1113–1127.

Guldenmund, F.W. 2007. The use of questionnaires in safety culture research—an evaluation. Safety Science, 45, 723–743.

Guldenmund, F.W., Hale, A.R., Goossens, L.H.J., Betten, J., Duijm, N.J. 2006. The development of an audit technique to assess the quality of safety barrier management. *Journal of Hazardous Materials*, 130(3), 234–241.

Salvi, O., Debray, B. 2006. A global view on ARAMIS, a risk assessment methodology for industries in the framework of the SEVESO II directive. *Journal of Hazardous Materials*, 130(3), 187–199.

Villa, V., Paltrinieri, N., Khan, F., Cozzani, V. 2016. Towards dynamic risk analysis: A review of the risk assessment approach and its limitations in the chemical process industry. *Safety Science*, 89, 77–93.

Risk Analysis and Management – Trends, Challenges and Emerging Issues – Bernatik, Huang & Salvi (Eds)
 ISBN 978-1-138-03359-7

Comparison of the ARAMIS risk assessment methodology with traditional approach

Kazimierz Lebecki
Główny Instytut Górnictwa, Katowice, Poland

Adam S. Markowski & Dorota Siuta
Lodz University of Technology, Lodz, Poland

ABSTRACT: The typical process plant consisting of storage tank, pump system and collecting tank for processing of ethanol was analyzed by two separate approaches:

1. ARAMIS methodology and,
2. Traditional approach based on the HAZOP and LOPA application.

The data comparison was made concerning all aspects of the risk assessment indicating the differences and uncertainties.

Risk Analysis and Management – Trends, Challenges and Emerging Issues – Bernatik, Huang & Salvi (Eds)
© 2017 Taylor & Francis Group, London, ISBN 978-1-138-03359-7

The importance of the ARAMIS approach outside the chemical industry: Application of MIMAH methodology to biogas production

Valeria Casson Moreno & Valerio Cozzani
Laboratory of Industrial Safety and Environmental Sustainability, University of Bologna, Bologna, Italy

ABSTRACT: Biogas produced from anaerobic digestion is currently one of the most emerging industrial sectors for energy production from renewable sources and current government funding are favoring its exploitation and development of the biogas market. From a process safety standpoint, Seveso Directive does not generally regulate biogas facilities since production plants are predominantly small to medium scale. However, recent literature has shown the rising trend of major accidents in the biogas supply chain, revealing the need for specific hazard identification. This paper describes the application of ARAMIS, more specifically the Methodology for the Identification of Major Accident Hazards (MIMAH), to a reference process scheme representing a typical small-size biogas production plant via anaerobic digestion. Bow-ties for the possible critical events related to the loss of containment of biogas from the anaerobic digester, and the most relevant hazardous piece of equipment were developed and their results are discussed.

Keywords: Biogas; Risk assessment; Bow-tie analysis; MIMAH; Emerging technology; ARAMIS

1 INTRODUCTION

Over the last 15 years, the opportunity of producing sustainable fuel from biomass has driven the attention to biogas production from Anaerobic Digestion (AD) (EurObserv'ER, 2014; REN21, 2015). This implied a fast growth of the biogas sector and consequent proliferation of AD plants and during 2014, more than 14,500 biogas production facilities were operated in Europe (REN21, 2015).

This trend has been associated with an even faster increase in the number of accidents and incidents as proven in recent literature (Casson Moreno et al., 2016; Casson Moreno and Cozzani, 2015), accompanied by severe consequences in terms of human and environmental losses.

On one hand, the majority of biogas production facilities are medium to small scale, therefore falling below the thresholds for the application of Seveso III Directive (2012/18/EU). ATEX (2014/34/UE) and D.lgs 81/2008 are a requirement only. On the other hand, biogas production takes place in a processing plant with complex biological reactions (Findeisen, 2015) entailing several hazards (e.g. fire, explosion, toxicity) but is not usually run by expert personnel (Saracino et al., 2016). The overall result is a non-negligible risk profile for the sector (Casson Moreno et al., 2016) revealing the need for ad hoc hazard identification.

For all the above-mentioned reasons, this paper presents the results of the application MIMAH to a reference process scheme representing a typical small-size biogas production facility involving anaerobic digestion of livestock slurry. This case was chosen as a representative of most of the biogas production facilities widespread in Europe (EurObserv'ER, 2014).

Specific bow-ties were identified using MIMAH (which was historically derived from petrochemical industry expertise (Delvosalle et al., 2006; Salvi and Debray, 2006)) and customized on the basis of (i) information collected in literature, (ii) a detailed accident analysis, and (iii) the elicitation of experts of the biogas sector.

The core importance of the results obtained is their broad validity for most of the existing biogas production facilities from anaerobic digestion.

2 ANAEROBIC DIGESTION BIOGAS: A TYPICAL PROCESS SCHEME

Biogas is the product of anaerobic digestion of organic materials, such as manure, sewage sludge, the organic fraction of household and industry waste, and energy crops. All types of biomass can be used as substrates for biogas production as long as they contain carbohydrates, proteins, fats, cellulose, and hemicelluloses as the main components. Only strong lignified organic substances, e.g., wood are not suitable due to the slowly anaerobic decomposition (FNR, 2013).

Based on the type of substrate, the bacteria species involved (mesophilic or thermophilic) and the final requirements of the resulting biogas (to be burned in cogeneration units or to be upgraded to biomethane), the configuration of anaerobic digestion plants changes. Despite this, the main steps for biogas production are (G.E. Scarponi et al., 2016):

1. Anaerobic Digestion (AD);
2. Hydrogen Sulfide (H_2S) removal;
3. Drying.

The process itself (1) takes place in the digester, which is the core component of the plant. AD can be carried out in a single stage or in two stages depending on the biomass used and to economic considerations about the desired yield in biogas and the retention time (Scarponi et al., 2015). The digester is usually integrated with a gasometer that is necessary to store the biogas produced. The gasometer basically is a plastic membrane able to handle the fluctuation in both biogas demand and production.

The desulfurization step (2) is usually carried on within the digester itself but, in some cases, this is not feasible (due to the characteristics of the digester and the H_2S content). In such cases, a dedicated scrubbing column is used. Biogas drying is generally achieved by cooling it in a chiller (3).

According to previous considerations, a reference scheme for biogas production was defined and is presented in Figure 1, representing most of the existing biogas production plants in Europe. In this configuration, the substrate consists of livestock slurry and is degraded by anaerobic bacteria resulting in the formation of digestate, which is removed, and biogas that is stored in the membrane gasometer placed on top of the digester. Here, a colony of desulfurizer bacteria reduces the content of hydrogen sulfide under 200 ppm_{vol}. Then, a blower carries the biogas stream (300 Nm^3/h, 60% vol of methane—CH_4- and

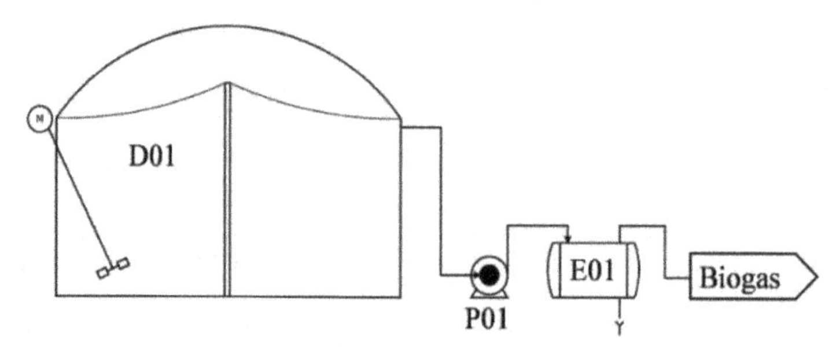

Figure 1. Reference process scheme for biogas production from anaerobic digestion of livestock slurry (G.E. Scarponi et al., 2016). D01: anaerobic digester with gasometer and internal desulfurization system; P01: a blower; E01: dryer.

40% vol of carbon dioxide—CO_2) to a dryer that lowers the water content and makes the biogas ready for either being burned in a cogeneration unit or being upgraded to biomethane (G.E. Scarponi et al., 2016).

3 RESULTS OF THE APPLICATION OF MIMAH

The bow-tie analysis was carried out with the approach present in the Methodology for the Identification of Major Accident Hazards (MIMAH), as described by Delvosalle et al., 2006; Salvi and Debray, 2006. A summary of the main steps of MIMAH is reported in Figure 2.

The first step consists in collecting all the information required for the analysis, e.g., the layout of the plant, the description of the process, equipment and pipes, operating conditions, substances and their properties, etc. In step 2, based on the information collected, a list of the hazardous substances present in the plant and a list of equipment containing these substances are drawn up. The hazards come from the flammability and toxicity of the biogas produced, a mixture of methane, carbon dioxide and hydrogen sulfide in traces. Details are displayed somewhere else in literature (G.E. Scarponi et al., 2016) and a summary is proposed in Table 1. In step 3, relevant hazardous equipment are selected, as summarized in Table 1.

Performing step 4, the possible Critical Events (CEs) initially associated with the equipment are those reported in Table 2. However, several CEs and associated bow-ties were found to be coincident or very similar:

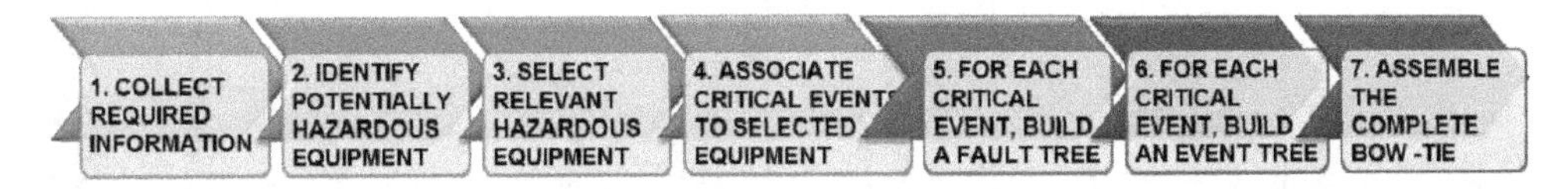

Figure 2. Summary of the steps of MIMAH methodology. Adapted from Delvosalle et al., 2006.

Table 1. Summary of the results of MIMAH's steps 1 to 3.

	Digester + Gasometer	Blower	Chiller	Pipes
State of the stream	Gas	Gas	Gas	Gas
Volume [m³] or Diameter [mm]	500	–	–	50 mm
Pressure [barg]	0.002	0.12	0.002	0.002
Temperature [°C]	39	39	15	15–39
$CH4/CO_2$ [%v]	60/40	60/40	60/40	60/40
H_2S [ppm]	200	200	200	200
Q [Nm³/h]	48	48	48	48

Table 2. Critical events associated with relevant hazardous equipment of Table 1.

	Critical Events (CE)						
	Breach on the shell in vapor phase (Small, Medium, Large)			Leak from gas pipe (Small, Medium, Large)			
Equipment	Small	Medium	Large	Small	Medium	Large	Catastrophic rupture
Digester + Gasometer	X	X	X				X
Chiller	X	X	X				X
Blower							X
Pipes				X	X	X	

UE		DDC		DC		NSC	CE	SCE	TCE	DP
Excessive conditions created by the environment	or	Aging	or	Seal, joint loss of effectiveness	or	Functional opening	Breach on the shell (medium size) in vapour phase	Gas jet	Gas dispersion	VCE
Excessive conditions created by the process										Flash fire
Lacking or defective maintenance										Toxic cloud
Wrong material delivered	or	Improper material								Environmental damage
Wrong material used										
Wrong dimension	or	Bad design							Gas jet ignited	Jet fire
Wrong material										Toxic cloud
Not replaced like with like	or	Bad installation or maintenance								Environmental damage
Bad installation or maintenance procedure										
Normal use/storage of aggressive chemical	or	Physical or chemical aggression								
Contamination										
		Normal functioning of the safety valve	or	Safety valve, safety relief device						
Lacking or defective maintenance	or	Too sensitive safety valve								
Design error										
Installation/calibration error										
Lacking or defective maintenance	or	Leaking isolation equipment	or	Fail to clear out contents before opening containment						
Design error										
Manufacturing error										
Installation error										
Human error	or	Hazardous contents removal procedure failed								
Blocked outlets										
Incorrect sensor signal	or	Lacking or wrong information about containment	and	Operation started when containment open						
Interpretation error										
Transmission error										
Human error										
Incorrect command and/or control signal	or	Containment closing procedure failed								
Incorrect sensor signal										
Human error										

Figure 3. Bow-tie of the CEs "medium size breach on the shell in vapor phase" for the anaerobic digester and gasometer shown in Figure 1.

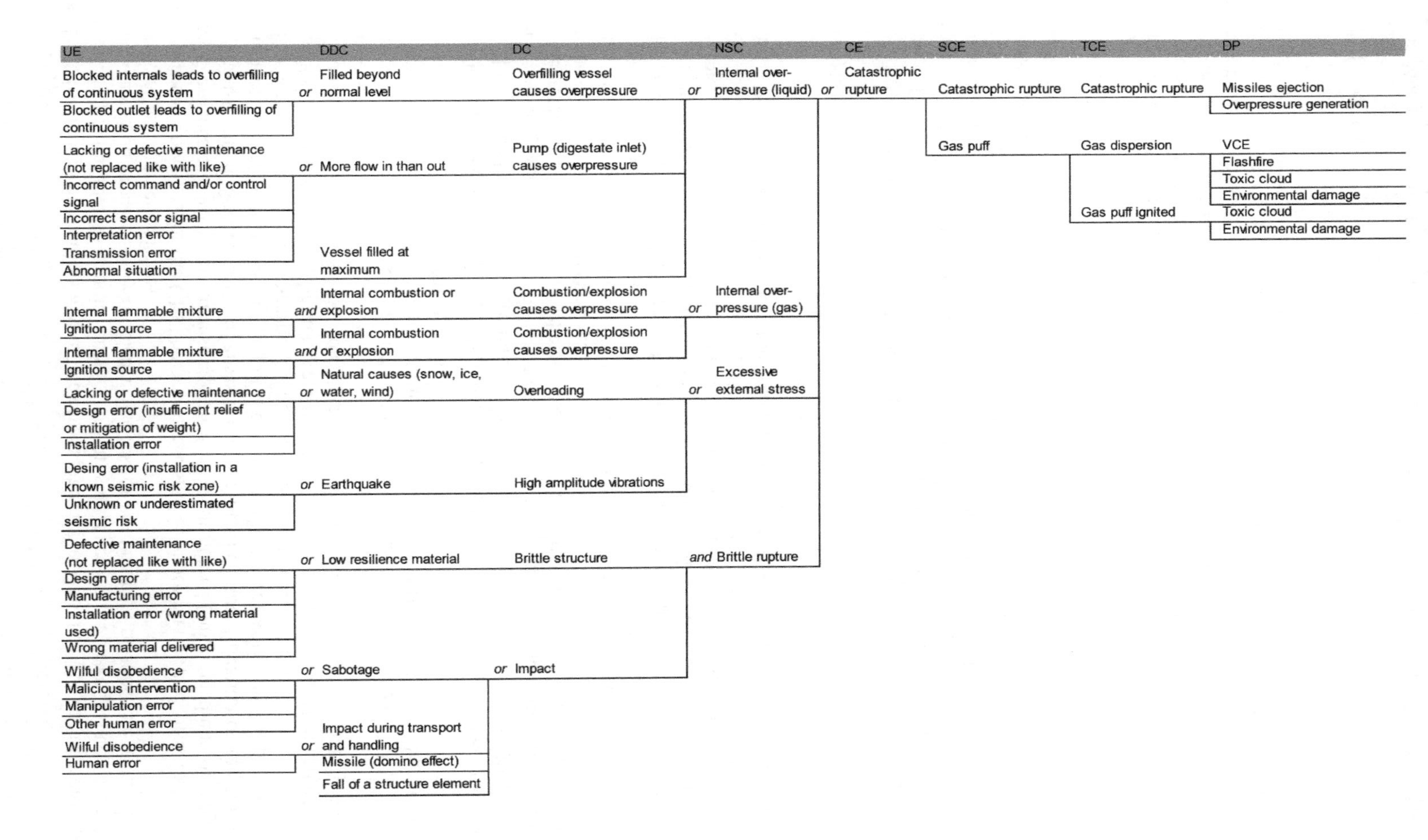

Figure 4. Bow-tie of the CEs "catastrophic rupture" for the anaerobic digester and gasometer shown in Figure 1.

- The CEs "breach on shell" as well as "leak from pipe" referred to the liquid phase were discarded since the analysis handles only hazardous substances in the gaseous state, focusing the analysis on process safety problems. The loss of containment of substrate is mainly a cause of environmental issues.
- The generic bow-ties for the CEs "breach on the shell in vapor phase" and "leak from gas pipe" are identical when the same size of rupture is considered (small, medium, or large). Obviously, the difference in the analysis is significant when quantification of the consequences is carried out, which falls out of the scope of the present work and was performed elsewhere (Giordano Emrys Scarponi et al., 2016; Scarponi et al., 2015).
- Small size rupture fault tree branches are included in medium size rupture fault tree that, in addition, is able to describe functional openings related to valves and safety relief devices.
- Large size rupture fault tree branches are included in catastrophic rupture fault tree.
- The anaerobic digester is the most critical item with respect to the quantities of hazardous substances handled, as shown in Table 1.

Furthermore, the gasometer is not included in standard equipment considered by the original MIMAH methodology. According to a previous study (Casson Moreno et al., 2016), the rupture of the gasometer seems to be a quite frequent CE. Since the gasometer and digester are integrated into the reference scheme considered, the fault trees of the digester include both the digester itself and the gasometer.

This allows circumscribing our analysis to two main critical events: "catastrophic rupture" and "medium size breach on the shell in vapor phase" of the anaerobic digester.

The application of MIMAH procedure as shown in Figure 2 provided the bow-ties for the CEs "medium size breach on the shell in vapor phase" and "catastrophic rupture" of the anaerobic digester and gasometer reported in Figures 3 and 4. Reading the fault tree from the CE to the left, the Necessary and Sufficient Causes (NCS), the Direct Causes (DC), the Detailed Direct Causes (DDC), and the Undesirable Events (UE) of the CE can be found.

4 CONCLUSIONS

The MIMAH methodology was born based on expertise acquired in the petrochemical industry. This is one of the first attempts of application outside the conventional chemical process industry. Biogas production from anaerobic digestion is a widespread industrial biotechnological process, in which hazards such as fire, explosion, and toxicity are present. The advantage of using MIMAH is its generality and orderliness that makes it flexible when used in a new case study like the presented one. With respect to the previous literature review, accident analysis, and the elicitation of experts, the anaerobic digester with integrated gasometer was confirmed to be the most relevant hazardous equipment. Bow-ties for CEs related to it were built and integrated on the basis of knowledge acquired by literature review, accident analysis, and the elicitation of experts of the biogas sector.

The main limitation found within the analysis was the lack of homogeneity in available information for the biogas sector due to the relative novelty of this industrial sector and the consequent limited acquired experience. Still, the importance of obtained results is their broad validity for most of the biogas facilities present in Europe and worldwide. A future step to be performed is the application of MIRAS, devoted to safety barriers identification and a complete Quantitative Risk Assessment.

REFERENCES

Casson Moreno, V., Cozzani, V., 2015. Major accident hazard in bioenergy production. J. Loss Prev. Process Ind. 35, 135–144. doi:10.1016/j.jlp.2015.04.004.

Casson Moreno, V., Papasidero, S., Scarponi, G.E., Guglielmi, D., Cozzani, V., 2016. Analysis of accidents in biogas production and upgrading. Renew. Energy 96, 1127–1134. doi:10.1016/j.renene.2015.10.017.
Delvosalle, C., Fievez, C., Pipart, A., Debray, B., 2006. ARAMIS project: a comprehensive methodology for the identification of reference accident scenarios in process industries. J. Hazard. Mater. 130, 200–19. doi:10.1016/j.jhazmat.2005.07.005.
EurObserv'ER, 2014. Biogas Barometer 12.
Findeisen, C., 2015. The importance of safety standards, risk assessment and operators training for a successful biogas market development Content, in: Biogas for Productive Uses, Industrial and Mobility Applications. Wein.
FNR, 2013. Biogas an introduction. Gulzow.
REN21, 2015. Renewables 2015 Global Status Report.
Salvi, O., Debray, B., 2006. A global view on ARAMIS, a risk assessment methodology for industries in the framework of the SEVESO II directive. J. Hazard. Mater. 130, 187–199.
Saracino, A., Casson Moreno, V., Antonioni, G., Spadoni, G., Cozzani, V., 2016. Application of a self-assessment methodology for occupational safety to biogas industry. Chem. Eng. Trans. 53, 247–252. doi:10.3303/CET1653042.
Scarponi, G.E., Guglielmi, D., Casson Moreno, V., Cozzani, V., 2016. Assessment of inherently safer alternatives in biogas production and upgrading. AIChE J. 62. doi:10.1002/aic.15224.
Scarponi, G.E., Guglielmi, D., Casson Moreno, V., Cozzani, V., 2016. Assessment of Inherently Safer Alternatives in Biogas Production and Upgrading. AIChE J. 62, 2713–2727. doi:10.1002/aic.15224.
Scarponi, G.E., Guglielmi, D., Casson Moreno, V., Cozzani, V., 2015. Risk Assessment of a Biogas Production and Upgrading Plant. Chem. Eng. Trans. 43, 1921–1926. doi:10.3303/CET1543321.

Risk Analysis and Management – Trends, Challenges and Emerging Issues – Bernatik, Huang & Salvi (Eds)

From ARAMIS methodology to a "dynamic risk" monitoring system

Emmanuel Plot
Ineris Developpement Sas, Verneuil-En-Halatte, France

Zoe Nivolianitou
DEMOKRITOS, Greece

Chiara Leva
DIT, Ireland

Vassishtasaï B.P. Ramany
MEEM, SNOI, France

Christophe Coll
DEKRA, France

Frédéric Baudequin
INTERACTIVE, France

ABSTRACT: The ARAMIS[1] project is believed to be able to address several goals, in particular: (a) the use of state-of-the-art methods to study processes to predict potential hazardous events and their likelihood; (b) to achieve "transparency" of processes that allows both the users and the regulating authorities to understand, validate, and comment on risks consistently. The ARAMIS methodology first introduced the concepts of safety barriers and bow-ties, which, nowadays, are used regularly by the European Industry and are considered as a valuable means to perform risk assessment and to share the results with stakeholders. However, in order to address a risk assessment usable for real-time safety management, a further step needs to be accomplished, namely the dynamic monitoring of risk, that is, how the actual status of the equipment and/or conditions in a moment in time can be taken into account to update the risk assessment and therefore estimate the risk exposure of the installation toward the accidental scenarios identified. This step was developed thanks to further EU-funded project TOSCA[2] that built on ARAMIS achievements. The actual risk level of an installation with respect to hazardous phenomena is in fact a property that changes over time taking into account the actual status of equipment and their management. In this paper, we explain the progress of this specific goal as well as present an applied case study.

1 INTRODUCTION

The ARAMIS methodology addresses several risk-related industrial needs: (a) the need for a methodology to identify, assess, and reduce the risk and demonstrate the risk reduction as

---

[1]Accidental Risk Assessment Methodology for Industries in the framework of SEVESO II directive—accepted for funding in the 5th Framework Program of the European Commission, which was started in January 2002.
[2]TOSCA Total Operations Management for Safety Critical Activities accepted for funding in the 7th Framework Program of the European Commission, which was started in December 2012 project ID 310201.

required by the SEVESO directive. This should be a state-of-the-art methodology and must also provide useful information about the ways to reduce the existing risk level and to manage it regularly; and (b) the need for a "reference" methodology, whose analysis results are accepted by the competent authorities. The latter can also use it to assess the safety level of the plant.

A reasonable question to be asked is "How the owner could monitor his plant safety level over time?" A plausible answer to this question is by using the risk assessment methodology proposed by ARAMIS as a basis of an IT monitoring performance system. In our opinion, this system should be seen as a mandatory part of the SEVESO directive requirements[3]. However, both ARAMIS methodology and other existing solutions are not yet able to address completely this need. They are generally focused on a mere safety barrier monitoring (see, for instance, the DNV tool, 2016) or on "simplified scenario" monitoring (see, for instance, the Petrotechnics, 2016). The idea is to extend the ARAMIS methodology by incorporating the time dimension so as to address the plant "dynamic risk". The authors of this paper have tried to develop a monitoring system able to address completely this need through several research projects and real-time applications (Leva et al 2010, Demichela et al. 2014, Monferini et al. 2013, Leva et al. 2012). The IT tool developed (using Interactive platform[4]) supports the continuous assessment of the safety barrier status and the automatic recalculation of the actual risk level, together with its comparison with the target levels of the accidental scenarios. The developed risk model and the relevant barriers selected can be connected to an overall plant model; then, all data coming from both the equipment and the critical task performing inflow as an input to the plant risk model and update the actual risk perception about plant running. In addition, information can be visualized with appropriate tools and visualized to relevant stakeholders/operators through online decision support tools. In the following, the whole methodology and procedure is presented in detail. At each step of the implementation process, the links between the ARAMIS methodology and the new product will be explained.

2 DESCRIPTION OF THE METHODOLOGY

This paragraph presents the links between the already developed ARAMIS methodology and the newly formulated dynamic risk assessment methodology.

2.1 *STEP 1: Bow ties*

The new methodology begins with the same first step as in ARAMIS, namely the Identification of the Major Hazards in the installation and the construction of the plant model in the form of bow-ties (BT, without safety barriers), which are designed using the MIMAH method[5]. Then, these BTs are introduced into the IT Tool. The analyst has also the possibility to use directly the IT tool for the initial design of bow-ties, making the following steps, as depicted in Figure 1:

a. Draw the bow-tie using the graph editor of the IT Tool by creating in a sequence:
   - The central event and its estimated frequency;
   - The initiating event;
   - The dangerous phenomena; and
   - The safety barriers and their estimated level of confidence.
b. Activate the frequencies (probabilities) propagation calculations by taking into account the estimated initiating event frequencies and safety barrier levels of confidence.

The calculations are done according to the ARAMIS proposed approach:

[3]SEVESO III.

[4]See: www.interactive.fr.

[5]ARAMIS's Methodology for the Identification of the Major Hazards, 2001.

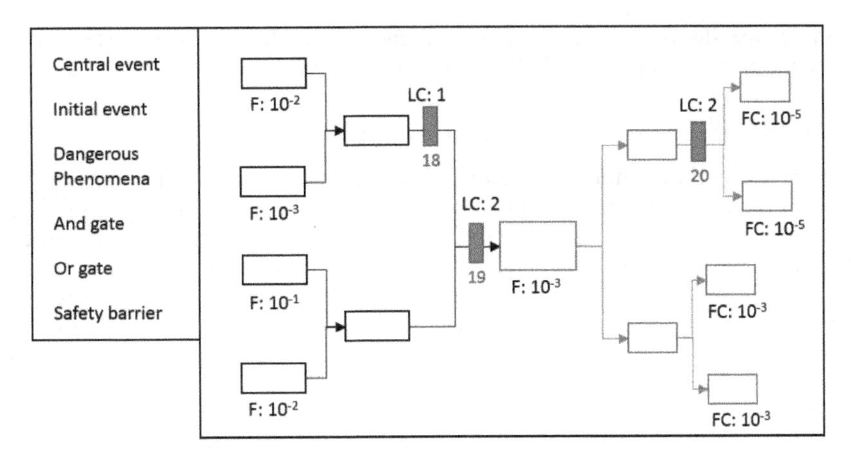

Figure 1. Bow-ties graphical editor and calculator (source: the Internet).

c. The analysis is made by a gate-to-gate calculation and taking into account the safety barriers on the fault tree. Briefly, the gate-by-gate method starts with the initiating events of the fault tree and proceeds upward to the critical event. All inputs to a gate must be evaluated before calculating the gate output. All the bottom gates must be computed before proceeding to the next higher level. In parallel, the influence of safety barriers on the accident scenario is taken into account. The prevention and control barriers decrease the transmission probabilities between two events in the fault tree and influence the critical event frequency. Indeed, if the level of confidence of a barrier on a branch is equal to n, then the frequency of the downstream event on the branch is reduced by a factor 10^{-n}.
d. Calculate the occurrence frequencies of the dangerous phenomena. The objective is to proceed step by step in the event tree to obtain, as output, the frequency of each dangerous phenomenon. First of all, in the generic event trees built with MIMAH, there is no AND/OR gate explicitly drawn. In fact, these gates are implicitly included in the event trees. AND gates are located between an event and its simultaneous consequences. OR gates appear downstream, and an event of one of the consequent events may occur and the others not. Second, when OR gates appear in the event tree, figures for the transmission probabilities linked with these gates is assessed. The transmission probabilities can be the following ones: the probability of rain-out and leakage, the probability of immediate ignition, the probability of delayed ignition, or the probability of VCE. Finally, safety barriers related to the event-tree side are taken into account, in terms of both consequences and frequency of dangerous phenomena. Briefly, it can be pointed out that the prevention and control barriers decrease the transmission probability between two events by their level of confidence and influence of the dangerous phenomena frequency. The limitation barriers reduce the consequences of dangerous phenomena in limiting the source term or in limiting their effects. In the event tree, when a limitation barrier is met, two branches must be built: (1) if the barrier fails with a probability equal to the Probability of Failure on Demand (PFD) and (2) if the barrier succeeds with a probability equal to (1-PFD). The PFD of a safety barrier is equal to 10^{-n}, with n being the level of confidence of the barrier. Both branches are kept in the event tree, because they will lead to different dangerous phenomena, one with less severe consequence but at a higher frequency, and the other one with a more severe consequence but a lower frequency.

2.2 *STEP 2: Equipment and bow-ties*

To be able to identify the criticality of equipment according to the criticality of their related dangerous phenomena, links between bow-ties and equipment have to be input into the IT Tool. However, bow-ties are often generic, abstract scenarios, designed for several equipment of the same type. Because of that, specific equipment is not directly linked to bow-ties; in the data model, specific equipment is linked to generic equipment linked to bow-ties.

2.3 *STEP 3: Safety barriers and equipment and critical tasks*

In this step, the level of confidence is estimated for a whole safety barrier (and not for a single device), including the different subsystems comprising the barrier (detector, safety system, action). For each subsystem, the level of confidence, effectiveness, and response time will be estimated and combined to calculate a global level of confidence of the barrier. A proper acquisition of relevant information about a system and a task in a safety critical environment is the foundation of every sound human factor analysis. The scope of the analysis may cover a human reliability assessment, an evaluation of a human–machine system as a whole, the writing of a procedure, or the preparation of a training program. When this foundation is correctly set, the conclusions of the analysis will be already addressed toward a useful and reliable direction.

More and more studies have highlighted that this critical first step of the analysis has been taken for granted and not given the attention required for collecting and structuring the information about the tasks and contexts. Task analysis is the process of gathering data about the tasks people perform, acquiring a deep understanding of it, and representing it. Traditionally, the main steps for achieving a task analysis relevant also in the context of a bow-tie are:

a. Preliminary data collection about the task to be modeled (especially if this is a safety barrier).
b. Update collected data through interviews or observations about the actual way the task is performed.
c. Representation of the information collected.
d. Evaluation of task reliability (i.e., what is the reliability of a task if it is to be considered a safety barrier and what are the performance shaping factors influencing it, which includes an evaluation of the task demands against the operators' capabilities or an evaluation of specific safety issues related to it).

The estimation of safety barrier level of confidence therefore also needs:

- To monitor all equipment and critical task identified as a part of safety barriers; therefore, it is mandatory to list these equipment and tasks and to enter them in the IT Tool.
- For each equipment and task, to know the impact of a failure on the safety barrier level of confidence; therefore, the qualitative safety barrier structures (detection–treatment–action systems and subsystems) have to be described into the IT Tool.

2.4 *STEP 4: Critical tasks and incidents*

The monitoring of the real-time level of confidence for a safety barrier is then based on two types of information:

- The delay in a critical task realization (those directly involved in the safety barrier performances and those indirectly involved, such as schedules in the preventive maintenance of the equipment directly involved in the safety barrier performances), which indicates "non-confidence".
- The incidents that indicate the unavailability of equipment or of person in charge of human barrier.

On the basis of these two types of information, it is possible to re-estimate the true safety barrier level of confidence.

As an example, we could take the following: If a detector is unavailable (or if there is a low-confidence indicator) for more than 10% of time on a given period, then it is known that the level of confidence of the related safety barrier is null (inexistent).

For being able to monitor the level of confidence of any given safety barrier, we should take measurements daily and insert them into the IT Tool.

On the basis of this, the tool is able to recalculate periodically the criticality of the dangerous phenomena.

2.5 *STEP 5: Changes*

If the hypothesis on which the bow-ties designed is no longer valid, then the monitoring of the level of confidence and the periodic recalculation of the dangerous phenomena criticality is no longer legitimate. Therefore, the IT Tool has to monitor through time the validity of these hypotheses. A set of related information has to be daily inserted into the tool, namely:

- The quantity of hazardous substances handled;
- The type of hazardous substances handled; and
- The frequency of initiating events.

3 CASE STUDY

The methodology and IT tool presented above have been developed within the European TOSCA project (the tool is referred to as the Computerized Barrier Manager System—CBMS, TOSCA, 2014, Konstantinidou et al. 2015). The innovation lies in the fact that the latter switches from research to business case with the SNOI case study (a national pipeline system managing petroleum facilities and depots). In 2015, the SNOI officers decided to base their monitoring management on the TOSCA methodology and tool. This was an excellent opportunity for the steps of the approach and the IHM of the tool to be reviewed, improved, and redesigned so as to suite the daily industrial needs. Three major partners have played significant role in this process: INERIS, SNOI, and DEKRA.

For this case study, it has been decided to select a prevalent accidental scenario. Having selected the one, four safety barriers have been positioned. At first glance, it may have seemed too simple. Though, when we started to link in the IT tool, the 88 storage tanks of the 15 SNOI sites concerned with the accidental scenario and to link the equipment involved in their safety barriers, the real complexity appeared. We realized, at the end, that we have to monitor 314 equipment, proving how comprehensive a risk assessment could be. For each equipment, the key data to be taken into account were been identified and their storage, format, and repetition patterns were provided.

It has been decided then to collect the data every 6 months for updating the criticality calculation of the dangerous phenomena of the petroleum tanks concerned. The first run discovered several "mistakes" or missing data. For instance, analysis performed by subcontractors on a given type of equipment was not usable because it was neither conclusive nor performed in the same time interval and with the same measurement units. This was not known beforehand unless one tries to reuse the data to recalculate anew the risk level on the basis of a different risk assessment method; then, the real problems in collected data appear. An additional example is the delays in preventive maintenance. The former were considered from the risk management perspective, but, nevertheless, adopted by the maintenance team because of organizational constraints and because no one in the field knew what was the actual risk assessment requirement to take into account.

It takes more than a year for the management to correct these inconveniences or the missing data. After that period, practices on field are on line, which are patently highlighted throughout the monitoring system put in place.

The methodology and IT tool has been presented to the regulating authorities. They considered it as a step ahead in the "transparency" of processes that increases the confidence in the daily work of the industrialist according to what has been validated on risk management requirements.

4 CONCLUSION

On the basis of this case study, it seems that the ARAMIS methodology has a better managerial impact when it is used to support a continuous monitoring.

As expected by the TOSCA researchers, it seems that this approach supports the development of a COP (Common Operational Picture) within a company. This approach seems as a kind of sting for bridging the gap between "actual practice" and "official work systems". It is a support for human factor and organizational improvements, addressing the fact that in complex process control industries, the different stakeholders, regulatory bodies, contractors, managers, and operating teams may have their own idiosyncratic "concept" or "picture" of the conditions that give rise to risks. Even within the same operational departments, the term "risk" means different things to different team members who may have different baselines and priorities. At the end, this approach tries to establish a common framework of safety performance, which is very important because these mutual understanding and interrelating mechanisms ultimately determine the level of system risk.

There are situations in the industry where the human actions are the main safety barriers to abnormal or accidental conditions. In order to maximize the reliability of good human and organizational barriers, we need to ensure that the action plans generated are based on a valid risk assessment of the situation to be addressed and informed by a relevant human factor analysis. This implies that the process needs to be participatory in nature, thus involving end user all the way through. For the example proposed, we involved the end user also in suggesting possible improvement actions. The one to be selected were rated on the basis of their impact and the difficulty/cost of implementation, the impact was informed by not only the risk assessment effects on the reliability of the barrier but also what priority the action was assigned during a focus group with the end users. Examples of suggested actions are not reported in this paper.

The evaluation of this work is based on the:

- coherence in the proposals themselves;
- coherence among the proposals; and
  a. the bow-tie analysis;
  b. the task analysis informing the bow-tie and the list of the performance shaping factors selected.

The benefit of the approach reflects not only the quality of the background information provided for the risk assessment but more importantly the involvement of the main end users of the system in assessing their own work performance and being proactively called to identify the way of improving its reliability and safety. Therefore, even if a further level of automation is identified as a further safety barrier, it will be designed in a way that will keep the user proactively in the loop.

The research team believes that an extension of this approach could also support binding a plant risk model with advanced process control based on emerging developments in cloud storage and computing so as to achieve operational optimization. We know that Overall Equipment Effectiveness (OEE) can be significantly increased by networking various isolated solutions with the help of software agents in automation technologies. This will allow bottlenecks, cost drivers, and process upsets to be better defined and energy consumptions precisely assigned. The data from a networked, integrated system can be used to optimize fuel usage and schedules and prevent plant trips or unwanted downtime. One of the industrial partners, an energy generation company, has estimated that a software agent able to better monitor trends and aggregate data from their DCS, PLCs, and a risk model of the plant can help them save over 5 M euro per year in unwanted process upsets and trips across their plants. Operations managers will only need to handle a uniform engineering tool system wide. These innovations have, on the contrary, increased the amount of data operations that the manager needs to handle to achieve a complete overview of plant performance.

The next step should be to set up a new project to build on these results and overcome these difficulties by:

- providing a real-time framework to connect process data from SCADA, DCS, and PLCs to enable real-time intelligence for operations control;

- proposing an overall plant risk forecast model to be used in conjunction with predictive control techniques, to achieve production efficiency and downtime minimization; and
- offering a novel empirically proven human–machine interface to provide task support and training to increase human reliability and situational awareness.

REFERENCES

Demichela, M., Pirani, R., Leva, M.C. 2014. Human factor analysis embedded in risk assessment of industrial machines: Effects on the Safety Integrity Level. *International Journal of Performability Engineering 10* (5), 487–496.

DNV, 2016 decision support tool for dynamic barrier management.

Konstantinidou, M., Plot, E., Leva, M.C, Mavridis, G., Aneziris, O., Nivolianitou, Z. 2015. Effective identification and management of critical tasks for safer performance, EMChIE 2015.

Leva, M.C., Bermudez, Angel C., Plot, E., Gattuso, M. 2013. When the Human Factor is at the core of the safety barrier, Chemical Engineering Transactions 2013, VOL. 33 pp. 439–444.

Leva, M.C., Cahill, J., Kay, A., Losa, G., McDonald, N. 2010. The advancement of a new human factors report—'The Unique Report'—Facilitating flight crew auditing of performance/operations, as part of an Airline's Safety Management System. Ergonomics 53(2), 145–148.

Leva, M.C., Kontogiannis, T., Balfe, N., Plot, E., Demichela, M. 2015. Human factors at the core of total safety management: The need to establish a common operational picture, Contemporary Ergonomics pp. 163–170.

Leva, M.C., Kontogiannis, T., Plot, E., Demichela, M. 2014. Total Safety Management: what are the main area of concern in the integration of best available methods and tools? Chemical Engineering Transactions, VOL. 36 pp. 559–564.

Leva, M.C., Pirani, R., De Michela, M., Clancy, P. 2012. Human Factors issues and the risk of high voltage equipment. Chemical Engineering Transactions, 26, 273–278.

Monferini, M., Konstandinidou, M., Nivolianitou, Z., Weber, S., Kontogiannis, T. Kay, A.M., Leva, M.C., Demichela, M. (2013). A compound methodology to assess the impact of human and organizational factors impact on the risk level of hazardous industrial plants. Reliability Engineering & System Safety 119, 280–289.

Proscient tool, 2016 Petrotechnics.

TOSCA (February 2012 – February 2016; Funding provider: EU-FP7).

Risk Analysis and Management – Trends, Challenges and Emerging Issues – Bernatik, Huang & Salvi (Eds)

Connection of ARAMIS methodology approach with APELL programme approach in the Czech Republic

Petra Ruzickova
Fire Rescue Service of Moravian-Silesian Region, Ostrava—Zabreh, Czech Republic

Ales Bernatik
Faculty of Safety Engineering, VSB—Technical University of Ostrava, Ostrava—Vyskovice, Czech Republic

ABSTRACT: The need to adequately assess and communicate major accident risks in connection with dangerous chemicals and chemical mixtures has been proved by many accidents and circumstances of accidents occurring in production and storage facilities in various branches of industry. In spite of the fact that statistical records exhibit a slight downward trend in the frequency of major accidents in the EU countries, historical experience shows that the occurrence of these events and their dangerous manifestations cannot be avoided wholly. In the last few years, intense expert discussions concerning the way of analysis processing and major accident risk assessing, including interpretation and communication of outputs of the analysis with the public have taken place in the Czech Republic. Here, a number of studies dealing with the problem of major accident prevention have been performed. The objective of this paper is risk assessment in small and medium-sized enterprises. The risk assessment is done using the ARAMIS methodology with a view to verifying its applicability to SMEs.

Keywords: Major accident prevention, ARAMIS methodology, LPG, APELL, risk communication

1 INTRODUCTION

Major accident prevention is a multi-criteria discipline and particular branches of the discipline are very closely interconnected and often follow one another. It also depends on the knowledge of modern science, technological trends and socio-economic development in individual countries. With reference to this fact, in the year 2005 we tried to make risk analysis and risk assessment for a selected industrial establishment in the Czech Republic and subsequently to communicate ascertained data with the public and also rescue services and local authorities and public administration. For this purpose we used two methodologies, namely the ARAMIS methodology ("Accidental Risk Assessment Methodology for IndustrieS in the Framework of the SEVESO II Directive") and the APELL programme (Awareness and Preparedness for Emergencies at Local Level), which was unique at that time. Outputs and benefits of connection of these two apparently different approaches that have been used and improved in the framework of valid legislation in the Czech Republic up to now are described below.

2 ARAMIS METHODOLOGY CASE STUDY

The ARAMIS (Accidental Risk Assessment Methodology for IndustrieS in the framework of the SEVESO II directive) methodology was developed in the framework of an EU 5FP project. A harmonised methodology for risk assessment, aimed especially at reducing uncertainties and variability in results and at including the evaluation of risk management efficiency into the analysis, was proposed. It is necessary to regard ARAMIS as a comprehensive tool for efficient implementation of risk identification and analysis with many pre-prepared and recommended steps (ARAMIS, 2004).

Risk analysis and assessment utilizing the ARAMIS methodology were carried out in an industrial establishment using the LPG for powering forklifts and having a built-up LPG filling station on its premises. The filling station consists of two cylindrical horizontal tanks; the total maximum amount of LPG is 4.9 t (see Figure 1).

In the vicinity of the filling station, which is located on the boundary of the industrial establishment, there are a housing estate and a public road. In case of a major accident, a threat to both the employees of the establishment and the population can be expected. For the purpose of simplification, other risk sources in the establishment are not assessed.

The procedure of the ARAMIS methodology can be divided into three basic steps, the outputs of which are relevant indexes:

1. assessment of consequence severity (S—severity index),
2. evaluation of risk management efficiency (M—management index),
3. assessment of surrounding environment vulnerability (V—vulnerability index).

All the indexes can be evaluated separately, but above all the indexes S and M are considerably interconnected in the selection of reference accident scenarios and consequence severity determination, when efficient measures to reduce the risks can affect the frequency of accidents or to limit the consequences of them.

Furthermore, the procedure for determining the index S-assessment of consequence severity will be presented. In the first phase, a list of sources of risks of major chemical handling accidents is made up.

- LPG filling station, 4 900 kg of propane-butane, extremely flammable liquefied gas.

For risk source identification, 16 types of equipment (selected EQ equipment) are defined in the methodology. In our case, it is EQ4-Pressure storage type of equipment.

To each selected risk source, a critical event has to be assigned (CE—Critical Event). The critical event is defined as a release of liquid content (LOC—Loss of Containment) from equipment. The method assumes 12 critical events. For our purposes, a critical event CE10 —Catastrophic rupture was selected.

Figure 1. LPG filling station in the establishment.

Further, fault trees and event trees that are connected to a so-called bow-tie (diagram) are built. Bow-ties are to be understood as major accident scenarios without considering installed safety systems.

The aim of the next phase of ARAMIS methodology is to select a reference scenario of accidents from scenarios identified in the first part. It is based on the study of influence of safety elements and risk management on selected scenarios. The Reference Accident Scenarios (RAS) represent a real hazardous potential of equipment after considering safety systems (including management).

The result of this part of the methodology is the frequencies of critical event after considering safety barriers in the fault trees.

- For CE10 —catastrophic rupture of LPG storage tank — $2.1.10^{-6}$/year.

The aim of the next step is to determine the frequencies of all dangerous phenomena of selected critical events. The procedure is based on considerations concerning the safety barriers in the event trees that can decrease the frequency or consequences of the dangerous phenomena.

Results for the selected critical event are illustrated in Figure 2. The methodology again offers rough values of probability of immediate ignition, probability of delayed ignition and probability of VCE from specialized literature (Delvosalle et al., 2004, Hourtolou et al., 2004).

In the next step, it is necessary to carry out the rough evaluation of consequences of dangerous phenomena. This qualitative evaluation of consequences is based on classifying the dangerous phenomena into 4 classes of consequences (C1-4), when the class C4 means the most serious consequences on human health and/or the environment. For individual dangerous phenomena the methodology offers pre-defined classes of consequences that can be modified according to the efficiency of barriers limiting the released amount of the substance or the impact of the dangerous phenomenon. The final classes of consequences are given in Table 1.

Reference scenarios are selected by means of a tool—risk matrix (see Figure 3).

It is necessary to state that the risk matrix in this phase of the methodology does not decide about the acceptability of risks but that it merely selects the reference accident scenarios that are further modelled for the purposes of severity determination.

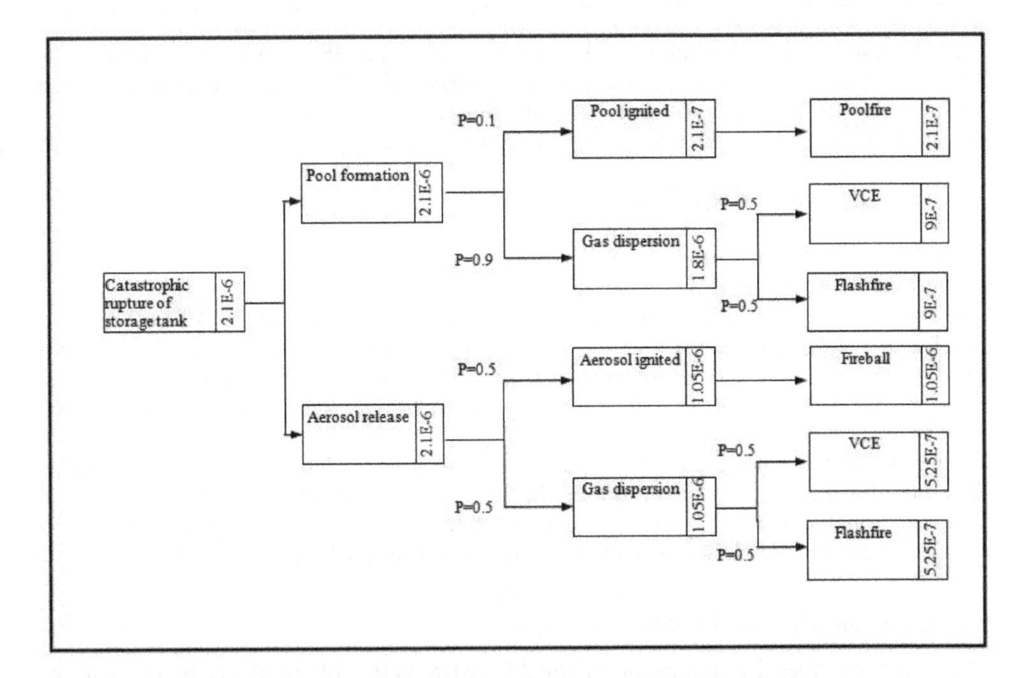

Figure 2. Event tree with the frequencies of dangerous phenomena—rupture of LPG storage tank.

Table 1. Frequencies and classes of consequences of dangerous phenomena.

Number	Dangerous phenomenon	Frequency	Class of consequences
1.	1a) Poolfire	2.1×10^{-7}/year	C2
	1b) VCE	1.4×10^{-6}/year	C3
	1c) Flashfire	1.4×10^{-6}/year	C3
	1d) Fireball	1×10^{-6}/year	C4

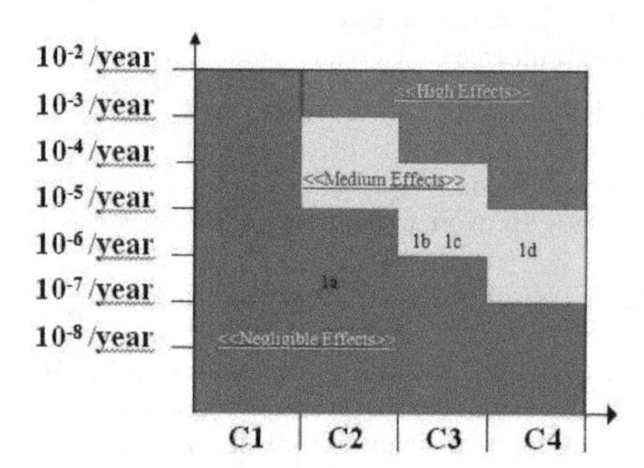

Figure 3. Risk matrix with results (Bernatik 2005).

Table 2. Distances for effects of dangerous phenomena for CE10.

Critical event		Frequency	2,1E-06				
Dang.phenomena	CE10	d1	d2	d3	d4	Frequency	Type
VCE	1b	260	165	75	45	1,4E-06	Overpressure
Flashfire	1c	72	72	68	35	1,4E-06	Thermal
Fireball	1d	430	330	255	195	1,0E-06	Overpressure

In the last phase of evaluation of index S by the ARAMIS methodology, the severity indexes of reference accident scenarios are determined using the proposed parameters. It is just the proposal of threshold values for individual effects of accidents that is another significant benefit of the methodology, because in the European Union uniform recommended values do not exist yet.

For the determination of individual distances, any mathematical model can be used. With regard to the fact that the use of computing module is entirely independent of the methodology for the determination of index S, the user can employ any mathematical model for the evaluation of accident effects.

Distances for particular levels of effects were calculated using the Dutch EFFECTS model (see Table 2). Table 2 states the results in the form of distances (in metres) for particular levels of consequences d1–d4 and for each dangerous phenomenon. The type of consequences expresses one of four possible serious effects of accidents (thermal radiation, overpressure, missiles, toxic effects).

The results of the study show the highest threat to the population and the employees of the establishment in case of a rupture of the LPG storage tank (fatal consequences within distance of about 195 m). Nevertheless, one can state that the level of safety in the establishment is, thanks to a whole series of technical and organizational measures, high.

2.1 *Discussion about results of the case study*

This paper describes the application of the ARAMIS methodology, especially of its introductory part, MIMAH and MIRAS, to the industrial establishment unclassified under the

Seveso II Directive. The objective was to test the applicability of the methodology to SMEs. The following conclusions may be drawn:

- ARAMIS enables the assessment of unclassified sources of risks.
- By pre-defined data it facilitates the procedure of detailed risk assessment.
- The making of the analysis has, however, high demands on time and professionalism.

For this reason, the further work will focus on considering the possibility of simplifying the ARAMIS methodology for SMEs (unclassified sources of risk).

3 APELL PROGRAMME

APELL (Awareness and Preparedness for Emergencies at Local Level) is an international programme that is one of the most comprehensive approaches in the area of increasing emergency preparedness. This program was created by the UNEP (United Nations Environment Programme) organization in cooperation with governments and industrial establishments to reduce the frequency of occurrence and consequences of industrial accidents and natural disasters (UNEP, 1996).

As is clear from the very name, the strategy of the APELL programme puts main emphasis on the local level of emergency preparedness. A good many of examples from history prove that the level of accident severity depends above all on immediate measures of relevant local authorities. It offers an integrated methodology for coordination of activities necessary to achieve the sufficient preparedness of the population and intervening units for an undesirable event for the given area. In ten basic steps, it describes in detail a recommended procedure for the implementation of the whole process.

In the Czech Republic, the application of the APELL methodology was carried out in the year 2005; it was the first project of such great extent that focused on the issue of major accident prevention. The whole study was performed in the area where a significant quantity of dangerous substances was handled without awareness of the population living in the vicinity of the premises about risks associated with the activities of the operator and without risk communication. Implementation of this study influenced significantly the further development, especially of emergency preparedness and risk communication with the public. Thanks to this application, some well-proven communication tools for risk communication with the public are used in the Czech Republic (Nevrla et al., 2005).

3.1 *Risk communication with the public*

One important communication tool, created thanks to this application, is a leaflet. This communication medium intended for communication with the public is used in major accident prevention for establishments that are obliged to inform the public about, among other matters, risks and following appropriate behaviour in case of a major accident.

At present, this idea of processing so-called "Information intended for the public in the emergency planning zone" is still preserved and firmly anchored in legislation of the Czech Republic (The Czech. Act no. 224/2015 Coll.). Experts however keep working on increasing the level and quality of communicated information, including the ensuring of integration of form and content of communicated data, because the achievement of these objectives will again contribute to improvement of levels of information on risks caused by selected dangerous substances and mixtures as well as appropriate behaviour in case of an accident. The main producer of Information intended for the public is a local competent regional authority. To support effective communication a methodology dealing with the risk communication strategy and proposing the form of communicated information was prepared in the framework of the project KOMRISK. Although the proposed procedure is designed especially for employees of regional authorities and operators of risk sources, it is also suitable for representatives of other institutions dealing with risk communication with the population as a significant stakeholder of the whole process of major accident prevention. As can be seen in two

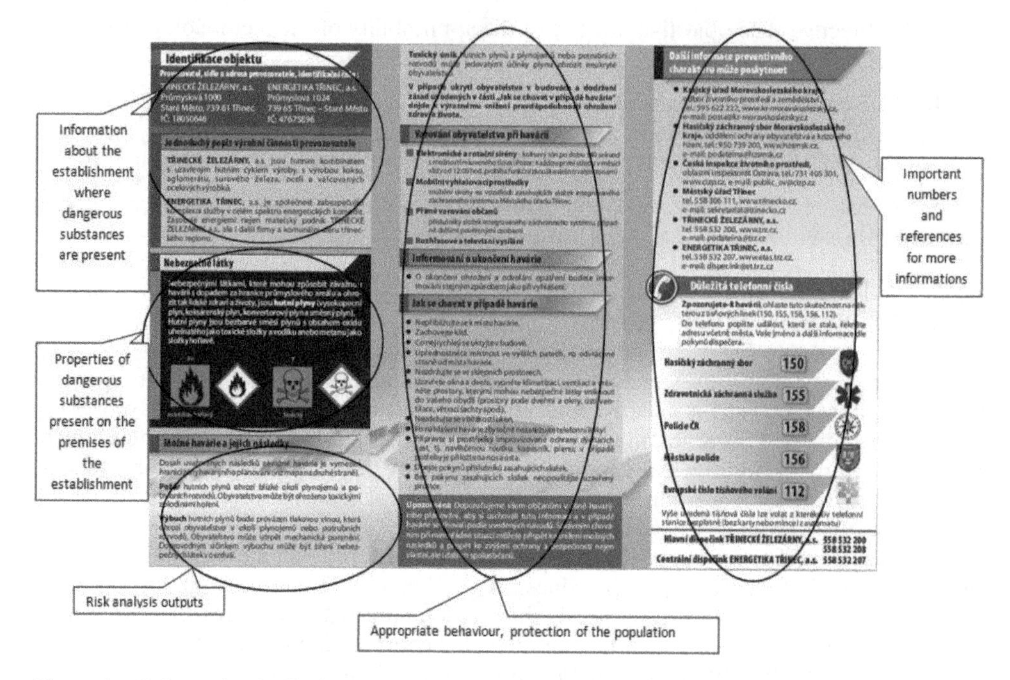

Figure 4. Information leaflet intended for the public living in the emergency planning zone.

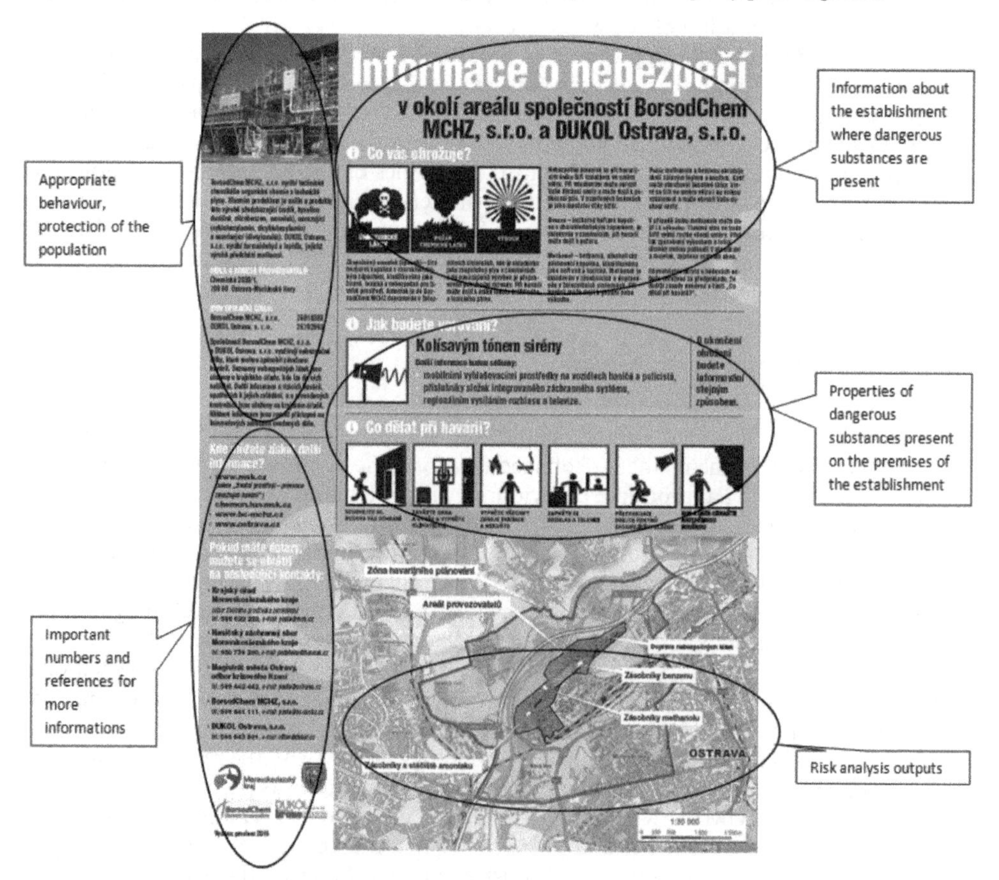

Figure 5. Information poster intended for the public living in the emergency planning zone.

figures given below (Figures 4 and 5), one of the forms of processing Information intended for the public has been preserved almost without change in the original design since the year 2005; the design was prepared thanks to the application of the APELL programme. Another form already uses the recommendations and new elements of the methodology created in the framework of the project KOMRISK. As new elements of Information leaflet that have been evaluated as the most suitable for both making risk communication more effective and creating altogether higher intelligence communication of desirable information, pictograms are used (Ministry of Environment, 2007). Pictograms are used, in addition to symbols for dangerous substances and mixtures, also for representation of appropriate behaviour and risks in connection with activities and handling dangerous substances and mixtures by the given establishment.

3.2 *Emergency preparedness*

Another valuable output of APELL programme application is the creation of a system of regular testing and training of units of Integrated Rescue System. These partial steps have a fundamental influence on increasing the emergency preparedness of not only rescue units, but also operators, local authorities and public administration, including important establishments, such as school facilities that may occur in the vicinity or in the emergency planning zone. For ensuring the regular testing, an emergency card system of the Integrated Rescue System has turned out to be best; the system has been gradually improved to its present form. For this purpose, risk analysis and assessment are carried out. Furthermore, distances for the effects of dangerous substances and mixtures are determined. Based on these outputs, procedures for all involved stakeholders, i.e. putting individual rescue units into action, correct procedure of operator's action upon accident detection, and steps leading to population protection are processed.

Emergency cards prepared like that are regularly tested by training. In the training, important establishments occurring in the zone of consequences of the possible accident are involved in the majority of cases. Thanks to this approach, school facilities are involved, and thus emergency preparedness of the general public is enhanced as well. Thus, the problems of population protection and appropriate behaviour in case of an extraordinary event become to be well known.

4 CONCLUSION

Only a consistent approach to enhancing emergency preparedness, regular informing the public and continuous maintaining risk communication between all stakeholders is in this sense a key part of the whole process of risk management. For these purposes, results of the ARAMIS study seem to be a promising tool not only for risk communication with the public, but also for effective sharing information about risks in the framework of professional community and competent institutions (Nevrly, V. 2004). Outputs of the risk analysis and their interpretation for the given area with the use of tools of the ARAMIS methodology contribute significantly just to better risk communication in the case of establishments unclassified under Seveso III. The implementation of the above-mentioned APELL programme has contributed to improving the integral view on the problems of emergency preparedness. It has helped to identify stakeholders interested in the process of major accident prevention and to formulate the communication strategy in relation to the public, units of the Integrated Rescue System, establishments handling dangerous substances and local authorities and public administration concerned. Verified partial procedures of individual methodologies can be understood as activities leading unambiguously to enhancing public awareness of risks of accidents and appropriate behaviour in case of their occurrence, which is one of the primary tasks concerning the minimization of possible consequences of a possible extraordinary event.

REFERENCES

ARAMIS. 2004, Accidental Risk Assessment Methodology for IndustrieS in the Framework of the SEVESO II Directive, User Guide, contract number: EVG1-CT-2001-00036, December 2004, Website: http://aramis.jrc.it.

Bernatik, A., Bris, R., Horehledova, S. Application of the ARAMIS Methodology to Small and Medium-Sized Enterprises A., *In European Safety and Reliability Conference ESREL 2005*, Tri City, Poland, 27.-30.6. 2005, ISBN 0415 38340 4, pp. 179–185.

Delvosalle C. & Fiévez C. & Pipart A. 2004. ARAMIS Project, Deliverable D.1.C.

Hourtolou D. & Salvi O. 2004. ARAMIS project: Achievement of the Integrated Methodology and Discussion about Its Usability from the Case Studies Carried Out on Real Test Seveso II Sites, Loss Prevention and Safety Promotion; Proc. intern. symp., Prague, 1–3 June 2004.

KOMRISK, VG20132015131, Document Zefektivnění komunikace o rizicích pro zvýšení bezpečnosti obyvatel v rámci novelizace zákona o prevenci závažných havárií (Making risk communication more effective to enhance population safety in the framework of amendment to act on the prevention of major accidents) supported by the Ministry of the Interior from the Security Research Programme of the Czech Republic in the Years 2010–2015.

MŽP ČR. ODBOR ENVIRONMENTÁLNÍCH RIZIK. Metodický pokyn odboru environmentálních rizik Ministerstva životního prostředí při zpracování dokumentu Analýza a hodnocení rizik závažné havárie podle zákona 59/2006 Sb., o prevenci závažných havárií. (Guideline of the Department of Environmental Risks of the Ministry of the Environment for preparation of the document Analysis and assessment of major accident risks according to Act No, 59/2006 Coll., on the prevention of major accidents). Věstník Ministerstva životního prostředí, 2007, 17 (3), pp. 1–15.

Nevrlá, P.; Komunikace mezi stakeholdry při prevenci závažných havárií a katastrof, diplomová práce,(Communication between stakeholders in the prevention of major accident hazards and catastrophic events), VŠB-TUO, 2004.

Nevrlá, P.; Kanichová, K. APELL a POkR—možné přístupy při informování veřejnosti v oblasti prevence závažných havárií (APELL and POkR—possible approaches to informing the public in the area of prevention of major accidents). In M. Šenovský (Ed.) Fire Protection 2005 (collection of papers). 14–15 Sept. 2005. Ostrava: Association of Fire and Safety Engineering, 2005. ISBN: 80-86634-66-3.

Nevrlý V.: Srovnání metod pro hodnocení rizik závažných havárií, diplomová práce, (Comparison of Methods of Major Accident Risk Assessment, Diploma thesis), VŠB-TUO, 2004.

UNEP. Industry and Environment. Management of Industrial Accident Prevention and Preparedness: A training resource package, first edition, 1996, 195 pp.

Zákon č. 224/2015 Sb., o prevenci závažných havárií způsobených vybranými nebezpečnými chemickými látkami nebo chemickými směsmi a o změně zákona č. 634/2004 Sb., o správních poplatcích, ve znění pozdějších předpisů (Act No. 224/2015 Coll., on the prevention of major accidents caused by selected dangerous chemicals and chemical mixtures and on amendment to Act No. 634/2004 Coll., on Administrative Fees, as amended). In Collection of Laws of the Czech Republic. 2015, Part 93, ISSN 1211–1244.

Risk Analysis and Management – Trends, Challenges and Emerging Issues – Bernatik, Huang & Salvi (Eds)
 ISBN 978-1-138-03359-7

The methodologies used in France for demonstrating risk control of a major accident: A heritage of the ARAMIS project?

Agnès Vallee, Bruno Debray, Valérie De Dianous & Christophe Bolvin
INERIS, Accident Risks Division, Verneuil-en-Halatte, France

ABSTRACT: French regulations regarding risk prevention and risk management were mainly reinforced by the law, introduced on July 30, 2003, which defines both prevention measures and repair conditions for the damage caused by industrial and natural disasters. Since then, regulations have been made considerably tighter and the entire approach towards risk assessment has changed. This law has developed very interesting tools for risk assessment and risk management (some of which are unique worldwide) and has initiated the use of frequency and probability in the French system. Better information to the public, stronger regulations, new methodology for safety reports, over-hauling of land-use planning and improved accident analysis are some of the mainstays of the law. Regarding the introduction of frequencies and probabilities, as operators in France are free to choose the methodology of probability assessment, it is interesting to review the different methodologies used by operators, with their advantages and disadvantages. Some of these methodologies are based on the ARAMIS project which has defined a methodology for risk assessment. This article aims to present major different methodologies used by operators.

Keywords: probability, land-use planning, frequency, safety barrier, risk assessment, bow-tie, major accident

1 INTRODUCTION

Co-funded under the 5th EC Framework Programme, ARAMIS ("Accidental Risk Assessment Methodology for IndustrieS") was a three-years project that started in January 2002. ARAMIS overall objective was to build up a new "Accidental Risk Assessment Methodology for IndustrieS" that combines the strengths of both deterministic and risk-based approaches, and aimed at becoming a supportive tool to speed up the harmonised implementation of SEVESO II Directive in Europe (Salvi et al., 2006).

This project took place in a particular context in France after the AZF accident (on September 21, 2001, in Toulouse). Discussions were carried out at various levels (competent authorities, local authorities, industry, experts...) to elaborate solutions aimed at improving the prevention and management of risks generated by hazardous industrial sites. A new law was introduced on July 30, 2003, which defines both prevention measures and repair conditions for damage caused by industrial and natural disasters (Loi, 2003).

The entire approach towards risk assessment has changed. Before 2003, risk assessment was based on the worst-case scenario. It was a deterministic approach. The law of July 30, 2003 now requires that the risk analysis, made in the context of the safety report, takes into account the probability of occurrence, the kinetic and the gravity of potential accidents (Lenoble et al., 2011). It specifies precisely and objectively the conditions under which an industrial site can be operated in a build environment. The assessment takes into account reducing the risks at the source, by cutting down the hazard (i.e reducing the quantity of used / stored dangerous substances...), or reducing the probabilities of occurrence of the potential accidents and limiting the consequences through organizational and technical safety measures. Nevertheless, it should be noted that the French regulation does not make compulsory a specific risk assessment methodology, as long as the operator justifies his or her choice.

The location of AZF factory in a very urbanized environment, which suffered catastrophic damages, strongly marked the minds and revealed the limits of land-use planning tools that existed until then. All these tools were intended to avoid aggravating the existing situations by preventing the increase of land use around industrial sites. It was necessary to go further by adopting a policy of gradual reabsorption of excessive promiscuity between hazardous installations and inhabited areas. The Technological Risk Prevention Plan (PPRT) is one of the flag-ship measures of the law of July 30, 2003. The aim of the PPRT is to protect people by acting on the existing urbanization and also by controlling the future land-use planning in the vicinity of the existing top-tier Seveso establishments.

Among the different features of the ARAMIS project, this paper focuses on those that have inspired some aspects of the regulation and that are currently commonly used in France for demonstrating risk control of major accident in the new regulation framework, especially:

- the use of bow-tie diagrams for the identification of the major accidents scenarios;
- the assessment of the probability of occurrence of dangerous phenomena, taking into account the efficiency and reliability of the safety barriers;
- the demonstration of risk control based on the combination of gravity and probability of occurrence of the accidental scenarios;
- and the interest of decisions taken by local authorities to reduce the global risk level by reducing the surrounding vulnerability (whereas the plant operator only can act on the potential hazard or risk of the installation).

The regulatory approach currently in force in France on which this article is based was wraped-up in 2010 in a regulatory guideline (circulaire du 10 mai 2010).

2 IDENTIFICATION OF THE MAJOR ACCIDENT HAZARDS USING BOW-TIES

The identification of major accident hazards likely to occur on an industrial site is the first step of the risk analysis. This work is made possible by the methodology MIMAH (Meth-

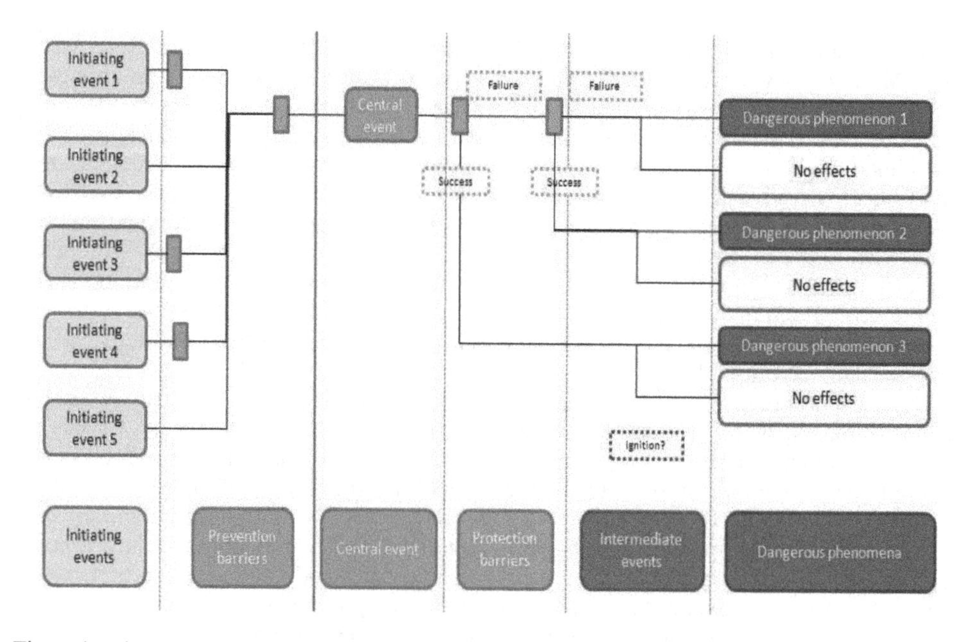

Figure 1. An example of a bow-tie diagram.

odology for the Identification of Major Accident Hazards) developed in the ARAMIS project.

The methodology is based mainly on the use of bow-tie diagrams (Figure 1), centred on a critical event and composed of a fault tree on the left and an event tree on the right. Initiating events include human errors such as a mistake in a filing procedure, technical failures as the failure of a control sensor, external aggressions such as an impact of a vehicle... A critical event is generally defined as a loss of containment or a loss of physical integrity. Then, central event leads to several dangerous phenomena such as BLEVE, boil-over, pool fire, toxic cloud dispersion...

Safety systems, technical or organisational, can be placed on the different branches of the bow-tie. Safety barriers may operate for prevention (before the central event) or for protection (after the central event).

The bow-tie concept is gaining in popularity and is believed to offer a good overview of the different accident scenarios considered. All causes and consequences are clearly identified on the bow-ties.

In the safety report, this tool is often used to describe more precisely the major accidents having consequences beyond the limits of the industrial site. It is used as a complementary tool to other risk analysis methodologies, like Preliminary Risk analysis, HAZOP... (methods based on the use of tables). The use of bow tie allows for a more complete analysis of every incident/accident that can happen.

3 EVALUATION OF THE PROBABILITY OF DANGEROUS PHENOMENA

Classical risk analysis methods propose to assess the probability of major accident. But, during the ARAMIS project, this calculation of the probability was shown not to be an easy task. An inventory of the probabilistic data sources was carried out. It showed that very generic frequency ranges are usually used in these data sources, both for critical events and causes. These generic figures have to be considered very cautiously; indeed, they may have been averaged from different kinds of plants and substances, the safety systems are not clearly identified in figures and the global level of safety of the considered plants is unknown. Eventually, the use of generic figures does not underline the efforts made by the industrialists on their specific site both in prevention and mitigation, and in their safety management system.

An alternative method was also proposed, which really takes into account the safety barriers implemented on the industrial site (De Dianous et al., 2006). Indeed, the ARAMIS project proposes the assessment of the frequency of the accident scenarios starting from the original frequency of occurrence of the deep causes and by reducing it taking into account the probability of failure of the safety functions identified on each scenario. An evaluation of the barriers is performed to validate that they are relevant for the expected safety function and to assess their probability of failure.

The calculations of the safety functions probabilities of failure are carried out according to the principles derived from the Safety Integrity Level concept (SIL) introduced by IEC 61508 (NF-EN 61508, 2002) and IEC 61511 (NF-EN 61511, 2005) standards (extended to all active barriers) and according to the known reliability of the safety barriers.

In order to be considered, a safety barrier (technical or human) has to meet the following requirements. It has to be:

- independent: the safety barrier must be independent of the causal events of the scenario and independent of other safety barriers (or in case of dependence with other safety barriers it must be taken into account when assessing the reliability);
- effective: able to fulfil the safety function that it was chosen for, in its usage context, for a period of process operation;
- with a response time in accordance with the kinetic of the scenario;
- testable;
- covered by preventive maintenance designed to guarantee that performance levels are maintained over time.

These developments were the basis of a methodology described in the Ω 10 and Ω 20 reports (INERIS, 2005 and 2006) (see Figure 2). The assessment is based on the evaluation of each component of the barrier with regards to three criteria: Effectiveness (Eff), Response Time (RT) and Level of Confidence (LC). The level of confidence is then converted into a "Risk Reduction factor" (RR). The risk reduction factor represents the amount by which the presence of a given barrier would divide the frequency of a given scenario. For example, if a scenario (defined as the sequence of events initiated by an initiating event and leading to a given hazardous phenomenon) has an initial frequency of 10^{-1} occurrence per year, adding a safety barrier with a risk reduction factor of 10 to prevent its occurrence would result in a frequency of the final hazardous phenomenon of 10^{-2}/year. In order to assess the risk reduction factor of the whole barrier, an aggregation of the data related to its components is realised. The assessment of each component is guided by a questioning process as described in the Ω 10 and Ω 20 reports.

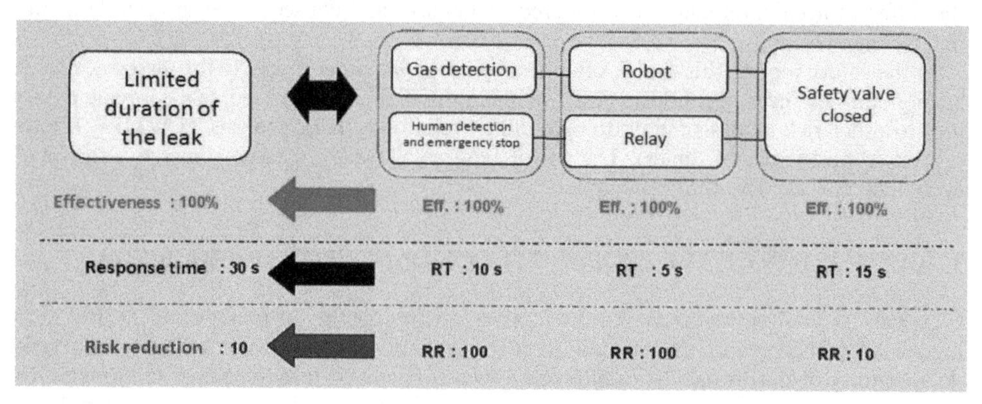

Figure 2. An example of the assessment of a barrier according to the Ω 10 methodology.

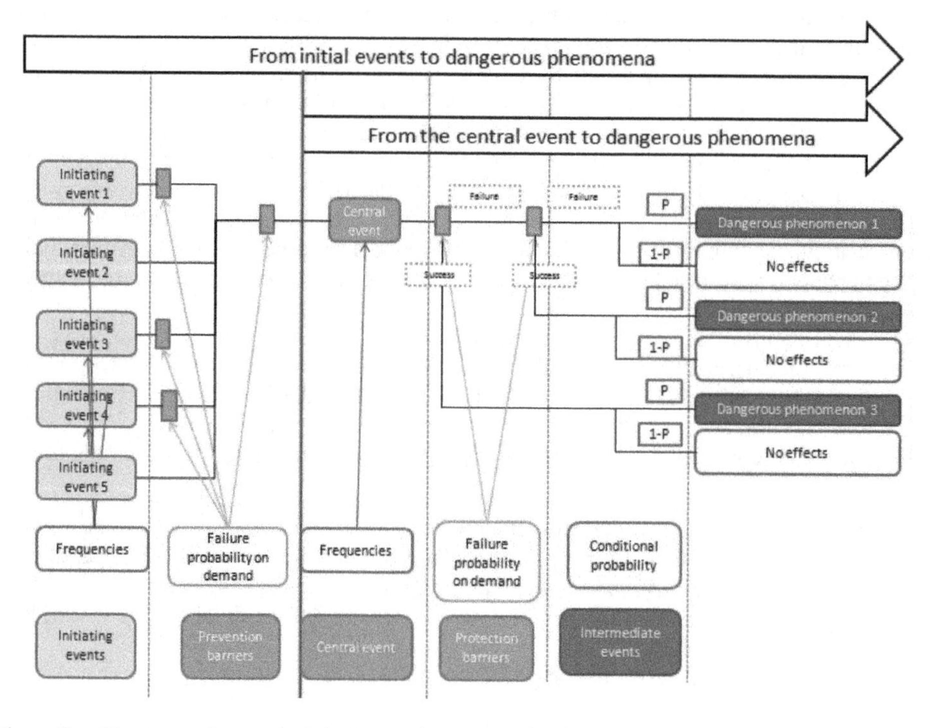

Figure 3. The two mains methodologies used in France for the assessment of dangerous phenomena occurrence probability.

The safety management has a strong influence on the capacity to control the risk, and hence for the evaluation of barriers performance. The approach in ARAMIS consists in devising a process-oriented audit protocol focusing on the activities relating to the life cycle of the safety barriers including design, installation, use, maintenance and improvement activities.

As operators are free to choose the methodology to be used in the safety report for assessing the occurrence probability of a dangerous phenomenon, the methodologies can be very different. Since 2005, two main methodologies have been used for major accident probability assessment:

- Quantitative evaluation "from initiating events to dangerous phenomena" (as proposed by the ARAMIS project);
- Quantitative evaluation "from the central event to dangerous phenomena".

This later methodology uses generic frequencies for central events that are independent from the actual implementation of prevention safety barriers. Only the control or protection safety barriers, which become active after a central event has occurred are taken into account.

These two methodologies are approximately equally used. Their principle is presented in Figure 3.

4 DEMONSTRATION OF RISK CONTROL AND ACCEPTABILITY

The risk analysis has the purpose to demonstrate that the industrial site has a good level of risk control. This risk control is built on the reduction of the frequency of occurrence of the major dangerous phenomena taking into account the safety barriers, so that the dangerous

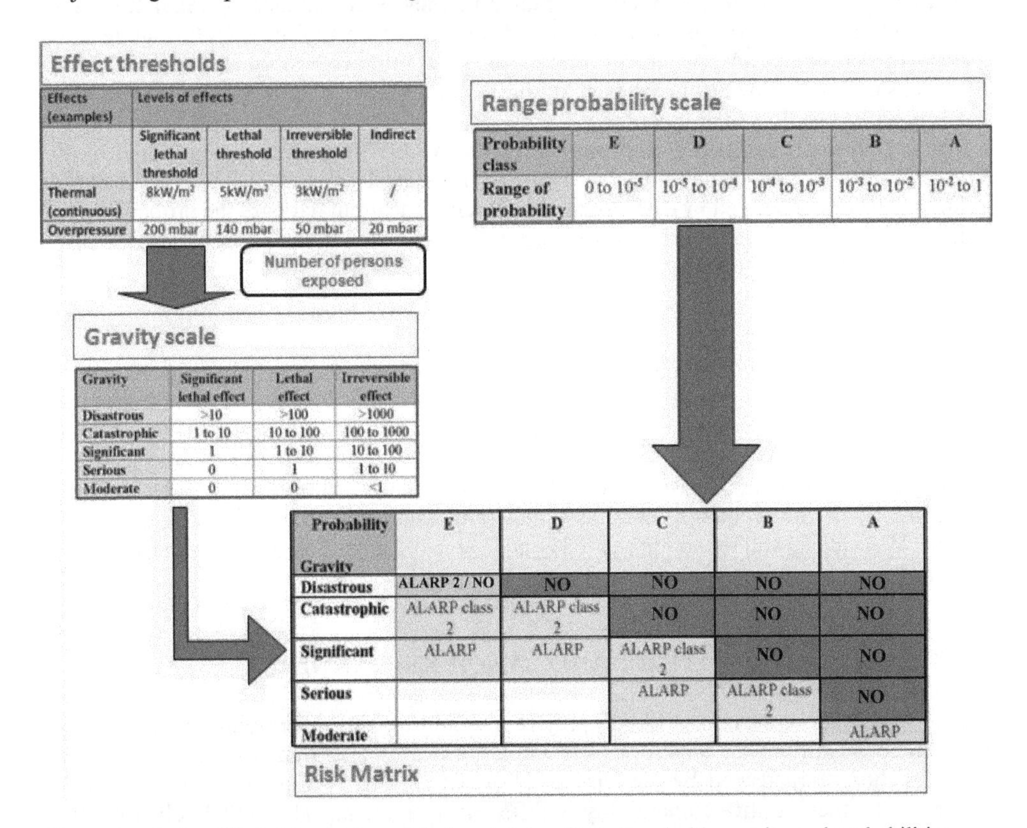

Figure 4. A simplified risk matrix and its input data: effects thresholds, gravity and probabilities.

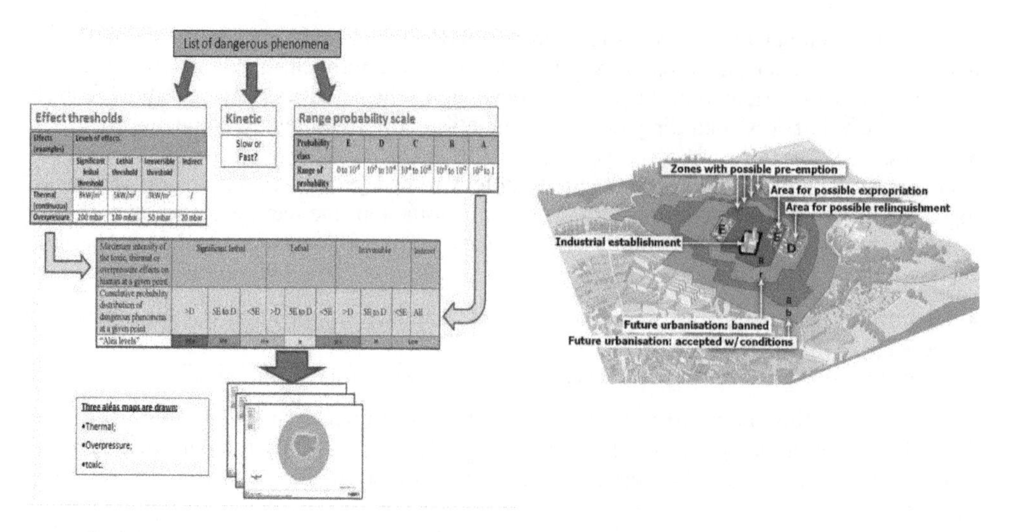

Figure 5. Simplistic scheme presenting the elaboration process of a PPRT.

phenomena are finally assessed with an acceptable combination of gravity of their consequences and frequency of occurrence. The consequences are defined by combining the intensity of the hazardous phenomenon (i.e. distance to regulatory effect thresholds) with the potentially impacted population.

The dangerous phenomena are then ranked according to their classes of probability and consequences in a risk matrix where acceptable risk levels are defined by regulation. This approach is interesting because it encourages the plant operators improving the management of risks by defining clear risk reduction targets. It takes into account the existing risk reduction measures, and provides recommendation for additional safety measures in case some scenarios have insufficient risk control.

The concept of risk matrix, as promoted in the ARAMIS project, was adopted by the French government, which has indeed developed a risk matrix for assessing the societal acceptability of the risk generated by a Seveso establishment. The input data are the probabilities of major accidents and the number of people potentially exposed to their consequences.

The matrix (see Figure 4) defines three levels of accidental risk:

- Acceptable (in white on the matrix): the risk is acceptable because both its consequences and probability are low;
- ALARP (As Low As Reasonably Practicable, in yellow): the risk has to be reduced in order to become "as low as reasonably practicable". The operator has to add additional safety measures in order to reduce the risk as low as possible, given an acceptable cost/effectiveness ratio;
- Unacceptable (in red): the risk is too high. The safety must be improved in order to reduce the gravity or the probability and, as a matter of fact, move the accident to the yellow zone. If accidents still remain in the red zone in spite of an improvement of the safety, the establishment is liable to be closed.

5 VULNERABILITY MAPPING AND LAND-USE PLANNING

An innovative attempt from AMARIS project is to address the vulnerability of the environment independently of the hazardous site. On a given spot of the environment, the vulnerability is thus characterised by the number of potential targets (human, environmental or material) and their relative vulnerability to different phenomena. The global vulnerability is a linear combination of each target vulnerability to each type of effects for the various types of impacts (overpressure, thermal radiation, gas toxicity or liquid pollution). The ARAMIS project also showed the value of using GIS for mapping vulnerable areas.

These elements contributed to the discussions around the creation of PPRT, new tool for the land-use planning introduced by the law of July 30, 2003.

The most important stages of PPRT are the following:

- Determination of the aléa[1] by combining the probability of a dangerous phenomenon and the potential intensity of its effects. Aléas are calculated for each point of the territory and for each type of effects (thermal, overpressure and toxic effects);
- Analysis of the stakes in the vicinity of the establishment. The types of construction and public buildings are distinguished;
- Cross-reference of the *aléas* and the stakes, allowing to draw a zoning map.

When the PPRT is finalized, it delineates a risk exposure perimeter, at the heart of which regulated zones are established. These zones can be either:

- "ban zones", within which future constructions are banned. Inside this zone, areas can be defined for expropriation or relinquishments;
- or "limitation zones", within which protective measures on the future or existing buildings can be compulsory.

6 CONCLUSION

The law of July 30, 2003 has introduced the use of probabilities and frequencies into the French legislation and regulations regarding industrial risks. The risk analysis, at the heart of the safety report, has now to take into account the probability of occurrence, along with the kinetic and the gravity of the potential accidents. Combined with the gravity of potential accidents, the probability is used to assess, using a regulatory risk matrix, the acceptability of an industrial establishment in its environment and the demonstration of risk control. Then, for land-use planning, probabilities and intensities of hazardous phenomena are used for the calculation of the aléas. Combined with an analysis of the stakes, they are used for the elaboration of the PPRT, which constitutes the basis for land-use planning decision, including the possibility to reduce existing land use if the risk cannot be further reduced at the source.

To respond to these changes concerning the identification, assessment and prevention of risks, the results of the ARAMIS project have been very useful.

Points are still open to discussion and questions:

- The authorities and the operators have faced some difficulties in this new context for safety report elaboration requiring the quantification of the probability of occurrence of major accidents, due to the lack of available data. Uncertainties exist for the calculation of these probabilities. To overcome this difficulty, some operators have put in place feedback analysis on their initiating events, on the failure of their barriers to obtain probability values closer to the experience of their installations, and thus to take them into account during the upcoming revisions of the safety reports.
- It can be noted that the gravity of major accidents is currently evaluated in the safety reports exclusively by taking into account the human consequences outside site (exposed persons). It might also be interesting to integrate the assessment of environmental consequences of major accidents into the risk assessment process.
- There is an advantage to having a P/G risk matrix setting national criteria for risk acceptability, as it helps the authorities to prioritize actions to be requested from the operator to reduce risk and hence promotes a homogeneous treatment on French territory. Yet it turns sometimes difficult to apply the ALARP principle, as the assessment of what is

[1]Aléa: Probability that a dangerous phenomenon creates effects of a given intensity, and over a determined period of time at a given point of the territory.

"reasonably practicable" may be different between the operators and the authorities in charge of enforcing the regulation. Differences have even been observed between regional offices of the national authorities. And setting the limit of what can be imposed to the plant operator is difficult, when the risk exists of closing the industrial site for economic reasons.

- Although the interest of the PPRT in evident, in order to avoid the recurrence of accidents with major consequences, the introduction of this new tool is felt unfavorably at local level and for the large public. The elaboration procedure is long and difficult to implement, as it involves many actors. The dissatisfaction is also due to the measures resulting from the PPRT that directly impact the local residents and economic activities (expropriation/relinquishment, reinforcement of building structure...). Actions are still under way to promote the effective implementation of the PPRTs.

REFERENCES

Circulaire du 10 mai 2010 récapitulant les règles méthodologiques applicables aux études de dangers, à l'appréciation de la démarche de réduction du risque à la source et aux plans de prévention des risques technologiques (PPRT) dans les installations classées en application de la loi du 30 juillet 2003.

De Dianous, V. & C. Fievez, C. 2006. ARAMIS project: A more explicit demonstration of risk control through the use of bow-tie diagrams and the evaluation of safety barriers performance, *Journal of Hazardous Materials*, 130 220–233.

INERIS (2008) Ω10, « Évaluation des Barrières Techniques de Sécurité », available on the website www.ineris.fr.

INERIS (2009), Ω20 « Démarche d'évaluation des Barrières Humaines de Sécurité », available on the website www.ineris.fr.

Lenoble, C. & Durand, C. 2011. Introduction of frequency in France following the AZF accident, *Journal of Loss Prevention in the Process Industries, Elsevier*, 24 (3), pp 227–236.

Loi n°2003-699 du 30 juillet 2003 relative à la prévention des risques technologiques et naturels et à la réparation des dommages.

NF-EN 61508 (March 2002), parts 1 to 7. Functional safety of electrical, electronic and programmable electronic safety-related systems.

NF-EN 61511 (March 2005), parts 1 to 3. Functional safety—Instrumented systems for the industry sector.

Salvi, O. & ARAMIS, B.D. 2006. An integrated risk assessment methodology for SEVESO plants, Chemical Engineering Transactions, Volume 9, 421–426.

Risk Analysis and Management – Trends, Challenges and Emerging Issues – Bernatik, Huang & Salvi (Eds)

ISBN 978-1-138-03359-7

Author index